*Whitman the Political Poet*

# Whitman the Political Poet

## Betsy Erkkila

*New York    Oxford*
OXFORD UNIVERSITY PRESS
*1989*

*In memory of my parents*
*Edwin Sulo Erkkila (1917–1957)*
*and Jean Troup Erkkila (1919–1959)*

Oxford University Press

Oxford   New York   Toronto
Delhi   Bombay   Calcutta   Madras   Karachi
Petaling   Jaya   Singapore   Hong Kong   Tokyo
Nairobi   Dar es Salaam   Cape Town
Melbourne   Auckland

and associated companies in
Berlin   Ibadan

Published by Oxford University Press, Inc.
200 Madison Avenue, New York, New York 10016

Oxford is a registered trademark of Oxford University Press

Library of Congress Cataloging-in-Publication Data
Erkkila, Betsy, 1944–
Whitman the political poet/Betsy Erkkila.
p.   cm. Includes index.
ISBN 0-19-505438-5
1. Whitman, Walt, 1819–1892—Political and social views.
2. Political poetry, American—History and criticism. 3. Poets,
American—19th century—Biography. 4. United States—Politics and
government—19th century. I. Title.
PS3242.P64E74 1989
811'.3—dc19                                      88–3651 CIP

Printing (last digit): 9 8 7 6 5 4 3 2 1

Printed in the United States of America
on acid-free paper

# Preface

*Whitman the Political Poet.* The title is at once a statement of subject and a challenge. It is an attempt to restore a series of linkages—Whitman, political, poet—that have been torn asunder in the wake of Modernist, Formalist, and New Critical strategies. I use the term *political* both in its traditional Aristotelian sense as a specific concern with the governance of *polis*, or state, and in its more recent—and still vigorously contested— sense as the entire network of power relations in which texts and authors are implicated. In reading Whitman the political poet, I shall be concerned not only with his overtly political postures and designs but also with the more subtle and less conscious ways his poems engage on the level of language, symbol, and myth the particular power struggles of his time—power struggles that came to center on the issue of state governance as well as on the issues of race, class, gender, capital, technology, western expansion, and war.

In *Democracy in America,* Alexis de Tocqueville observed: "The relations that exist between the social and political condition of a people and the genius of its authors are always numerous; whoever knows the one is never completely ignorant of the other." In tracing Whitman's development from his early work as a journalist and popular writer to his later work in poetry and prose, I shall be concerned with the interactive relation of writing and history, politics and art. I intend *Whitman the Political Poet* both as a reading of Whitman's life and work within and against the sociopolitical context of his time and as a meditation on the problem of America itself—its history, its paradox, its best, and its worst—written by a critic who is herself no doubt reflecting and engaging the concerns of her own post–Vietnam era generation. I am interested, that is, in the ways that the social and political conditions of nineteenth-century America might serve as an enriching context in which to read Whitman's poems; I am also interested in the ways

that the text of the poems opens onto and into the complexity and contradictions of American history—how the formal and symbolic text of Whitman's
poems might itself serve as an enriching context in which to read the past as
it enters and shapes the present and the future.

I would like to thank the Woodrow Wilson International Center for Scholars in Washington, D.C., for providing fellowship support, stimulating
colleagues, and an ideal setting in which to begin work on this book. I
would also like to thank the American Council of Learned Societies for
awarding me a grant to help complete this study. I am grateful to the many
colleagues and friends—especially Gay Wilson Allen, Roger Asselineau,
Nina Auerbach, David DeLaura, Daniel Hoffman, Jerome Loving, Robert Regan, Carroll Smith-Rosenberg, and Maureen Quilligan—who offered support and advice at various stages in this project. I owe a particular debt of gratitude to Larzer Ziff for his friendship, his exemplary work
as a scholar and teacher, and his generous support over many years.
    Through several visions and revisions of this book, I have, as always,
been particularly blessed by the love, good humor, and constant encouragement of my husband, Larry Stuelpnagel.
    This book is dedicated to my parents, Jean Troup Erkkila and Edwin
Sulo Erkkila, who died when I was young but who left me the memory of
love and possibility.

*Philadelphia*                                                              B.E.
*February 1988*

# Contents

A dance of walls agitates the meadows
and America drowns itself in machines and lament
I want the strong air of the most profound night
to remove flowers and words from the arch where you sleep,
and a black boy to announce to the gold-minded whites
the arrival of the reign of the ear of corn.

—FEDERICO GARCÍA LORCA, "Ode to Walt Whitman"

*Whitman the Political Poet*

FIGURE I. Engraved frontispiece for *Leaves of Grass* (1855), by S. Hollyer. Courtesy of MA 6069 The Pierpont Morgan Library, New York.

# 1

# *A Revolutionary Formation*

> Every page of my book emanates Democracy, absolute,
> unintermitted, without the slightest compromise, and the
> sense of the New World in its future, a thoroughly revolution-
> ary formation. . . .
>
> —WALT WHITMAN, *Notebooks*

The publication of *Leaves of Grass* on or about July 4, 1855, was an act of revolution, an assault on the institutions of old-world culture that was as experimental and far-reaching in the artistic sphere as the American revolt against England had been in the political sphere. Everything about the book, including its date of publication, was revolutionary. Designed by Whitman and printed at his own expense, the volume was quarto sized, with clusters of leaves embossed on its dark green cover; the title, which was printed in gold, sprouted lush roots and leaves, suggesting the motifs of growth, fertility, luxuriance, and regeneration that figure throughout the poems. With its visual iconography of roots taking shape as *Leaves of Grass* and its verbal pun on leaves as pages and pages as grass, the title challenged traditional notions of poetic taste by suggesting the common, local, and democratic ground out of which American literature emerged.

The title page bears no author's name, only the title *Leaves of Grass* and the date and place of publication: Brooklyn, New York: 1855. Whitman inscribed his authorial signature not in a name but in an engraved frontispiece of himself dressed as a day laborer in workingman's trousers, a shirt unbuttoned to reveal his undershirt, and a hat cocked jauntily on his head (Figure I). This working-class figure was part of Whitman's revolt against the profession of authorship, his attempt to equalize the traditional hierarchical relationship between writer and reader by present-ing the author as a democratic presence, a common man who speaks as and for rather than apart from the people. His workingman's dress and informal pose represented a radical departure from the conventions of literary portraiture, in which the artist sits starched, buttoned, and hatless in coat, high collar, and tie.

At a time when the growth of the literary marketplace was altering the relationship between writer and reader, Whitman's substitution of a picture of himself for a name was an attempt to transgress the bounds of both literature and the print medium itself in order to make personal contact with his audience. In the as-yet-untitled 1855 poem "A Song for Occupations," he says:

> I was chilled with the cold types and cylinder and wet paper
>     between us.
> I pass so poorly with paper and types . . . . I must pass with the contact
>     of bodies and souls.[1]

By placing a picture of himself at the beginning of his first and future editions of *Leaves of Grass,* Whitman sought to project the sense of personal presence and magnetism he admired in such orators as Elias Hicks and George Fox. "Camerado, this is no book," he says in his 1860 poem "So Long": "Who touches this touches a man."[2]

But Whitman's 1855 frontispiece was not intended to be a literal presentation of himself. Engraved from a daguerreotype made in 1854 by Mathew Brady's cameraman, Gabriel Harrison, the portrait is a construct, an invention of the poet as the representative American who emerges in the poems:

> I celebrate myself,
> And what I assume you shall assume,
> For every atom belonging to me as good as belongs to you.
>                     (*LG* 1855, p. 25)

The daguerreotype figure projects the proletarian energies and sympathies of Whitman's first volume of poems, in which he said he "deliberately and insultingly ignored all the other, the cultivated classes as they are called, and set himself to write 'America's first distinctive Poem,' on the platform of these same New York Roughs, firemen, the ouvrier class, masons and carpenters, stagedrivers, the Dry Dock boys, and so forth."[3]

The difference between Whitman's actual self and his invented self is suggested by the difference between Walter Whitman, whose name appears in the copyright notice on the verso of the title page, and the figure who names himself in his poems:

> Walt Whitman, an American, one of the roughs, a kosmos,
> Disorderly fleshy sensual . . . . eating drinking and breeding.
>                     (*LG* 1855, p. 48)

Like his poet as common man, Whitman's self-naming is an assault on literary decorum and the Puritan pieties of the New England literary establishment. The name Walt Whitman refuses the first, middle, and last name formality of Ralph Waldo Emerson, John Greenleaf Whittier, and Henry Wadsworth Longfellow. In using his nickname "Walt," Whitman once again breaks literary rank, insisting on a familiarity and intimacy with his audience that defied the conventional distance, reserve, and formality that governed relations between writer and reader.

In a twelve-page, two-column preface, Whitman sounds the cry of revolt that is implicit in the design of the 1855 *Leaves of Grass*. Like the American Revolution, the entire thrust of the preface is to relocate authority inside rather than outside the individual; it is a declaration of self-dependence, a radical assertion of liberation and release—personal, political, and artistic. Turning traditional notions of gentility and decorum on their head, Whitman declares: "The old red blood and stainless gentility of great poets will be proved by their unconstraint. A heroic person walks at his ease through and out of that custom or precedent or authority that suits him not" (*LG* 1855, p. 13).

More than any other American writer of the nineteenth century, Whitman realized that a truly democratic American literature would require not merely a revolution in content but also a revolution in literary form and in the traditional conceptions of literature itself: "Of the traits of the brotherhood of writers savans musicians inventors and artists nothing is finer than silent defiance advancing from new free forms. . . . The cleanest expression is that which finds no sphere worthy of itself and makes one" (*LG*, 1855, p. 13). The elimination of commas as marks of distinction and division in this passage is—like the use of suspension points, the lack of subordination, and the violation of grammatical rules and linear logic throughout the preface—a fitting introduction to the formal revolution of the poems.

The twelve untitled poems that follow the preface make good Whitman's declaration of literary independence. Defying the rules of rhyme, meter, and stanza division and breaking down the distinction between poetry and prose, Whitman's verse rolls freely and dithyrambically across the page in what one 1855 reviewer called "a sort of excited prose broken into lines without any attempt at measure or regularity, and, as many readers will perhaps think, without any idea of sense or reason."[4] The poems are untitled, and individual poems, separated only by two horizontal bars, appear to flow together as part of a single florid growth entitled *Leaves of Grass*.

Even the reviews were revolutionary. Thumbing his nose at uncomprehending critics and the privileged mechanisms of literary reviewing, Whitman wrote no fewer than three rave reviews of himself: "An American bard at last!" he exclaimed in the *United States Review,* "One of the roughs, large, proud, affectionate, eating, drinking and, breeding, his costume manly and free, his face sunburnt and bearded, his postures strong and erect, his voice bringing hope and prophecy to the generous races of young and old." He draws attention to the revolutionary sources of his creation: "We shall start an athletic and defiant literature. . . . The interior American republic shall also be declared free and independent." And he stresses the antiliterary nature of his revolt: "Self-reliant, with haughty eyes, assuming to himself all the attributes of his country, steps Walt Whitman into literature, talking like a man unaware that there was ever hitherto such a production as a book, or such a being as a writer."[5]

In seeking to explain the transformation of the political journalist Walter Whitman into the American bard who emerged in the 1855 *Leaves of Grass,* critics have been mystified by the apparent split between the politician and the poet. But this split has been at least partly the construction of critics who, under the influence of the Modernist and New Critical insistence on the separation of politics and art, have been eager to rescue Whitman's poems from the charge of political contingency in order to save them for the universality of art. Stressing Whitman's nonpolitical sources, critical studies—particularly in the last forty years—have emphasized the formal, mystical, and psychological dimensions of his art, and most studies dwell on his Emersonian and transcendental sources.

Unable to reconcile Romantic and Modernist notions of the purity of art with the fact of Whitman's early and active political engagement, critics have advanced a number of elaborate and sometimes contradictory theories to explain his emergence as a poet: a reading of Emerson or George Sand or some other writer whose influence was determining; a love affair with a New Orleans octoroon; a mystical experience; or, more recently, an oedipal or some kind of sexual crisis.[6] Few critics have sufficiently stressed, and no one has rigorously examined, the obvious political passion and struggle that were at the very foundation of Whitman's democratic songs.

Considered in the political context of his times, however, Whitman's emergence in 1855 as the poet of democracy seems neither mystifying nor particularly disconnected from his life during the 1840s as radical Democrat and party journalist. In fact, the critical tendency to discuss Whitman not in relation to the revolutionary traditions on which he was raised and the political battles in which he engaged but in relation to a line of poetic development that runs from Edward Taylor to Ralph Waldo Emerson to

Robert Frost and Wallace Stevens reflects the privileging of New England and the "New English" sensibility in America that Whitman challenged.

Whitman is one of America's most overtly political poets. Having said this, one is immediately struck by the fact that the Whitman we have learned to read in the academy and among the critics comes to us curiously purified of his political designs. We have learned that Whitman is best and most interesting as a personal rather than as a political poet; indeed, that he is at his worst and most problematic as an artist when he is being most political. What is at stake here is an American canon, a particular way of reading and interpreting literature, and a literary profession grounded in the assumption that aesthetic value is an indwelling essence detached from ideological interest and the messiness of history.[7] This putative opposition between what Raymond Williams has called "attention to natural beauty and attention to government, or between personal feeling and the nature of man in society" had its roots in the works of the Romantic poets themselves.[8] Whitman, however, would not have understood the opposition. Even among American Romantic writers, he stood out in consistently avowing the specifically political project of his work.

But in the final years of his life and in the years immediately following his death, the split between Whitman the politician and Whitman the poet was fostered by his most ardent supporters in America. Deploying an eternalized vocabulary of absolute transcendental value in his book *Cosmic Consciousness* (1901), Richard Maurice Bucke commented on the "case" of Whitman in which "writings of absolutely no value were *immediately* followed by pages across each of which in letters of ethereal fire are written the words ETERNAL LIFE."[9] Reflecting on "this instantaneous evolution of the Titan from the Man," Bucke was the first to suggest that a mystical experience explained the supposed gap between the political journalist and the visionary poet.

In stressing the religiospiritual nature of his achievement, Whitman's inner circle of friends was in part reflecting his own increased emphasis on religion in the last two decades of his life. But like William Douglas O'Connor's earlier defense of Whitman against charges of immorality, in *The Good Gray Poet* (1866), the religiospiritual and ultimately transcendental focus of Bucke and others has tended both to remove Whitman's work from the historic specificity of his time and to deflect attention from the more radical political posture of a poet who challenged the traditional hierarchies of power and domination; who celebrated the liberation of male and female, sex and the body, workers and poor persons, immigrants and slaves; and who placed the values of liberty and equality, comradeship and solidarity at the heart of his democratic songs.

In the complex landscape of literary Modernism, Whitman emerged as a figure of contest. He was at the center of the aesthetic debate between the William Carlos Williams school and the T. S. Eliot school—between those who, in Williams's terms, remained rooted in the American locality and those who went abroad physically and intellectually, like Henry James, Ezra Pound, and T. S. Eliot. For all his emphasis on literary tradition as an embodiment of values that transcends ideological conflict, Eliot could not stomach Whitman's poetic democracy, for obvious political reasons. As a self-confessed "classicist in literature, royalist in politics, and anglo-catholic in religion," Eliot "had to conquer an aversion to his form as well as to much of his matter" before he could read Whitman at all. Although Williams was theoretically more sympathetic, he, too, enforced a formalist separation between politics and art and criticized Whitman for allowing his democratic message to interfere with the main matter of his technical innovation. "He had seen a great light but forgot almost at once after the first revelation everything but his 'message,' the idea which originally set him in motion, the idea on which he had been nurtured, the idea of democracy—and took his eye off the words themselves which should have held him."[10]

To Ezra Pound, Whitman was a similarly problematic figure—not Homer, Dante, or Shakespeare, he regretted in "What I Feel About Walt Whitman" (1909), but "America's poet" nevertheless, bearing the "exceeding great stench" of America itself. "I come to you as a grown child/ Who has had a pig-headed father," Pound finally conceded in his 1915 poem "A Pact." But the renewed "commerce" between them was made possible only by dint of the knife of aestheticism: "It was you that broke the new wood,/Now is a time for carving."[11]

While Eliot and Pound were experiencing an almost physical revulsion against the political content of Whitman's verse, Vernon Parrington was lionizing Whitman in specifically political terms as a heroic democrat—"a great figure, the greatest assuredly in our literature." But for all his interest in the political and economic sources of literature in *Main Currents in American Thought* (1927–30), Parrington treats Whitman abstractly and from afar, idealizing him as transhistorical bearer of American national ideology and a keeper of the democratic flame amid the "breakdown and disintegration" of the post–World War I period. Newton Arvin is more critical in *Whitman* (1938), a book that reflects the literary leftism of the 1930s. Examining Whitman from a socialist point of view, Arvin presents what is still one of the best studies of the intellectual origins of Whitman's social thought.[12]

In *American Renaissance: Art and Expression in the Age of Emerson and*

*Whitman* (1941), F. O. Matthiessen seeks to reconcile a New Critical emphasis on aesthetic form with his political sympathies as a Christian socialist. But even though he is interested in the common "devotion to the possibilities of democracy" shared by the five writers he discusses— Emerson, Thoreau, Hawthorne, Melville, Whitman—his focus is New Critical, concentrating, as he says in his introduction, "entirely on the foreground, on the writing itself." As translated into his final book on Whitman, this means that Matthiessen emphasizes Whitman's linguistic experiments and his organic form, tracing the analogues of his work in oratory, opera, painting, and nature rather than in American historical experience.[13]

Matthiessen's magisterial study set the critical frame and interpretive and evaluative terms for future readings of Whitman. In *Walt Whitman Reconsidered* (1955), which is still one of the liveliest critical discussions of Whitman's work, Richard Chase treats Whitman as a comic, albeit neurotic, genius, a master of New Critical wit and irony. Other critical studies of the time, including Charles Feidelson's *Symbolism and American Literature* (1953), R. W. B. Lewis's *The American Adam* (1955), and Howard Waskow's *Whitman: Explorations in Form* (1966) focus on the symbolic, mythic, and formal dimensions of Whitman's art.[14]

Although Gay Wilson Allen and Roger Asselineau consider the social background of Whitman's thought in their major centennial biographies, Allen stresses the psychological origins of Whitman's poetic creation in *The Solitary Singer* (1955), and Asselineau stresses the personal and essentially mystical sources of Whitman's art in *The Evolution of Walt Whitman* (1954; trans. 1960, 1962). James E. Miller's New Critical reading of the poems in *A Critical Guide to Leaves of Grass* (1957), a reading that emphasizes the mysticospiritual basis of Whitman's art, is still one of the standard critical texts on *Leaves of Grass*.[15]

Joseph J. Rubin's *The Historic Whitman* (1973), which concludes before Whitman published *Leaves of Grass* in 1855, and Justin Kaplan's *Walt Whitman: A Life* (1980) have begun the work of reexamining Whitman in the political context of his times. However, the emphasis in studies of Whitman is still largely psychological and transcendental, as indicated by Paul Zweig's celebratory *Whitman: The Making of the Poet* (1984) and David Cavitch's *My Soul and I: The Inner Life of Walt Whitman* (1985). In "Pastoralism and the Urban American Ideal: Hawthorne, Whitman, and the Literary Pattern," James Machor reflects a prevalent attitude toward Whitman when he says that the "creation of a separate, subjective, poetic world is precisely the primary purpose of *Leaves of Grass* in general and his poetic treatment of the city in particular."[16]

Simply stated, my own aim is to repair the split between the private and the public, the personal and the political, the poet and history that has governed the analysis and evaluation of Whitman's work in the past. I shall attempt to present Whitman in the thickness of his times, yoking body, work, and body politic in a sustained analysis that will open, I hope, into a reading of America itself. My own overriding assumption is that works of art have a particular history that is not merely biographical but social and political in the broadest sense of the terms as well, for as Walter Benjamin has taught us, "the uniqueness of a work of art is inseparable from its being imbedded in the fabric of tradition."[17] Indeed, what makes Whitman unique as an artist, and perhaps also most interesting and valuable, is his embeddedness in his time rather than his transcendence of it.

To say this, however, is not to limit the significance of Whitman's work to a mirror reflection of his times. On the contrary, as I hope to show, it is by reading Whitman in the context of nineteenth-century American historical experience that we discover him at his most revolutionary. Whitman lived at a crossroads of history, at a time when the twin revolutions in politics and technology, democracy and industry were bringing the radical transformations that would alter the landscape of America and the world. As a poet Whitman positioned himself at the center of the diverse and contradictory energies of American culture, seeking through his poems to order and shape these energies into a harmonious democratic world. In the symbolic order of his poems, the real contradictions in the American marketplace— contradictions that center in particular on the issues of race, class, gender, and capital—continually collide with and threaten to explode Whitman's democratizing designs. In attempting to manage symbolically the facts and contradictions of history at a point when the republican values and artisanal economy of the past were eroding, Whitman's poems engage the particularity and difference of their time, at the same time that they reveal a commonality of interests and desires, experiences and forms that link the struggles of the past with those of the present and the future. Ultimately, Whitman's poems may be of value not as pointers to some transcendent spiritual realm but as visionary markers of where we have been, where we are, and where we might still go.

Whitman himself repeatedly stressed the historical embeddedness of his work. As he said in "A Backward Glance o'er Travel'd Roads" (1888), *Leaves of Grass* originated from "a feeling or ambition to articulate and faithfully express in literary or poetic form, and uncompromisingly, my own physical, emotional, moral, intellectual, and aesthetic Personality, in the midst of, and tallying, the momentous spirit and facts of its immediate days, and of current America" (*LGC*, p. 563). In a note on his *Complete*

*Poems and Prose,* he wondered whether his work might not be "in one sort a History of America, the past 35 years, . . . coming actually from the direct urge and developments of those years, and not from any individual epic or lyrical attempts whatever, or from pen or voice, or any body's special voice."[18] The political and capitalist transformation of America that we receive as a matter of history, Whitman experienced firsthand, on the level of the body and the senses: as abundance, energy, desire, and possibility; as deprivation, anxiety, conflict, and dislocation. This experience was no mere background to his work but the substance from which his poems were constituted.

Freudian critics have taught us to see beneath the mask of the bard of democracy the anguished visage of an American neurotic, but they have read Whitman's neurosis as a merely personal affair. I am interested in the ways that Whitman's family romance was bound up with the romance of America—in the ways the signs of personal neurosis and crisis we find in his poems are linked with disruptions and dislocations in the political economy. I am interested, finally, in the ways that Whitman the poet and America the polis reflect and refract each other.

Like the Romantic project in general, Whitman's poems are a response to and an attempt to manage the disintegrative forces of both democracy and technology in the nineteenth century. Like Wordsworth's vision of the poet as "an upholder and preserver, carrying every where with him relationship and love," Whitman's grammar of reconciliation and union responds to a world of rupture and dislocation.[19] His assertion in the 1855 preface that poets rather than presidents will become the "common referee" of the nation is—like Shelley's assertion that poets will become the legislators of mankind—uttered at a time when the poet was becoming increasingly marginalized. But as Raymond Williams has shown us in his study *Culture and Society* (1958), it was not until the late nineteenth century that the split between poet and politician was institutionalized in the "New Aesthetics" of "art for art's sake" and the Modernist aesthetics of *la poésie pure.* At midcentury Whitman was still writing in a revolutionary and postrevolutionary tradition in which poetic creation was integral to rather than at odds with the nation's political creation.

Living at street level in Brooklyn and Manhattan, Washington, D.C., and later in Camden, New Jersey, Whitman did not distinguish between private interest and public interest. At the time Whitman was writing, the American republic was less than a century old. The state of the nation was everybody's business; in particular, it was the business of the writers who during the revolutionary and postrevolutionary period had played a strategic role in naming and defining what the American republic would or would not be. Self, family, society, world: poet, preacher, orator, politician. These were part of a common culture, a sequence of interlocking interests that began to unravel under the pressure of market capitalism and the politics of democracy.

In the year of Whitman's birth, 1819, the United States was entering the first of the periodic depressions that would characterize the modern commercial-industrial world. In the same year, James Tallmadge, Jr., a Republican representative from Whitman's home state of New York, introduced an amendment to prohibit the extension of slavery into Missouri and to provide for the emancipation of those slaves who were already there. Although the immediate issue of slavery in Missouri was resolved by the Missouri Compromise of 1820, the debate over the Tallmadge amendment provoked a conflict between North and South so bitter and far-reaching that even Thomas Jefferson was led to predict disaster for the Union: "All, I fear, do not see the speck on our horizon which is to burst on us as a tornado, sooner or later. The line of division lately marked out between different portions of our confederacy is such as will never, I fear, be obliterated."[20] The conflict between North and South was intensified by the conflict between rich and poor, capital and labor, brought by the Panic of 1819. Emerging in the year of Whitman's birth, these dark spots on the horizon of America were destined to become sources of major struggle in the development of both poet and nation.

In 1823 Whitman's father, Walter Whitman, Sr., moved his family from West Hills, Long Island, to Brooklyn, where he sought work as a carpenter (Figure II). Driven to the city by economic hardship, Whitman's family became part of a more general uprooting of the populace from country to city in the nineteenth century as America was transformed from an agricultural into a manufacturing society. For Whitman it was the first of a number of moves from country to city and back again that he and his family would be forced to make as a result of periodic economic downturns. Speaking of the houses his father built in Brooklyn between 1824 and 1826, Whitman said: "We occupied them, one after the other, but they were mortgaged, and we lost them" (*PW*, I, 13). Experiencing a general sense of dispossession and impotence in regard to modern economic conditions, Walter Whitman, Sr., was one of the common men, the masses of day laborers, to whom Andrew Jackson appealed when he won a majority of the popular vote in his bid for the presidency in 1824 and when he was elected president in 1828.

The politics of Walter Whitman, Sr., reflected the movement among American workers to reshape republican ideology to protest the inequality

FIGURE II. Whitman's West Hills, Long Island home. Courtesy of The Library of Congress.

and dependence fostered by the growth of market capitalism. The Jeffersonian ideology of popular sovereignty and inalienable right, as it was deployed by Jackson and his supporters, gave to Whitman, Sr., a means of ordering his experience and redeeming his sense of lost power and status in the eroding artisanal culture of early-nineteenth-century America. In his study of the rise of the working class in New York between 1788 and 1850, Sean Wilentz states: "Above all, the notion of independence, central to both republican politics and the order of the artisan system, propelled their critique of proletarianization."[21]

At a time when the independent artisanal economy of the past was being transformed into a society of wage laborers subordinated to capital, Whitman, Sr., embraced the revolutionary ideals of self-sovereignty and individual right, not as a matter of party politics only, but as articles of religious faith. It was from his father that Walt Whitman received his earliest training in the Enlightenment and revolutionary ideology that became the template of his life and work. "They all espoused with ardor the side of the 'rebellion' in 76," Whitman said of his father's side of the family. "I remember when a boy hearing grandmother Whitman tell about the times of the revolutionary war." In his daybooks Whitman recorded instances of patriotism and personal bravery on both the maternal and paternal sides of his family during the British occupation of Long Island.[22]

✓Raised among brothers whose very names—Andrew Jackson (b. 1827), George Washington (b. 1829), and Thomas Jefferson (b. 1833)— reflected their father's democratic ideals, Whitman soon realized that these ideals were both a political and a family matter. He later called attention to his "political" family in a story entitled "My Boys and Girls":

> What would you say, dear reader, were I to claim the nearest relationship to George Washington, Thomas Jefferson and Andrew Jackson? Yet such is the case, as I aver upon my word. Several times has the immortal Washington sat on my shoulders, his legs dangling down upon my breast, while I trotted for sport down a lane or over the fields. Around the waist of the sagacious Jefferson have I circled one arm, while the fingers of the other have pointed him out words to spell. And though Jackson is (strange paradox!) considerably older than the other two, many a race and tumble have I had with him— and at this moment I question whether, in a wrestle, he would not get the better of me, and put me flat.[23]

Racing, tumbling, and wrestling with his brothers as if they were the political heroes of the republic, Whitman began to develop a sense of self that was bound up with the political identity of America.

The political ideals of Whitman's family were complemented by his first experiences as a journalist's apprentice between 1831 and 1832 in the offices of the Long Island *Star* and Long Island *Patriot.* At the *Patriot* Whitman learned the printer's trade, published a few "sentimental bits," and listened "with a boy's ardent soul and eager ears" to the tales of the printer William Hartshorne, an old revolutionary character, who described in detail the "personal appearance and demeanor of Washington, Jefferson, and other of the great historical names of our early national days."[24] Supporting the interests of the Democratic workers against the Whig manufacturers, the *Patriot,* like other newspapers of the time, was openly partisan in its policy and appeal. In addition to learning the printer's trade and writing a few pieces, Whitman received at the *Patriot* his first apprenticeship in the political power of the printed word. Although he attended public school between 1825 and 1830, journalism and politics became—like whaling for Ishmael—his Harvard and Yale.

Whitman was also raised on the radical political and religious philosophies of Tom Paine, Frances Wright, and Constantin Volney, who were part of what Henry May has called the "revolutionary enlightenment."[25] Whitman's father was a friend of Tom Paine, and he may have been among the freethinkers who began on January 29, 1825, to gather annually to celebrate Paine's birthday. Copies of the major freethinking texts—Volney's *The Ruins* (1791), Paine's *The Age of Reason* (1794), and Wright's *A Few Days in Athens* (1822)—were cherished books in the Whitman household. Whitman also absorbed the radical social philosophy of the *Free Enquirer*, which he told Horace Traubel, "my daddy took and I often read."[26] Edited by Frances Wright and Robert Dale Owen, the *Free Enquirer* sought through the rhetoric of a "war of class" to unite the grievances of New York City workers in an anticapitalist and anticlerical platform.[27]

"I swore when I was a young man," he told Traubel,

> that I would sometime—I could not say when but as the opportunity appeared—do public justice to three people—three of the superber characters of my day or America's early days who were either much maligned or much misunderstood. One of them was Thomas Paine: Paine, the chiefest of these: the other two were Elias Hicks and Fanny Wright. I determined that I would bear witness to them—true witness where the great majority have borne false witness—in thick and thin, come what might to me. (*WWC*, II, 205–6)

Whitman carried out his early resolve when in 1877, at a gathering to celebrate Paine's one hundred fortieth birthday, he delivered a speech entitled "In Memory of Thomas Paine" at Lincoln Hall in Philadelphia. His main source of personal information about Paine was Colonel John Fellows, whom he had met at Tammany Hall in the early 1840s. "I liked to draw him out reminiscently," Whitman recalled. "He was an intimate associate of Thomas Paine. . . . From him I learned the truth about Paine—how literally nothing true was at the bottom of all the vile slanders."[28] Like the artisan freethinkers who idolized Paine in the 1820s, Whitman believed that these slanders had been perpetuated by a conservative theological establishment fearful of Paine's deist appeal to reason and natural law. "Woe be to the man," Whitman said, "who invokes the antagonism of priests and property!" (*WWC*, I, 278).

Setting himself against a conservative tradition that had attempted to discount Paine's more radical influence on the founding ideology, Whitman stressed the powerful role that *Common Sense* (1776), *The Rights of*

*Man* (1791), and *The Age of Reason* (1794) had played in the birth and growth of the American nation:

> That he labor'd well and wisely for the States in the trying period of their parturition, and in the seeds of their character, there seems to me no question. I dare not say how much of what our Union is owning and enjoying to-day—its independence—its ardent belief in, and substantial practice of, radical human rights—and the severance of its government from all ecclesiastical and superstitious dominion—I dare not say how much of all this is owing to Thomas Paine, but I am inclined to think a good portion of it decidedly is. (*PW*, I, 141)

In 1829 Whitman was taken by his parents to hear the radical Quaker preacher Elias Hicks. As a boy of ten, Whitman could not follow Hicks's lecture, but later in *November Boughs* (1888), he remembered the affective and transforming power of his oratory: "A pleading, tender, nearly agonizing conviction, and magnetic stream of natural eloquence, before which all minds and natures, all emotions, high or low, gentle or simple, yielded entirely without exception, was its cause, method, and effect" (*PW*, II, 637–38). He also remembered his doctrine of the "inner light": "More definitely, as near as I remember (aided by my dear mother long afterward,) Elias Hicks's discourse there in the Brooklyn ball-room, was one of his old never-remitted appeals to that moral mystical portion of human nature, *the inner light*" (*PW*, II, 644).

The "inner light" of religious spiritualism and the "outer light" of the revolutionary enlightenment—the doctrines of the soul and the doctrines of the republic—became the early and potentially self-contradictory poles of Whitman's thought. Long before he had any contact with the self-reliant and transcendental philosophy of Ralph Waldo Emerson, Whitman was trained in the Hicksite doctrine that the "ideals of character, of justice, of religious action . . . are to be conform'd to no outside doctrine of creeds, Bibles, legislative enactments, conventionalities, or even decorums, but are to follow the inward Deity-planted law of the emotional soul" (*PW*, II, 639). Whitman stressed the political implications of Hicks's "mystical and radical" doctrines: "I wanted to write of Hicks as a democrat," he said of his essay, "the only real democrat among all the religious teachers: the democrat in religion as Jefferson was the democrat in politics" (*WWC*, II, 36). In the post–Civil War period, as democratic word and world grew farther apart, Hicks's "mystical and radical" democracy of the soul provided Whitman with an easy and blessed escape from the tawdry and self-serving democracy of the American marketplace. But in Whitman's early years, at least, Hicks's concept of a divine law written in the heart of every rational creature and his practical emphasis on the Christ-like life rather than on the worship of Christ complemented and justified the rational and democratic philosophies of Paine and Jefferson, Wright and Volney.

In the mid-1830s, Whitman was in the habit of frequenting the anti-slavery halls in New York. "It was there," he said, "I heard Fanny Wright. . . . She spoke in the old Tammany Hall there, every Sunday, about all sorts of reforms. Her views were very broad—she touched the widest range of themes—spoke informal, colloquially. . . . She has always been to me one of the sweetest of sweet memories: we all loved her: fell down before her: her very appearance seemed to enthrall us" (*WWC*, II, 204–5). Although Whitman never fulfilled his youthful resolve to bear "true witness" to the woman whom the *New York Courier and Enquirer* had called the "great Red Harlot of Infidelity," he always spoke of Wright as a model of female heroism.

He was particularly fond of *A Few Days in Athens* (1822), Wright's defense of Epicureanism and the pleasure principle, which was based on the materialist philosophy of Jeremy Bentham. "Her book on Epicurus was daily food to me," Whitman said. "I kept it about me for years" (*WWC*, II, 445). His own political thinking, particularly his ideas on the regenerative power of education and his negative attitude toward mon-eyed wealth and the clergy, bears traces of Wrights' radical philosophy.[29] Like Paine and Volney, Wright looked upon organized religion as the root cause of personal unhappiness. "Imagine the creature man," says Epicurus, "not bending the knee of adulation to visionary beings armed by fear for his destruction, but standing erect in calm contemplation of the beautiful face of nature. . . . Thus considered, he is transformed into a god of his present idolatry."[30]

Drawing on the "philosophy of the Garden" expounded by Epicurus and his disciples, Whitman included scenes from *A Few Days in Athens* in an early sketch for *Leaves of Grass* entitled "Pictures":

And here Athens itself,—it is a clear forenoon,
Young men, pupils, collect in the gardens of a favorite master, waiting
    for him.
.  .  .  .  .  .  .  .  .
Here and there, couples or trios, young and old, clear-faced, and of
    perfect physique, walk with twined arms in divine friendship, happy,
Till, beyond, the master appears advancing—his form shows above the
    crowd, a head taller than they,
His gait is erect, calm and dignified—his features are colossal—he is
    old, yet his forehead has no wrinkles,

Wisdom undisturbed, self-respect, fortitude unshaken, are in his
   expression, his personality;
Wait till he speaks—what God's voice is that, sounding from his mouth?
He places virtue and self-denial above all the rest,
He shows to what a glorious height the man may ascend,
He shows how independent one may be of fortune—how triumphant over
   fate.

                                          (*LGC*, pp. 643–44)

At a time when the emergence of sweatshops and outwork cellars was
commodifying individuals and the relations among them, Whitman may
have been drawn to Epicurus as a model republican, showing "how independent one may be of fortune—how triumphant over fate." As one of his
earliest poetic sketches for *Leaves of Grass,* the passage is striking in
suggesting Whitman's debt to the writing of Frances Wright and the
philosophy of Epicurus, as well as to the ideals of virtue, independence,
and cooperation fostered by artisan republicanism. Although virtue and
self-denial are not values one normally associates with Whitman's verse,
the passage provides early evidence of the ways that Whitman's poetic
celebration of a balanced and self-regulating body intersects with and
affirms a utopian republican discourse of self-mastery and "divine friendship," independence and social community, bodily control and political
happiness.

    Speaking of *A Few Days in Athens,* Whitman said: "I always associated
that book with Volney's *Ruins,* which is another of the books on which I
may be said to be raised" (*WWC,* II, 445). Written amid the fervor of the
French Revolution, Volney's *Ruins, or Meditations on the Revolutions of
Empires* (1802) argues against aristocratic and monarchical privilege in
favor of personal sovereignty and representative democracy. Citing religion as the primary source of human bondage, in the second part of his
political treatise Volney undertakes a materialist critique of revealed religion, proposing to replace it with a "religion of evidence and truth" based
on natural law.

    In his early notebooks, Whitman made several notes from *The Ruins,*
which he later incorporated into the poetry of *Leaves of Grass.* More
important than his specific borrowings from *The Ruins,* however, is the
enlightened vision of human history that Whitman found in Volney: "I will
ask of the ashes of legislators," says Volney, "by what secret causes do
empires rise and fall; from what sources spring the prosperity and misfor-
tunes of nations; and on what principles can the peace of society, and the
happiness of man be established."[31] Locating the source of human misery
neither in the state nor in divine providence but in the ignorance and

cupidity of the individual, Volney proposes a prudent and enlightened love of self as the source of personal and political transformation. In Volney's concept of natural law and self-love as the source of happiness among individuals and nations and in his hopeful vision of human progress toward an international order of liberty and equality, peace and harmony, Whitman found the enlightened perspective on humanity and the millennial view of history that would frame his own political philosophy and literary work.

Translated by Thomas Jefferson and Joel Barlow in 1802, Volney's *Ruins*, like the political writings of Paine and Wright, complemented the Jeffersonian principles on which Whitman was raised. He owned a nine-volume set of *The Writings of Thomas Jefferson*, which was published in 1853–54, but his primary Jeffersonian document was the Declaration of Independence, which became the sacred text of his political philosophy and the mythical text of his poems. Jefferson's political ideals, as they were reinterpreted and deployed in the age of Jackson, were reinforced for Whitman during the 1830s and 1840s by the writings of William Leggett, whose editorials for the *New York Evening Post* and the *Plaindealer* educated readers in the Jeffersonian premises of Jacksonian democracy.[32]

As America entered another economic depression in 1837 and Whitman himself became one of those who, like his protagonist in the story "Tomb Blossoms," was forced to move to the country in search of a livelihood, to escape being "racked" in the city "with notes due, or the fluctuations of prices, or the breaking of banks," he came to see a return to Jefferson's principles as a cure for his own as well as the nation's social ills (*EPF*, p. 88).

At the end of the 1830s, Whitman entered local politics, actively committing himself to the Democratic party. In an editorial written for the Long Island *Democrat* in September 1839, he invoked the revolutionary principles of equality and inalienable rights in support of the Democratic candidates in the upcoming election:

> Democracy has its foundations in the very broadest notion of good to our fellow creatures and to our countrymen. It is based on the doctrine of equality in political rights and privileges; it overlooks the distinctions of rank and wealth; it comprehends in its protection all classes and conditions of society, nor allows that the refined and rich shall receive more consideration in its decrees than the poor and lowly born.[33]

Whitman regarded the Democratic party as the safeguard of the ideals of the American Revolution: "Those once derided, but now widely worshipped doctrines which the great Jefferson and the glorious Leggett promulgated . . . form the best elements of the Democratic creed, and of the Democratic Party."[34] When in the late 1840s, Jefferson's doctrines ceased, in Whitman's view, to be synonymous with those of the Democratic party, he broke with the party and began his own campaign to recover the creed of Jefferson and the sacred fire of the revolutionary founders as a living heritage in America.

By the time Whitman came of age in 1840, he had acquired a reputation as a political figure in Queens County, where one of the local Whig papers described him as a "well-known locofoco of the town" and "champion of the Democracy."[35] Identified with the radical wing of the Democratic party, he wrote editorials supporting the Democratic platform and actively campaigned for the reelection of Martin Van Buren to the presidency. Although the Democrats lost nationally, they did win locally, and Whitman took partial credit for their victory. Whitman commented a few years later on the electoral state of mind in Queens: "It's a pity the people of Queens have not some one to stir them up in the columns of the *Democrat* as in the fall of 1840, by a certain young fellow from the eastern part of the island, who . . . contributed materially to the unexpected triumph of the party at that period."[36] The political effectiveness of his columns in the Long Island *Democrat* was the first sign to Whitman of his own power to reach the people through the written word and thereby bring about political change.

Whitman's active engagement with the party of Jefferson gave him the sense of political power and the coherent social mythology he needed in order to move away from the morbid and derivative lyrics that he began writing during the 1830s. Written during his own and the nation's depression, these early lyrics are characterized by a sense of uncertainty and powerlessness and a longing for death as a release from earthly care. His earliest known published poem, "Our Future Lot," is typical. In this poem, which appeared in the Long Island *Democrat* on October 31, 1838, Whitman expresses despair at not being able to pierce the veil of the future:

> O, powerless is this struggling brain
>   To pierce the mighty mystery;
> In dark, uncertain awe it waits
>   The common doom—to die!
>             (*EPF*, p. 28)

Subsequent stanzas of the poem, which Whitman later deleted, counter this gloomy vision with the hope of regeneration through death. In the spring and summer of 1840, the *Democrat* printed several of these early lyrics—including "The Inca's Daughter," "The Love That Is Hereafter," "The End of All," "We Shall All Rest at Last," and "My Departure"—all of which express a similarly mixed mood of "flashing hope and gloomy fear."

Whitman's move away from the darkly brooding *other* of his early poems to the confident and self-generating democrat of his journalism and prose illustrates on a broader level the ways that ideology works in society. His commitment to democratic ideology gave him the language, forms, rituals, and symbols through which to speak and act as an autonomous being with a significant relation to a rapidly transforming and potentially disruptive world.[37] At a time when traditional orders were eroding, democratic ideology gave Whitman a reason for being, a language of possibility, and a country to dream in. To maintain his political commitment—and his sense of self-significance—Whitman had to leave behind, or repress, the decentered, nay-saying alien of his early verse, and he also had to silence the fact of radical contradiction in the founding fathers, in the American marketplace, and at the very heart of democracy itself.

Coming of age during the age of Jackson, Whitman embraced the values of liberty, equality, mobility, and free enterprise promulgated by the Democratic party. Invested with a sense of personal power in and through democratic ideology, Whitman never really admitted what later historians have demonstrated to be true: that for all the fulsome political rhetoric about democracy and the common man, the age of Jackson was a period of social disruption and class conflict that widened rather than narrowed the gap between rich and poor, capital and labor.[38] What Whitman called the "widely worshipped doctrines" of Jeffersonian liberalism were in fact part of the problem of democracy in America.

At the same time that Whitman's lyrics of self-doubt were appearing in the *Democrat*, his journalism began to show the regenerative effects of his political involvement. In February 1840, the *Democrat* began publishing a series entitled *Sun-Down Papers from the Desk of a Schoolmaster*. Written in a chatty and informal style, the papers show Whitman's effort to establish an intimate rapport with the workers he envisioned as his readers. Anticipating his later assaults on gentility in *Leaves of Grass* and *Democratic Vistas* (1871), Whitman in *Sun-Down Paper* no. 2 contrasts the bluff and hearty manliness of a democratic character named Hom with the aristocratic manners and artifice of Tom Beprim. In no. 3, he urges working-class families to maintain their self-respect and independence by resisting the urge to be genteel and fashionable: "A working man's family . . . in its emulation after the style of those above them in riches and conventional breeding, runs a dangerous race over what is not only not worth having, but what is pernicious and vile."[39]

In *Sun-Down Paper* no. 7, Whitman makes his first known declaration of literary intention, announcing his ambition to write a book along the political lines of Volney's *Ruins:* "I would compose a wonderful and ponderous book. Therein should be treated on, the nature and peculiarities of men, the diversity of their characters, the means of improving their state, and the proper mode of governing nations." Like Volney, Whitman wants to make an "enlightened" critique of money and property the central focus of his book: "One principal claim to a place among men of profound sagacity, by means of the work I allude to, would be on account of a wondrous and important discovery, a treatise upon which would fill up the principal part of my compilation. I have found out that it is a very dangerous thing to be rich" (*UPP*, I, 37–38).

Written while Whitman was himself teaching in the Long Island schools, the *Sun-Down Papers from the Desk of a Schoolmaster* represent his first appearance in the guise of an enlightened instructor of the people. "Enlighten the people generally," Jefferson had said, "and tyranny and oppressions of body and mind will vanish like evil spirits at the dawn of day."[40] By 1840 Whitman had become convinced not only of the need to enlighten the people in the ideals of the revolutionary founders but also of the personal role he might play in this enlightenment.

"I may perhaps be the only one living today," he told Horace Traubel, "who can throw an authentic sidelight upon the radicalism of the post-Revolutionary decades." Having absorbed, through both his firsthand experience and his reading, the political radicalism of the postrevolutionary period, Whitman came to see himself as a repository of political traditions that must be passed on to future generations. In fact, his evolution from teacher to politician to journalist to fiction writer to would-be orator and American bard might be seen as stages in his attempt to find the most effective means of spreading to the people the original ideals of the American republic. "Whitman was undoubtedly convinced that he had a mission," Horace Traubel observed. "This conviction never assumed fanatic forms. Whitman was the most catholic man who ever thought he had a mission. But he did regard himself as such a depository. . . . He often asked himself: How am I to deliver my goods?"[41]

A few days before the election of 1840, Whitman delivered his goods in the form of a poem entitled "The Columbian's Song." Drawing on the rising glory rhetoric of such revolutionary writers as Joel Barlow and Philip Freneau, Whitman expresses confidence in America's future:

> O, my soul is drunk with joy,
>     And my inmost heart is glad,
> To think my country's star will not
>     Through endless ages fade,
> That on its upward glorious course
>     Our red eyed eagle leaps,
> While with the ever moving winds,
>     Our dawn-striped banner sweeps:
> That here at length is found
>     A wide extending shore
> Where Freedom's starry gleam,
>     Shines with unvarying beam.
>                 (*EPF*, pp. 12–13)

Here for the first time Whitman invokes the theme of political union that became the overarching figure of his life and work. Despite the divisions of party and faction, he expresses confidence in the strength of a "heart-prized union band" continually renewed by the memory and hope of the revolutionary fathers:

> Though parties sometimes rage,
>     And Faction rears its form.
> Its jealous eye, its scheming brain,
>     To revel in the storm:
> Yet should a danger threaten,
>     Or enemy draw nigh,
> Then scattered to the winds of heaven,
>     All civil strife would fly;
> And north and south, and east and west,
>     Would rally at the cry—
> 'Brethren arise! to battle come,
> For Truth, for Freedom, and for Home,
>     And for our Fathers' Memory!'
>                 (*EPF*, p. 13)

Conventional in language, image, and sentiment, "The Columbian's Song" is Whitman's first sustained effort to gather into a coherent poetic form the revolutionary ideology on which he was raised. As the first poem to express his new political commitment and one of the last poems he wrote in the 1840s, "The Columbian's Song" signals a turn away from the self-doubting figure of his earlier poems and a move toward the impassioned political lyrics of 1850.[42] In the guise of the Columbian, remembering the heritage of the revolutionary past as he envisions the sweep of the

union toward the democratic future, is the embryo of the bard who set out
in 1855 inscribed with the contradictions of his culture. He gives the "sign
of democracy," but beneath his grammar of reconciliation, he bears the
knowledge of "Faction," "civil strife," personal and economic depression,
and a land "Where the stern resolve for liberty/Was writ in gushing
blood" (*EPF*, pp. 12–13).

# 2

# The Paradox of the American Republic

> If we were asked the particular trait of national character from which might be apprehended the greatest evil to the land, we should unhesitatingly point to the strife for gain which of late years has marked, and now marks, the American people. This unholy spirit seems to have no bound or check. . . . Even the battle spots where our old soldiers fought and died, are not beyond the reach of this pollution. The very hill made sacred by the blood of freedom's earliest martyrs, is sold and trafficked for.
>
> —WHITMAN, *New York Aurora*, 1842

"We are battling for great principles—for mighty and glorious truths," Whitman told a Democratic rally of about fifteen thousand people in New York City in 1841. "I would scorn to exert even my humble efforts for the best democratic candidate that ever was nominated, in himself alone. It is our creed—our doctrine, not a man or set of men that we seek to build up. Let us attend then, in the meantime, to measures, policy and doctrine, and leave to future consideration the selection of the agent to carry our plans into effect." As the guiding spirit of the Democratic party, Whitman invoked the female genius of the republic that hovered over the revolutionary era of Thomas Jefferson: "The guardian spirit, the good genius who has attended us ever since the days of Jefferson, has not now forsaken us. I can almost fancy myself able to pierce the darkness of the future and behold her looking down upon us with the same benignant smiles she wore in 1828, '32, and '36" (*UPP*, I, 51). In the dark days of the 1840s, Whitman continued to "behold" this revolutionary genius as a source of personal and political renewal. But as the issues of slavery, territorial expansion, the Mexican war, sectionalism, free trade, states' rights, worker strife, and capitalist domination threatened to dissolve not only the Union but also the foundations of democracy itself, the "good genius" of the American republic came to seem more and more violent, bloody, and mired in contradiction.

In a comment on the significance of the American Revolution, Jefferson wrote: "May it be to the world, what I believe it will be, (to some parts sooner, to others later, but finally to all,) the signal of arousing men to burst the chains under which monkish ignorance and superstition had persuaded them to bind themselves, and to assume the blessings and security of self-government."[1] For Whitman as for Jefferson, the revolt against King George represented not only a revolt against the tyranny of a monarch; the revolt represented a relocation of sovereignty in the individual rather than the state. It was this translation of power from the political authority of the state to the inalienable rights of the individual that burst the chains of the past and made the American Revolution both a change of regime and the basis for a *Novus Ordo Seclorum.*

"It is the manners and spirit of the people which preserve a republic in vigour," Jefferson wrote in *Notes on the State of Virginia:* "A degeneracy in these is a canker which soon eats to the heart of its laws and constitution."[2] The republican basis of the new American order—its dependence on the will of the people—was a source of possibility and weakness. That is, republicanism required the moral virtue of its citizens, an ability to act in the public interest, and thus a revolution in political institutions and in human nature itself. The republican vision of the founders was, writes Gordon S. Wood, "so divorced from the realities of American society" as "to make the Revolution one of the great utopian movements in American history."[3]

In Jefferson's view, the nation's republican vigor would be ensured by an agricultural economy of virtuous and independent husbandmen who owned and cultivated the land. As he observed in his now-classic formulation of the agrarian myth:

> Those who labour in the earth are the chosen people of God, if ever he had a chosen people, whose breasts he has made his peculiar deposit for substantial and genuine virtue. . . . Corruption of morals in the mass of cultivators is a phaenomenon of which no age nor nation has furnished an example. . . . While we have land to labour then, let us never wish to see our citizens occupied at a work-bench, or twirling a distaff. Carpenters, masons, smiths, are wanting in husbandry: but, for the general operations of manufacture, let our work-shops remain in Europe. (*Notes,* pp. 164–65)

Aware of the dependence of a political republic on a virtuous citizenry, Jefferson proposed a program of mass education "to avail the state of those talents which nature has sown as liberally among the poor as the rich, but which perish without use, if not sought for and cultivated." It is this program of mass education that is the safeguard of republican government: "Every government degenerates when trusted to the rulers of the people alone. The people themselves therefore are its only safe depositories. And to render even them safe their minds must be improved to a certain degree" (*Notes*, p. 148).

Whitman shared Jefferson's faith in education as a means of safeguarding republican institutions. Although he did not write his "wonderful" book during the 1840s, he did, in his capacity as journalist, politician, and maker of tales, maintain the guise of a schoolmaster seeking to cultivate the republican vigor of the American people. The ideals he taught—independence, freedom, equality, local sovereignty, and minimal government—were an urban version of Jeffersonian republicanism as it was promulgated by the Democratic party in the 1830s and 1840s. But in Whitman's vision of a harmonious society of artisans, farmers, and laborers owning homesteads in fee simple, his association of virtue with the laboring classes, and his emphasis on the interactive values of independence and cooperation, freedom and community, Whitman's ideal republic also reflected the artisan republicanism of the city workers among whom he was raised.[4]

When in 1842 Whitman became the editor of the *Aurora*, one of New York's penny papers, he celebrated the capacity of the penny press to reach and educate the masses of mechanics, workers, and poorer classes who were its readers: "Among newspapers, the penny press is the same as common schools among seminaries of education. They carry light and knowledge in among those who most need it. They disperse the clouds of ignorance; and make the great body of people intelligent, capable, and worthy of performing the duties of republican freemen."[5] Founded in 1841, the *Aurora* described itself as politically independent but "democratic, in the strongest sense of the word. Whatever is contrary to Democracy, among Democratic Whigs or Democratic Republicans, will find little favor" (*Aurora*, p. 1).

The very name *Aurora* suggests an enlightened, new-world focus, and Whitman, as its editor, began to assume the dimensions of a composite, democratic persona. Dreaming of making the *Aurora* the most readable journal in the republic, he exclaimed: "Our *countrymen!* the phrase rolls pleasantly from our tongue. We glory in being *true Americans*. And we profess to impress Aurora with the same spirit. We have taken high American ground . . . based upon a desire to possess the republic of a proper respect for itself and its citizens, and of what is due to its own capacities, and its own dignity" (*Aurora*, p. 117). Despite the paper's professed neutrality, Whitman longed to make the *Aurora* the organ of the Democratic party: "The Tammany party want, here in New York, a newspaper bold, manly, able, and *American* in its tenor; a newspaper vigorous and original and fresh" (*Aurora*, p. 62). These same Tammany party terms—bold, manly, vigorous, original, fresh, American—would be translated into the grammar of national myth in the 1855 edition of *Leaves of Grass*.

Like the poet of democracy who emerged in 1855, Whitman as the editor of the *Aurora* had designs on his readers; he wanted to imbue the American people "with a feeling of respect for, and confidence in, *themselves*" and to educate them in the principles of Jeffersonian democracy. Protesting the growing concentration of political and economic power in the hands of the government, he reminded his readers that "the true democratic principle, the genuine principle of the American system—teaches that the 'best' governing power is that which puts its power in play 'least.' " Declaring himself "among the foremost of those who desire our experiment of man's capacity for self government, carried to its extreme verge," Whitman, in the language of a good Jacksonian, attributed the country's economic ills to legislative meddling: "Every time that congress or a state legislature meddles in matters of finance, they only plunge the interests of the people deeper and deeper in difficulty" (*Aurora*, pp. 82, 90).

While he was editor of the *Aurora*, Whitman's primary battle was waged against Bishop John Hughes, who sought state support for Catholic schools. The plan elicited Whitman's fiercest anticlerical rhetoric. Describing Hughes's supporters as "a gang of foreign outcasts and bullies, prompted by this fanatical wretch and his slaves," he wrote a series of articles that attacked foreign influence on American institutions and culture, boldly asserted the principle of separation of church and state, and scolded Tammany for being tempted to support the Catholic proposal: "The whole city—the whole state," Whitman declared, "ought to rise up as one man, and let these jesuitical knaves, and their apt satellites, know what it is to feel the blast from an injured and outraged country" (*Aurora*, p. 58). Although Whitman was accused of supporting the native American group, which advocated disenfranchisement of the foreign-born and exclusion of Catholics from office, he vigorously denied the charge in words that beckon toward the world-embracing persona of "Salut Au Monde!": "We have no antipathy or bigotted ill will to *foreigners*. God forbid! Our love is capacious enough, and our arms wide enough, to encircle all men, whether they have birth in our glorious republic, the monarchies of Europe, or the hot deserts of Africa—whatever be their origin or their native land" (*Aurora*, p. 63).

Whitman's position on the separation of church and state had its origins in the antiauthoritarian and anticlerical sentiments of the revolutionary enlightenment. Critical of party members who were willing to introduce sectarianism into politics in order to win the support of Catholics in local elections, he beseeched the Democrats to take a stand worthy of the principles of the revolutionary founders: "As they love the memory of Washington—as they adhere to the teachings of Jefferson—as they prize the safety, present and future, of our beloved republic—we implore them to speak out against the machinations of these reverend demagogues" (*Aurora*, p. 65). Whitman's words register some of the New York City workers' growing discontent with political parties and government corruption. The following week, when Democratic members of the New York senate passed a bill to allocate funds to Catholic schools, Whitman announced that "the democratic party wants regenerating" if it is to keep "any of its former purity, strength and power" (*Aurora*, p. 70).

Whitman's commitment to regenerating not only the Democratic party but also the American people in the purity, strength, and power of republican principle became the base of his political journalism and fiction during the 1840s, just as it would later become the base of his poetics. "Can I hope," he said in the introduction to his temperance novel *Franklin Evans, or the Inebriate* (1842), "that my story will do good? I entertain that hope. Issued in the cheap and popular form you see, and wafted by every mail to all parts of this vast republic; the facilities which its publisher possesses, giving him the power of diffusing it more widely than any other establishment in the United States" (*EPF*, pp. 126–27). Whitman's comment attests to the increasing power of the press and popular fiction as agencies of ideological diffusion. As in his political journalism, Whitman conceived of himself as a writer for the mass, seeking through the power of the print medium to disseminate seeds of republican virtue throughout the land.

In his first published tale, "Death in the School-Room (a Fact)," which appeared in *The United States Magazine and Democratic Review* in August 1841, Whitman protested the use of corporal punishment in the schools. In other tales of the 1840s, he assumed a proletarian stance worthy of Frances Wright, as he criticized capital accumulation, corporate power, the oppression of workers and women, the corruption of businessmen and lawyers, religious institutions, capital punishment, mental asylums, child labor, and child abuse. He attempted to resolve these modern social and economic problems on a personal rather than a corporate level, by invoking the self-regenerating power of the individual and by teaching the values of self-restraint, compassion, and social love.

"The Child and the Profligate," which appeared in the *New World* in
1841 while Whitman worked there as a compositor, sets the pattern for
these early tales. Originally published under the title "The Child's Champion," the tale registers, through a mother's concern for her laboring
child, a protest against the enslavement and impoverishment of the
worker: "The thought of a beloved son condemned to labor—labor that
would break down a man—struggling from day to day under the hard rule
of a soulless gold-worshipper; the knowledge that years must pass thus;
the sickening idea of her own poverty, and of living mainly on the grudged
charity of neighbors—thoughts, too, of former happy days—these racked
the widow's heart, and made her bed a sleepless one without repose." In a
direct address to his readers, Whitman calls attention to the worker's
sense of powerlessness, "the pangs of hunger—the faintness of the soul at
seeing those we love trampled down, without our having the power to aid
them—the wasting away of the body in sickness incurable—and those dull
achings of the heart when the consciousness comes upon the poor man's
mind, that while he lives he will in all probability live in want and wretchedness" (*EPF,* pp. 70–71). In its language of "want and wretchedness,"
hunger and bodily disease, the tale registers the loss not only of a provident father but of a provident system "of former happy days" that pro-
tected the individual from the "hard rule" of the capital marketplace.

Rather than propose a change of system, however, Whitman solves the
problem in the terms of the system. To the worker's sense of despair and
powerlessness, he opposes the fate of the Profligate, whose personal regeneration through love and temperance represents the possibility of social
transformation in the tale. Through the "wish to love and be loved, which
the forms of custom, and the engrossing anxiety for gain, so generally
smother," the Profligate is transformed by the child into a provider; countering the ethos of self-interest with an ethos of social love, the Profligate
liberates the child from his abusive master, provides for the widowed
mother of the child, and becomes himself the father of a family.

During the early 1840s the *Democratic Review,* which published several of Whitman's tales as well as stories by Nathaniel Hawthorne and
Edgar Allan Poe, became associated with "Young America," a group
that called for the creation of a distinctively American literature.[6] Although Whitman's early tales have usually been treated as anomalous to
the aims of the Young America group, they are in fact much more
nationalistic in focus than is commonly assumed. In "The Last of the
Sacred Army," which appeared in the *Democratic Review* in March 1842,
Whitman's aim is openly political: "The memory of the WARRIORS of
our FREEDOM!—let us guard it with a holy care," he says at the outset

of the tale as he tries to inspire his readers with a religious reverence for the heroes who embody the revolutionary ideals of the American republic. He presents George Washington and General Lafayette as "Soldiers of Liberty" who are part of a universal contest against "kingcraft and priestcraft" which "stalk abroad over fair portions of the globe, and forge the chain, and rivet the yoke" (*EPF*, pp. 95–96).

Reflecting on the appropriateness of such idolatry to a democratic people, Whitman responds in the guise of a philosopher, defending the worship of the revolutionary founders as a kind of civil religion: "No: it is well that the benefactors of a state be so kept alive in memory and in song, when their bodies are mouldering. Then will it be impossible for a people to become enslaved; for though the strong arm of their old defender come not as formerly to the battle, his spirit is there, through the power of remembrance, and wields a better sway even than if it were of fleshly substance" (*EPF*, p. 99). Whitman's words sum up the revolutionary impulse that is at the center of his own life and work. As the last of the revolutionary figures died, he experienced sadness at the passing of a heroic epoch and anxiety lest the ideals of the founders be forgotten amid the scramble for material gain. In his journalism and fiction, his unsuccessful forays into politics and oratory, and later in his poetry and social criticism, he undertook to keep alive through the power of remembrance the spirit of the revolutionary founders.

Unlike "The Last of the Sacred Warriors," Whitman's early tales characteristically focused on the domestic rather than the political sphere. These domestic tales have usually been read as superficial exercises in the literature of popular sentiment, but here again Whitman's concern with issues of domestic governance and family union is related to the larger issues of national governance and political union. Whitman was in fact quite self-conscious about the potential role of popular literature as a vehicle of national ideology. In the introduction to *Franklin Evans*, he commented on how "the earlier teachers of piety used parables and fables, as the fit instrument whereby they might convey to men the beauty of the system they professed" (*EPF*, p. 127). In Whitman's tales of the 1840s, the domestic fable became one of these "fit instruments" for conveying the beauties of the democratic system of government and the horrors that result when the system is violated.

These seemingly innocuous tales illustrate the complex, interactive, and frequently concealed relation between literature and ideology. As Georg Lukacs has observed, it is often those features of a text that seem most devoid of historical interest that are most deeply indicative of its historical embeddedness.[7] Just as American revolutionary writers, in literature ranging from broadsides and ballads to Jefferson's Declaration of Independence, presented their revolt against King George in emotionally charged images of sons seeking to save mothers and daughters from the abusive violations of a father, so Whitman's domestic tales of cruel and intemperate patriarchs, oppressing mothers and children and fracturing families, became a means of exploring issues of gender, power, and authority in the political sphere.

Several tales enact in the political language of tyrant and victim a revolutionary pattern in which an oppressed child revolts against the domestic reign of an unjust patriarch: in "Bervance, or Father and Son," the father's unjust treatment of his son leads to the son's revolt and subsequent madness. The father realizes that it is all his fault as he remembers "that terrible cursing which, the last time tyrant and victim stood face to face together, rang from the lips of the Son, and fell like a knell of death on the ear of the Father"(*EPF*, p. 87). In "Wild Frank's Return," the father's preference for one son over the other results in discord between brother and brother and revolt against the tyranny of the father. The moral of the story is at once personal and political: "Oh, it had been a sad mistake of the farmer that he did not teach his children to love one another. It was a foolish thing that he prided himself on governing his little flock well, when sweet affection, gentle forbearance, and brotherly faith, were almost unknown among them" (*EPF*, p. 64).

The temperance theme, which is central to "One Wicked Impulse!", "The Child and the Profligate," "Reuben's Last Wish," and the novel *Franklin Evans*, has a similarly marked political nuance. "It is almost incredible," says Thomas L. Brasher, the editor of *Early Poems and Fiction*, "that the man who wrote *Leaves of Grass* also wrote *Franklin Evans*" (*EPF*, p. 125). Whitman himself later claimed that he wrote his temperance novel "with the help of a bottle of port or what not" (*WCC*, I, 93). But as Sean Wilentz has shown, the issue of temperance was basic to the cause of artisan republicanism in New York City, particularly among the Workingmen's movement and the trade unions of the 1830s.[8] Even though the temperance fable may be, as Whitman once said, "damned rot," the masses he sought to reach and the republican "system" he sought to teach through the temperance form were not. Critical skepticism and Whitman's own self-parody notwithstanding, the need for regeneration and temperance as a means of preserving the republican vigor of individual and nation was a theme of all his writing early and late, poetry and prose.

In *Franklin Evans* the protagonist himself suggests the connection between personal temperance and republican virtue when he describes the Temperance movement in language that parallels the political rhetoric of "The Last of the Sacred Army." Franklin Evans's vision of a future America regenerated through temperance is informed by the political terms of manifest destiny:

> I saw from the tops of the fortresses, the Star-Flag—emblem of Liberty—floating gloriously abroad in the breeze!
> And how countless were the inhabitants of that country! On I went, and still on, and they swarmed thicker than before. It was almost without boundary, it seemed to me—with its far-stretching territories, and its States away up in the regions of the frozen north, and reaching down to the hottest sands of the torrid south—and with the two distant oceans for its side limits. (*EPF*, p. 220)

When in this free and prosperous land, the last of the "serfs of Appetite" signs the temperance pledge, the event is described in language that suggests the final realization of the millennial dreams of the American Revolution: "Now man is free! He walks upon the earth, worthy the name of one whose prototype is God! We hear the mighty victory chorus sounding loud and long. Regenerated! Regenerated! . . . Victory! Victory! The Last Slave of Appetite is free, and the people are regenerated!" (*EPF*, pp. 221–23). The entire sequence plots the ambiguous relation between the best and the worst in American ideology—between personal power and imperial design, reformist zeal and nationalist expansion, the dream of revolutionary freedom and the politics of territorial conquest.

In 1843, Whitman published "Lesson of the Two Symbols" in the first issue of *The Subterranean,* a working-class paper whose purpose was to protect the people from "political cliques, fanatical traitors, bigotted sectarians, swindlers, speculators, robbers, bankers and brokers with their truckling, cringing servile tools to defend them." One of the few new poems that Whitman published between 1840 and 1850, "Two Symbols" marks the growing gap between the rhetoric of republicanism and the actual conditions of Jacksonian democracy. Like the branch of peace and the cluster of arrows held by the eagle in the revolutionary seal of America, Whitman's "Two Symbols" signify the values of gentleness and strength, virtue and courage associated with Washington and the founding moment. But even as Whitman extols these republican symbols of the "proud structure that our fathers based," he ends with an uneasy glance toward the future, wondering whether the structure of the fathers "be yet to grow."[9]

The muted question at the end of "Lesson of the Two Symbols" developed into a major controversy over the direction and future of the American republic during Whitman's editorship of the *Brooklyn Daily*

*Eagle* between 1846 and 1848. As an active participant in the local politics of Brooklyn, where he gave an occasional speech at a political rally and served in the Democratic party structure, Whitman looked upon his editorship of the *Eagle* as a powerful means of influencing public opinion in the direction of moral and political reform. "Much good can always be done, with such potent influence as a well circulated newspaper," he said in an article entitled "Ourselves and the Eagle": "To wield that influence, is a great responsibility. There are numerous noble reforms that have yet to be pressed upon the world. People are to be schooled, in opposition perhaps to their long established ways of thought. —In politics, too, the field of improvement is wide enough yet; the harvest is large, and waiting to be reaped" (*UPP*, I, 116–17).

The 1840s was a decade of agitation for abolition, labor reform, women's rights, temperance, educational reform, health cures, sexual freedom, utopian socialism, and world peace. In his editorials for the *Eagle*, Whitman placed himself at the center of these reformist energies and political struggles. He supported women's rights, open immigration, free trade, hard money, free speech, and the preservation of the native American heritage. He advocated reforms in the relations of labor and capital, urging improvements in the factory system, wages, working conditions, and the treatment of women laborers. He agitated against the slave trade and the extension of slavery. He urged reforms in education and the prison system, and he spoke out against corporal and capital punishment. He also supported changes in the physical conditions of city life, recommending improvements in lighting, garbage disposal, and the water system, and he advocated the construction of public baths.

Although Whitman never questioned the relations of private property and free enterprise at the foundation of the American system, his editorials for the *Eagle* reveal the signs of dispossession, dehumanization, and degeneration in American life and the growing inequality between rich and poor that were the true legacy of Jacksonian democracy.[10] Recognizing that there was something wrong in America, Whitman attributed the problem not to a failure in the system but to an insufficiently radical commitment to Democratic ideology. "As a fact without reasonable question," he said when the Democratic ticket lost the state and local elections in November 1847, "we would mention that our party has not been, of late, sufficiently bold, open and radical, in its avowals of sentiment." He called upon the party to advance in the direction of Democratic radicalism:

> It is to this progressive spirit that we look for the ultimate attainment of the perfectest possible form of government—that will be where there is the *least*

possible *government,* so called—when monopolies shall be things that *were,* but are not—when the barbarism of restrictions on trade shall have passed away—when, (and *this* 'when' we transfer to the present tense,) the plague spot of slavery, with all its taint of freemen's principles and prosperity, shall be allowed to spread no *further;* and when the good old Democratic Party— the party of the sainted Jefferson and Jackson—the party, which, with whatever errors of men, has been the perpetuator of all that is really good and noble and true in our institutions—the time-honored Democratic Party shall be existing and flourishing. (*GF,* I, 218–20)

Despite Whitman's confident rhetoric, his political journalism reveals disruptions created by the new industrial order that exceed his desire to contain them within the terms of Democratic party ideology. These disruptions were already evident in his editorials for the *Aurora* in 1842. In a piece entitled "The Last of Lively Frank," he says of the sickness and destitution of a once-lively young man: "Now his ghastly features, and the surrounding circumstances of penury, told a tale of chances thrown away, industry contemned, extravagance indulged in, and utter desperation at last, which it was terrible to think of" (*Aurora,* p. 24). Representing the individualist ethos of the dominant culture—an ethos propagated by the numerous evangelical and charitable societies of the time—Whitman sympathizes with the plight of young Frank, but he also blames him. He presents him as the victim of his own negligence and self-indulgence rather than of the new urban-industrial order. And yet Whitman's ideological commitment to the republican myth of self-regeneration is in uneasy conflict with his sense of some deeper wound in the new order that cannot be healed through a cultivation of personal virtue. "As we came forth from the house," he asserts, "we very naturally thought of the similar events that are daily going on in this great city. If some potent magician could lift the veil which shrouds, in alleys, dark streets, garrets, and a thousand other habitations of want, the miseries that are every day going on among us—how would the spectacle distress and terrify the beholder!" (*Aurora,* p. 25). In his vision of "delicate women . . . working themselves to illness, . . . young boys forced by the circumstances wherein they are bred, to be familiar with vice and all iniquity, [and] girls, whom absolute starvation drives at length to ruin, worse than starvation," the democratic image of the individual as a producer of self and history gives way to the disquieting "spectacle" of the individual as a helpless victim of the brute "circumstances" of city life. The story suggests the knowledge of powerlessness, the terror of "alleys, dark streets, garrets, and a thousand other habitations of want" that lay just behind (and in some sense nourished) Whitman's later poems in praise of the city.

It was in the demon of personal gain, which was largely responsible for this destitution, that Whitman found the main threat to the American republic: "If we were asked the particular trait of national character from which might be apprehended the greatest evil to the land, we should unhesitatingly point to the strife for gain which of late years has marked, and now marks, the American people." The triumph of capital over republican virtue was symbolized for Whitman in the public willingness to desecrate the graves of the revolutionary fathers in the name of commercial progress: "Even the battle spots where our old soldiers fought and died, are not beyond the reach of this pollution. The very hill made sacred by the blood of freedom's earliest martyrs, is sold and trafficked for" (*Aurora*, p. 41).

In his editorials for the *Aurora*, Whitman struggled with the central paradox in the progressive ideology of the American republic, namely, that in the entrepreneurial, self-interested economy secured by the Constitution of 1787, the progress toward the future was becoming a progress away from the revolutionary ideals of the republic set forth in the Declaration of Independence. This paradox was already present in Jefferson's mind. In *Notes on the State of Virginia*, in an uncharacteristically negative assessment of American prospects, Jefferson expressed fear that the republican ideals of the Revolution would be lost amid the pursuit of material gain:

> From the conclusion of this war we shall be going down hill. It will not then be necessary to resort every moment to the people for support. They will be forgotten, therefore, and their rights disregarded. They will forget themselves, but in the sole faculty of making money, and will never think of uniting to effect a due respect for their rights. The shackles, therefore, which shall not be knocked off at the conclusion of this war, will remain on us long, will be made heavier and heavier, till our rights shall revive or expire in a convulsion. (*Notes*, p. 161)

For all his belief in social and scientific progress, Jefferson regarded the Revolution as a privileged political moment from which the American republic would inevitably decline.

In the late 1840s, as the political tensions in the nation mounted, Whitman became increasingly conscious of the gap between democratic ideology and the actual conditions of American life. His editorials for the *Eagle* between 1846 and 1848 are a record of personal and national crisis, as the issues of slavery, capital, labor, urban conditions, women's rights, territorial expansion, sectional division, and the growing concentration of political and economic power exposed major tensions and contradictions in the ideology of the republic.

As in his *Aurora* editorials, Whitman located the prime source of danger in the dominance of capital and the corresponding growth of corporate conglomerates.

> It is the feverish anxiety after riches, that leads year after year to the establish-ment of those immense moneyed institutions, which have so impudently practiced in the face of day, frauds and violations of their engagements, that ought to make the cheek of every truly upright man burn with indignation. Reckless and unprincipled—controlled by persons who make them complete engines of selfishness—at war with everything that favors our true interests—unrepublican, unfair, untrue, and unworthy—these bubbles are kept afloat solely and wholly by the fever for gaining wealth. . . . The same unholy wish for great riches enters into every transaction of society, and more or less taints its moral soundness. (*GF*, II, 132–33)

The victims of these "engines of selfishness" were the laborers, whose oppression, impoverishment, and growing conflict with the Northern capi-talists were the subject of several of Whitman's *Eagle* editorials. "What lots of cents have gone out of poor folks' pockets, to swell the dollars in the possession of owners of great steam mills!" he exclaimed in an article protesting high tariffs. "Our American capitalists of the manufacturing order, would *poor* a great many people to be rich!" (*GF*, II, 70–71). In another article Whitman cited poor wages as the primary cause of female criminality, particularly prostitution. Aware of women's particular oppres-sion in the new economic order, he wrote: "How many *poor young women* there are in Brooklyn and New York—made so by the miserably low rate of wages paid for women's work, of all kinds, and in all its various depart-ments, from that of the most accomplished governess to that of the washerwoman!" (*GF*, I, 148).

At the same time that Karl Marx was making his revolutionary critique of Western capitalism, Whitman began mounting his own attack on Ameri-can capitalism from the viewpoint of the laboring class. Like Marx, he recognized that the economics of capitalism "enters every transaction of society" and "taints its soundness," but by focusing on the problem of monopoly and corporate wealth, he avoided the potential contradiction between the free-enterprise society he lived in and the harmonious and egalitarian democratic society of his dreams. Like Herman Melville in his more optimistic moments, Whitman wrote from an essentially eighteenth-century view of commerce in which, as Joel Barlow said in *The Vision of*

*Columbus*, "the spirit of commerce is happily calculated . . . to open an amicable intercourse between all countries, to soften the horrors of war, to enlarge the field of science and speculation, and to assimilate the manners, feelings and languages of all nations."[11] Envisioning the commercial spirit as an essentially benign, civilizing, and unifying force, Whitman never carried his critique of capitalism to an attack on the concept of free enterprise itself.

Whitman also expressed an increasingly pained awareness of the crowded conditions of city life and the deadening effects of its mechanized culture. His editorial entitled "Philosophy of Ferries" might be read as a parable of the new economic order. The Fulton ferry, which "takes precedent by age, and by a sort of aristocratic seniority of wealth and business," moves "like iron-willed destiny." Whitman notes the ferry's clocklike regularity and its "passionless and fixed" inattention to the disasters it may trail in its wake: "Perhaps some one has been crushed between the landing and the prow—(ah! that most horrible thing of all!) still, no matter, for the great business of the mass must be helped forward as before" (*GF*, II, 160). This unsettling vision is important, for it represents the underside of Whitman's poetic celebration in "Crossing Brooklyn Ferry," in which the ferry is symbolically transformed into a vehicle of human connectedness and advance through time.

The ferry of Whitman's journalism suggests a malign impulse in the "business of the mass": "A moment's pause—the quick gathering of a curious crowd, (how strange that they can look so unshudderingly on the scene!)—the paleness of the more chicken hearted—and all subsides, and the current sweeps as it did the moment previously. How it deadens one's sympathies, this living in a city!" Atomized and automatized by city life, Whitman's crowd is Chaplinesque, moving with the clockwork rhythms of the machines to which they are tied: "But the most 'moral' part of the ferry sights, is to see the conduct of the people, old and young, fat and lean, gentle and simple, when the bell sounds three taps. . . . Now see them as the said three-tap is heard! Apparently moved by an electric impulse, two-thirds of the whole number start off on the wings of the wind! Coat tails fly high and wide! You get a swift view of the phantomlike semblance of humanity, as it is sometimes seen in dreams—but nothing more—" (*GF*, II, 160–61).

In another article on the Fulton ferry, "Ten Minutes in the Engine Room of a Brooklyn Ferry Boat," Whitman begins with a more sanguine view of the machine: "There are few more magnificent pieces of handiwork than a powerful steam-engine, swiftly at work!" But his description bears the traces of a Puritan jeremiad: "At one end is the fiery region of living heat, the roaring, glowing coals, which look like a small edition of the infernal regions. The draft rumbles with a mighty hissing sound between a few little interstices; and through the mica plates, as through glass, one beholds a powerful mass of *hotness* quite terrible to look upon. It is enough to make a sinful man feel any other feeling on earth than that of a pleasurable anticipation" (*GF*, II, 210–11). Like Henry Adams's vision of the disintegrative and dehumanizing effects of the dynamo in *The Education of Henry Adams* (1907), Whitman's vision of the Brooklyn ferry registers an anxiety about modern technological culture that would become a disruptive subtext in his later poetic celebrations of American progress.[12]

In the July 1845 issue of the *Democratic Review*, John O'Sullivan declared that it was America's "manifest destiny to overspread and to possess the whole of the continent which Providence has given us for the development of the great experiment of liberty and federated government entrusted to us."[13] Dismayed by the speed and disruption of Northern industrial development, Whitman, like O'Sullivan and other nineteenth-century celebrators of the westward expansion, came to see the "boundless democratic free West" as the ultimate site of America's democratic experiment. In "Where the Great Stretch of Power Must Be Wielded," Whitman expressed uneasiness with the luxury and wealth spawned by the commercial culture of the East, arguing that the states "need a balance wheel like that furnished by the agricultural sections of the West" (*GF*, I, 26). As editor of the *Eagle*, he supported the expansionist policies of President James Polk, which resulted in the annexation of Texas in 1845, the acquisition of Oregon in 1846, and the war with Mexico in 1846–47; and for the remainder of his life he envisioned Cuba and Canada as future American states.

The connection between the disruptions of the new urban industrial order, the pastoral dreams of the West, and the Mexican war is suggested by an article Whitman wrote on a visit to Fort Hamilton, where a military detachment was preparing to embark for California. Amid the ringing of popular jingoistic tunes enjoining the people to

> Clear the way for General Taylor
> To lick the Mexicans he's a whaler,

Whitman reflected on the relief that nature provides from the crowded conditions of city life: "No one can tell, except a citizen tired out with pavements and crowding houses, how truly glorious is the country (God's work, indeed!) in comparison with that Babel which makes the eyes ache,

and the soles so weary! For ourself, we sometimes feel a yearning greediness for the wide seawaters, and the open fields of the country" (GF, II,
168–69). What is suppressed in this classic formulation of the agrarian
dream—or at least barely visible in the militaristic figure of Mexicans
being licked by General Zachary Taylor—is the fact of conquest, violation, and blood that was the very condition of America's pastoral "greediness" for the "open fields of the country."

Commenting on the people as the only sure reliance of American
liberty, Jefferson wrote James Madison in 1787: "This reliance cannot
deceive us, as long as we remain virtuous; and I think we shall be that, as
long as agriculture is our principal object, which will be the case while
there remain vacant lands in any part of America. When we get piled upon
one another in large cities, as in Europe, we shall become corrupt as in
Europe, and go to eating one another as they do there."[14] By making the
American dream of freedom contingent on the unlimited availability of
land, Jeffersonian ideology became tied to a national policy of violation. As
people began to pile up in large cities in the wake of the capitalist transformation, the national government pursued a more aggressive policy of
territorial expansion in the name of preserving American democracy.

Even Whitman quailed at the increasingly militaristic direction of
Polk's expansionist policies. In a comment on the military drills at Fort
Hamilton during the Mexican war, Whitman's Quaker pacificism sur-
faced: "Not being the least bit military ourself—but on the contrary, a
man of peace—we don't know whether they did very well or ill" (GF, II,
169). He appears to recognize the contradiction between the "promise" of
the American republic and the regimentation of military life: "In some
apartments we saw members of the gentler sex—wives, we were told, of
the men, and accompanying them to the distant land of promise; likewise
children of both sexes, who appeared to be as happy as clams—although
we doubt the moral influence of developing the 'young idea' in a soldier's
camp" (GF, II, 170).

The issues of virtue and commerce, progress and violation, expan-
sionism and the need for an "agrarian" balance wheel, even in the city, are
linked in Whitman's editorial campaign to save Fort Greene. On July 2,
1846, the Eagle carried an "Ode—By Walter Whitman," which was to be
sung to the tune of "The Star-Spangled Banner" for the Fourth of July
celebration at Fort Greene. Fort Greene was the burial site of several
thousand soldiers killed in the battle of Long Island during the Revolutionary War, and Whitman's "Ode" grew out of the movement to save the
park from destruction by local authorities.

Like the demonstrations provoked by the movement to save the "people's park" in Berkeley, California, in 1969, the agitation over the Fort Greene park focused the political tensions of the decade. The desecration of the park was defended in the name of commercial progress: "Trade and commerce are an irresistible power," declared the New York *Tribune,* "and before their necessities nothing can stand. The requirements of the rapidly flourishing city for 'more room' are constant and clamorous; and her citizens are justly proud of the rapid growth, even while lamenting that in her progress a spot so haunted with lofty associations must be despoiled" (*GF,* II, 46). For Whitman, the impending destruction of Fort Greene came to symbolize the larger crisis of values in the nation: In the plan to destroy the park he saw the betrayal of the republic's revolutionary heritage by commercial and capital interests and the paradoxical equation of progress with despoliation: " '*Must* be despoiled,' quotha!—What for, pray? Is the Dollar-god so ruthless that he grudges a few poor acres . . . to the service of health, of refinement, of *religion?* Is nothing to be thought of on earth, but cash? . . . *No,* Sir Much-worm! whoever you are 'a spot so haunted by lofty associations, must' *not* 'be despoiled' " (*GF,* II, 47–48).

Whitman's campaign to save Fort Greene shows how the problem of overcrowded cities, the obsession with material wealth, and the disruptions of market capitalism became linked in mid-nineteenth-century America with the flight to nature, the emphasis on bodily health, and the quest for spiritual transcendence. Whitman's "Ode" stresses the interrelationship between the preservation of revolutionary traditions, the preservation of nature, and the cultivation of the soul:

> O, God of Columbia! O, Shield of the Free!
>    More grateful to you than the fanes of old story,
> Must the blood-bedewed soil, the red battle-ground, be
>    Where our fore-fathers championed America's glory!
> Then how priceless the worth of the sanctified earth,
> We are standing on now. Lo! the slopes of its girth
> Where the Martyrs were buried: Nor prayers, tears, or stones,
> Mark their crumbled-in coffins, their white, holy bones!
>                                                 (*EPF,* p. 34)

Conflating the rhetoric of religion and revolution, in which God becomes an insurrectionary spirit, the poem presents Fort Greene as a locus of revolutionary traditions and of a new civil-religious worship. Through his ode and his plea for the preservation of the "sanctified earth" of Fort Greene, Whitman tried to ensure that "the battle, the prison-ship, martyrs and hill" would remain alive in the memory of future generations of Americans.

As the last poem Whitman wrote before he began the literary experiments that led to the publication of *Leaves of Grass* in 1855, his "Ode" represents an early attempt to use poetry as a form of political action. And on the issue of Fort Greene, at least, he was successful. It was in part due to Whitman's editorials and his Fort Greene "Ode" that the creation of Fort Greene Park was authorized by Congress in 1847. Years later, on November 14, 1908, a monument to the Prison Ship martyrs was finally dedicated by President William Howard Taft.

"The power of the periodical press is second only to that of the people," wrote Alexis de Tocqueville in *Democracy in America*.[15] As editor of one of the largest and most popular newspapers in Brooklyn and New York, Whitman became increasingly conscious of the power of the press and especially the power of the pen to reach the masses. With the example of the revolutionary writers in mind, he celebrated the pen as a weapon of political transformation, a means of influencing public opinion and political policy: "Where is, at this moment, the great medium or exponent of power, through which the civilized world is governed?" he asked in one of his *Eagle* editorials.

> Neither in the tactics or at the desks of statesmen, or in those engines of physical terror and force wherewith the game of war is now played. The *pen* is that medium of power—a little crispy goose quill, which, though its point can hardly pierce your sleeve of broadcloth, is able to make gaping wounds in mighty empires—to put the power of kings in jeopardy, or even chop off their heads—to sway the energy and will of congregated masses of men, as the huge winds roll the waves of the sea, lashing them to fury, and hurling destruction on every side! (*GF*, II, 246)

As editor of the Brooklyn *Daily Eagle*, Whitman began to conceive a role for the pen that exceeded the bounds of the political journalist. In words that again take up his dream of writing a "wonderful" book, he speculates: "At this hour in some part of the earth, it may be, that the delicate scraping of a pen over paper, like the nibbling of little mice, is at work which shall show its results sooner or later in the convulsion of the social and political world. Amid penury and destitution, unknown and unnoticed, a man may be toiling on to the completion of a book destined to gain acclamations, reiterated again and again, from admiring America and astonished Europe!" (*GF*, II, 246–47).

Politically potent and perennial as the grass, the book that Whitman imagined would pass on to present and future generations the sacred fire of the revolutionary fathers. Conceiving of the pen as a scourge of political justice, destroying all that threatened the life of the American republic, he saw himself wielding a revolutionary power that he could not achieve in his capacity as private citizen or political journalist: "The sly rogue and the profligate, with brazen, hard face,—the betrayer of his trust, and the wily seducer—the monarch on the throne and the unprincipled legislator—the heartless parent and the unthankful child—the tyrannical captain of a crew, and the brutal whipper of a slave, all bend beneath the whirlwind of its wrath, and know its power to punish wickedness" (*GF*, II, 248).

Whitman's *Eagle* editorials were a prose dress rehearsal for the political text of his poems. As he ministered to the crisis of individual and nation, reminding Americans of their revolutionary heritage and the role they were destined to play in the "holy millennium of liberty," his prose began to loosen into the expansive, oratorical rhythms and visionary democracy of *Leaves of Grass*. "Swing Open the Doors!" he proclaimed in one editorial, striking a pose that projects his political views on free trade, open banking, open immigration, free soil, free men, and free women. "We must be constantly pressing onward," he declares, "every year throwing the doors wider and wider—and carrying our experiment of democratic freedom to the very verge of the limit" (*GF*, I, 10). His phrases roll with the participial rhythms of his later free-verse poems, and his open-door image anticipates the democratic challenge he hurls at his readers in "Song of Myself":

> Unscrew the locks from the doors!
> Unscrew the doors themselves from their jambs!
> (*LG* 1855, p. 48)

Communing daily with his readers and growing accustomed to thinking of himself not as an individual but a collectivity—an en masse—Whitman expanded toward the latitude and longitude of the democratic comrade and lover of his verse: "We really feel a desire to talk on many subjects to *all* the people of Brooklyn," he confessed to his readers. "There is a curious kind of sympathy . . . that arises in the mind of a newspaper conductor with the public he serves. He gets to *love* them. Daily communion creates a sort of brotherhood and sisterhood between the two parties" (*UPP*, I, 115). Written shortly before Whitman did his first known writing toward *Leaves of Grass*, this comment marks the shortness of the path between the political journalist and the poet of democracy.

# 3

## The Poet of Slaves
## and the Masters of Slaves

I am the poet of slaves, and of the masters of slaves
I am the poet of the body
And I am

—WHITMAN, *Notebooks*

For Whitman, as for the American people, it was the issue of slavery more than any other that tested the nation's moral and political bounds. As editor of the Brooklyn *Daily Eagle*, he faced for the first time the central paradox of the American republic: the contradiction between republican ideology and the reality of an economy of masters and slaves. This contradiction had roots in the foundation of the republic itself. In a letter to his friend John Holmes, Jefferson commented on his own and the national dilemma over slavery: "We have the wolf by the ears, and we can neither hold him, nor safely let him go."[1] Divided between his fears for the Union, property, and the safety of the race and a revolutionary ideology that committed him to the liberation of all slaves, Jefferson in his wolf image aptly reflected the dilemma of the American republic.

As a journalist and political activist during the 1840s and then as a poet, Whitman took Jefferson's and the nation's wolf by the ears. Like the controversy over slavery in the political sphere, Whitman's response to the issue of slavery in the 1840s and 1850s triggered a crisis of democratic faith. As he attempted to resolve the paradox of slavery in the American republic by drawing on revolutionary ideology, he adopted increasingly radical postures. Inscribed with the contradictions of an entire culture, slavery is in many ways the original text of America, and it was in response to this text, in the heat of the slavery controversy, that Whitman made the final passage from party journalist to political poet.

Although Whitman never fully accepted the tactics of the Abolition-

44

ists, as editor of the *Eagle* he came to share their position on the contradiction of slavery in the American republic. The existence of the slave-trading business in America was, he said in a March 1846 editorial "Slavers—and the Slave Trade," "a disgrace and blot on the character of our Republic, and on our boasted humanity!" (*GF*, I, 187). Detailing the fracture of familial and communal bonds as a result of the slave trade, in which "the negro is torn from his simple hut—from his children, his brethren, his parents, and friends—to be carried far away and made the bondman of a stranger," Whitman does not make clear the distinction between the abomination of the slave trade and the equally abominable practice of selling and bartering slaves already in America (*GF*, I, 187–88). In fact, the entire editorial might be read as a critique of the slave system. Although Whitman was bound as the editor of a Democratic journal to support the proslavery position of President Polk, his awareness of the "plague spot" of slavery on the body of the American republic leads to a momentary crisis of democratic faith: "Pah! we are almost a misanthrope to our kind when we think they will do such things!" (*GF*, I, 189).

The Mexican war and American expansion became linked with the issue of slavery when on August 8, 1846, in response to President Polk's request for an appropriation of $2 million to purchase land from Mexico, Congressman David Wilmot asked for an amendment forbidding the extension of slavery into the new territory. He proposed as "an express and fundamental condition to the acquisition of any territory" that "neither slavery nor involuntary servitude shall ever exist in any part of said territory."[2] Under Whitman's editorship, the *Eagle* was the first of the New York dailies to support the Wilmot Proviso. "Set Down Your Feet, Democrats!" he declared in the December 21, 1846, issue: "let the Democratic members of Congress, (and Whigs too, if they like,) plant themselves quietly, without bluster, but fixedly and without compromise, on the requirement that *Slavery be prohibited in them forever*" (*GF*, I, 194). To extend rather than to prohibit slavery in the territories, Whitman argued, would be to act against the commitment of the founding fathers and to widen rather than close the gap between ideology and reality in America.

The issue of slavery in the territory, he wrote in an article entitled "New States: Shall They Be Slave or Free?" raised the question of whether "the mighty power of this Republic ... shall be used to root deeper and spread wider an institution which Washington, Jefferson, Madison, and all the old fathers of our freedom, anxiously, and avowedly from the bottom of their hearts, sought the extinction of, and considered inconsistent with the other institutions of the land" (*GF*, I, 201). To those like John C. Calhoun who argued that slavery was written into the U.S. Constitution, Whitman pointed to its compromise and contradiction. Northern Democracy had been faithful to the " 'compromise of the Constitution' by protecting the institution of slavery (uncongenial as that institution is to all the instincts and sympathies of Democracy) within the limits that the Constitution found it" (*GF*, I, 206). It was not the compromised Constitution of America but the egalitarian ideals of the Declaration of Independence that Whitman invoked in support of his antislavery position. Describing Jefferson as the "apostle of liberty" and citing his plan for prospective emancipation, Whitman remarked that he was "no doubt anxious that the time should arrive when the words which he put forth in the Declaration of Independence, 'that all men are created free and equal,' should be as true in fact as self-evident in theory" (*GF*, I, 199–200). Anxious himself to reconcile revolutionary ideology with the fact of slavery in the Constitution of America, Whitman had to invent founding fathers and an American republic more committed to the cause of emancipation than was in fact the case.

The Wilmot Proviso sprang from and was supported by the same confusion of antislavery and anti-Negro motives one finds among the revolutionary founders. The main purpose of the proviso, said Wilmot, was to defend "the rights of white freemen. I would preserve for free white labor a fair country, a rich inheritance, where the sons of toil, of my own race and own color, can live without the disgrace which association with negro slavery brings upon free labor."[3] Whitman's defense of the Wilmot Proviso, which he referred to as the Jefferson Proviso, occasionally echoed Wilmot's rhetoric. In "American Workingmen, Versus Slavery," he described the debate over slavery extension as a conflict between "*the grand body of white workingmen, the millions of mechanics, farmers, and operatives of our own country*" and a "few thousand rich, 'polished,' and aristocratic owners of slaves at the South. . . . An honest poor mechanic, in a slave State, is put on a par with the negro slave mechanic" (*GF*, I, 208–9).

Citing such passages, Whitman's critics have argued that his real concern in his antislavery writings was with the cause of white labor rather than black slavery. Actually, Whitman's position lay somewhere in between. As the rhetoric of labor versus aristocracy suggests, Whitman, like Frances Wright and William Leggett, conceived of the struggle against slavery as a class struggle against the feudal institutions of Europe. The only persons who would be excluded from the new territory, Whitman contended, would be "the *aristocracy* of the South—the men who work only with other men's hands" (*GF*, I, 204). At a time when wage labor was becoming a new form of slavery, Whitman's antislavery editorials bear the traces of labor movement radicalism in stressing the danger of the slave system to the rights and dignity of all laborers. "The influence of the slavery institution is to bring the dignity of labor down to the level of slavery," he said, and he called upon workers to defend their rights so "that their calling shall *not* be sunk to the miserable level of what is little above brutishness—sunk to be like owned goods, and driven cattle!" (*GF*, I, 209–10).

Whitman presented the struggle against slavery in America in the millennial terms of the revolutionary enlightenment, comparing the Southern planters with "kings and nobles when arraigned for punishment before an outraged and too long-suffering people. . . . But the course of mortal light and human freedom, (and their consequent happiness,) is not to be stayed by such men as they" (*GF*, I, 213). Here, as elsewhere in his writings, Whitman's commitment to the revolutionary ideal of freedom led him to support the cause of the "long-suffering people," black and white. But in making his case, he, too, swung between the antislavery rhetoric of the American Revolution and the anti-Negro phobia of his age.

Whitman's position on abolition was similarly ambivalent. "The mad fanaticism or ranting of the ultra 'Abolitionists,' " he wrote, "has done far more harm than good to the very cause it professed to aid" (*GF*, I, 192). Although he has been criticized for his failure, unlike Emerson and Thoreau, to embrace the cause of immediate abolition, his policy was, if not "ultra-Abolitionist," at least considerably more "revolutionary" than might at first appear.

For Whitman, the main issue was, as it was during the American Revolution, the problem of sovereignty and authority. Although he supported the Abolitionists' ideals, commending their "love of impartial 'liberty, our nation's glory' " and defended their right of free speech, he believed that emancipation should come from below and not from above, from the people of the southern states and not from the national government. Emphasizing this idea in an editorial "Abolition," he wrote: "*It is to the discoveries and suggestions of free thought, of 'public opinion,' of liberal sentiments, that we must at this age of the world look for quite all desirable reforms, in government and any thing else*" (*GF*, I, 193). His support of the principle of popular sovereignty over a national act of emancipation reflects the ideological potency of the doctrine of states' rights at a time when the rights of the individual and the states against federal interference were still jealously guarded as the primary heritage of the American Revolution.

The issue of popular and state sovereignty was for Whitman a key issue in the slavery controversy, an issue that would lead to his fury over the tightening of the Fugitive Slave Law in 1850 and eventually to his with-drawal from party politics. As he became increasingly disillusioned with party politics, he began to experiment with the idea of using poetry as a form of political action. In his earliest extant notebook, dated 1847, he recorded a series of prose and poetic notes that adumbrate the form and substance of *Leaves of Grass*.[4] The notes grew out of Whitman's desire to reach the congregated masses of the people as a means of enabling republican regeneration. "Be simple and clear—Be not occult," the notes begin, and he then describes the essential attributes of an American:

> True noble expanded American Character is raised on a far more lasting and universal basis than that of any of the characters of the "gentlemen" of aristocratic life, or of novels, or under the European or Asian forms of society or government.  —It is to be illimitably proud, independent, self-possessed generous and gentle. It is to accept nothing except what is equally free and eligible to any body else. It is to be poor, rather than rich—but to prefer death sooner than any mean dependence.  —Prudence is part of it, because prudence is the right arm of independence. (*UPP*, II, 63)

Like his political journalism, Whitman's poetic notes speak the egalitarian and millennial language of the revolutionary enlightenment. They also bear the rhetorical traces of the proletarian, anticapitalist appeals of Frances Wright and William Leggett: "I wish to see American young men the working men, carry themselves with a high horse" (*UPP*, II, 64), he says; and he wants to invest poetry with the democratic energies of the people: "I will not descend among professors and capitalists—I will turn the ends of my trousers around my boots, and my cuffs back from my wrists, and go with drivers and boatmen and men that catch fish or work in the field. I know they are sublime" (*UPP*, II, 68–69).

These early notes are instructive in demonstrating how *Leaves of Grass*, like the work of other putatively ahistorical Romantic writers, simultaneously engages and resists its time. Whitman's attempt to create "an illimitably proud, independent, self-possessed, generous, and gentle" human being was not a flight from history but an act of poetic resistance that had fairly marked origins in working-class discontent—in the dependence, dispossession, commodification, and merchandizing of the self fostered by market capitalism. At the base of Whitman's democratic poetics is the desire to participate in the act of political (re)creation by creating a regenerated republican self:

Test of a poem
How far it can elevate, enlarge, purify, deepen and make happy the
attributes of the body and soul of a man.

(*UPP*, I, 75)

This enlarged and purified body and soul is the source of national renewal: "Vast and tremendous is the scheme!" he exclaims in the conclusion to his notes:

It involves no less than constructing a nation of nations . . . and the people of
this state instead of being ruled by the old complex laws, and the involved
machinery of all governments hitherto, shall be ruled mainly by individual
character and conviction. —The recognized character of the citizen shall be
so pervaded by the best qualities of law and power that law and power shall be
superseded from this government and transferred to the citizen. (*UPP*, I, 76)

Whitman's poet participates in the act of national creation by carrying
on the revolutionary task of transferring power from the government to
the individual, but he is also a unifying figure, who speaks the "divine
grammar" of democratic reconciliation: "No two have exactly the same
language, and the great translator and joiner of the whole is the poet. He
has the divine grammar of all tongues, and says indifferently and alike,
How are you friend? to the President in the midst of his cabinet, and
Good day my brother, to Sambo, among the hoes of the sugar field"
(*UPP*, I, 65).

Whitman's concern with the issues of self-sovereignty and unity arose
out of a political situation in which, despite the rhetoric of optimism,
individual and nation were in a state of uncertainty and fragmentation, a
state that was in part a product of increasing instability in the political and
economic sphere as a result of the slavery controversy. His *Eagle* editorials
reflect this growing anxiety about the state of the Union. Enumerating the
personal and political blessings "involved in the UNION of these United
States together into an integral Republic, 'many in one,' " he exclaims:
"This Union dissolved? Why the very words are murky with their own most
monstrous portent!" (*GF*, I, 229–30). In another essay, "Disunion," he
quotes Jackson: "The Union! it must and shall be preserved." He reflects
on the possibility of dissolution with a sense of apocalyptic doom: "Any
man, with ordinary judgment must know, that the disjunction of the Union
would sow a prolific crop of horrors and evils, like dreary night, compared
to which the others are but as a daylight cloud" (*GF*, I, 239). Here, as in his
poems of the late 1850s, the possibility of disunion is presented in images
that reverse the millennial vision of the revolutionary enlightenment: images of murk and cloud, horror and evil, darkness and night.

Critical of the heightened rhetoric of disunion in both North and South, particularly in Massachusetts and South Carolina, Whitman counsels restraint: "We hear these angry recriminations and threats with alarm which we will not disguise—in connection with a certain momentous question which is now before Congress [the extension of slavery in new territory], and the finale of which, when it comes, is supposed capable of shaking the foundation of our Republic through and through, as the tempest shakes a noble ship at sea" (*GF*, I, 231).

The composite persona that Whitman imagines in his notebook dated 1847 is in part a response to the rhetoric of disunion: The poet rehearses the nation's conflicting energies within the confines of the unitary self, seeking to affirm poetically the political viability of the democratic republic. In casting the shape of this unitary self who would travel the final road to *Leaves of Grass*, the slavery issue was central. When in his notebook Whitman breaks for the first time into lines approximating the free verse of *Leaves of Grass*, the lines bear the impress of the slavery issue:

> I am the poet of slaves, and of the masters of slaves
> I am the poet of the body
> And I am
>
> > > (*UPP*, II, 69)

The lines join or translate within the representative figure of the poet the conflicting terms of master and slave that threaten to split the Union. Essential to this process of translation are the strategies of parallelism and repetition, which, as in the democratic and free-verse poetics of *Leaves of Grass*, balance and equalize the terms of master and slave within the representative self of the poet. By balancing and reconciling the many within the one of the poet, Whitman seeks to reconcile masters and slaves within the larger figure of the E PLURIBUS UNUM that is the revolutionary seal of the American republic.

And yet even in these trial lines, Whitman's unifying act appears to be short-circuited by the irreconcilability of an economy of masters and slaves within the figure of the republican self. After the phrase "And I am," the lines break off. The poet tries again:

> I am the poet of the body
> And I am the poet of the soul
> I go with the slaves of the earth equally with the masters
> And I will stand between the masters and the slaves,
> Entering into both, so that both will understand me alike.
>
> > > (*UPP*, II, 69)

These lines move in a direction different from that of the initial ones: The poet will "go" with masters and slaves, stand between them, and enter into them, but he cannot "contain" them within the bounds of the republican self. This initial short-circuiting continues throughout *Leaves of Grass.* Although Whitman could and did identify with "the hounded slave," he could never balance, absorb, or contain the term *slave master* within the republic of the self. His grammar of democratic union is, like the political Union itself, continually ruptured by the fact of slavery in the body of the republic.

Whitman's refusal to compromise on the issue of slavery extension led to his termination as editor of the *Brooklyn Daily Eagle.* When in 1847 the state Democratic convention at Syracuse failed to support the Wilmot Proviso, the owner of the *Eagle,* Isaac Van Anden, who was an officer in the party organization, chose to go along with the party's proslavery extension position. Whitman, however, stood firmly on the principle of free soil, criticizing the Democratic party for failing to adopt "the principles of the Jeffersonian Proviso—for so it should be called. The immortal author of the Declaration of Independence is as much the originator of the proviso as Columbus was the discoverer of this continent" (*GF,* I, 222). In order to keep the faith of the American people, Whitman asserted, the Democratic party must be "true to the memory of the Revolutionary Fathers who fought for freedom, and not for slavery" (*GF,* I, 224). As the party split into proslavery extension Old Hunkers and free-soil Barnburners, Whitman tried to unite them under the Jeffersonian rubric. The *Eagle* was neither Old Hunker nor Barnburner, he asserted: "We are a *Democratic*-Republican," he said, echoing Jefferson's inaugural address in which he, too, sought to unite party factions: "We are all republicans—we are all federalists."[5]

Whitman's assertion of unity notwithstanding, the truth was that the *Eagle,* like the Democratic party, was split between the Old Hunker position of its owner and the increasingly radical position of its Barnburner editor. In an editorial on January 3, 1848, Whitman opposed the anti–Wilmot Proviso position of General Lewis Cass, who was to be the Democratic presidential candidate in the November election. This was Whitman's last editorial for the *Eagle.* Summing up the events that led to his termination as editor, he later said: "The troubles in the Democratic party broke forth about those times (1848–49) and I split off with the radicals, which led to rows with the boss and 'the party,' and I lost my place" (*PW,* I, 288).

Having split off with the radicals, Whitman intended to carry on his free-soil activities by editing a Barnburner newspaper. On January 21, 1848, the New York *Tribune* printed the following note:

> We are informed from the best authority that the Barnburners of Brooklyn
> are about starting a new daily paper, as, it is said, The Eagle has returned to
> Old Hunkerism again. Mr. Walter Whitman, late of The Eagle, is to have
> charge of the new enterprise. (*GF*, I, xxxiii)

The projected Barnburner paper was begun in September. In the interven-
ing months, Whitman, perhaps disillusioned with party politics, made a
trip to New Orleans, where he wrote for the *Crescent.*

   If Whitman's three-month stay in New Orleans was not the occasion,
as was once believed, for a "New Orleans romance," his trip south did
broaden his perspective and intensify his sense of the conflict between
ideal and reality in the American republic. On the one hand, the mixture
of black, native American, French, and Hispanic people in New Orleans
made him conscious of the racial and cultural pluralism of America. On
the other hand, the daily experience of writing for a paper that advertised
the sale of slaves and living in a city where slaves were sold at auction
deepened his opposition to the extension of slavery in America.

   Whitman's radical sentiments were further aroused when on May 24,
1848, the Barnburners from New York bolted the Democratic National
Convention in Baltimore, and the convention nominated Lewis Cass, the
highly vocal opponent of the Wilmot Proviso, to head the Democratic party
ticket. The following day, on May 25, Whitman resigned his post at the
*Crescent* and returned east to join in the creation of the Free-Soil party. On
June 28, 1848, the *Advertiser* announced that Whitman had returned to
New York, "large as life, but quite as vain, and more radical than ever."[6]

   In August 1848, Whitman was elected as a delegate to the Free-Soil
convention in Buffalo, where he joined Barnburners, antislavery Whigs,
and Abolitionists, including Frederick Douglass, in nominating Martin
Van Buren for the presidency on a platform of "Free soil, free speech,
free labor, and free men." To support what he called the "genial and
enlightened doctrines of the Free Soil" party, he founded the Brooklyn
Weekly *Freeman,* which, he announced in the first issue of September 9,
1848, would "oppose, under all circumstances, the addition to the Union,
in the future, of a single inch of *slave land,* whether in the form of state or
territory."[7] Beyond his immediate dedication to the goals of the Free-Soil
party, however, Whitman's principal commitment was still to the revolutionary ideals of the American republic as he believed they were embodied
in the philosophy of Jefferson: "How he hated slavery!" Whitman exclaimed in the first issue of the *Freeman.* "He hated it in all its forms—
over the mind as well as the body of man. He was, in the literal sense of
the word, an *abolitionist;* and properly and usefully so, because he was a

Southerner, and the evil lay at his own door."[8] As Whitman came to identify with the Abolitionist position, he tried to reconcile his position with Democratic ideology by inventing a mythical image of Jefferson as Abolitionist.

The office of the *Freeman* was destroyed by fire after the first issue, and although Whitman was able to resume publication in September 1849, his thoughts at that time seemed to be moving away from journalism and party politics toward poetry and oratory as the most effective means of reaching and radicalizing the American people. Commenting on Whitman's activities at this time, his brother George recalled that he "made a living now—wrote a little, worked a little, loafed a little," frequented libraries, and "wrote what mother called 'barrels' of lectures"—on politics, language, slavery, diet, exercise, physique, and the like. His notebooks and scrapbooks also indicate that he was engaged in an extensive program of self-education—reading, annotating, and taking notes on books and articles on literature, language, history, science, astronomy, geography, foreign countries, and world events.[9]

Whitman became even further disillusioned by party politics when in the summer of 1849 the Barnburners returned to the fold of the Democratic party. As the talk of Southern secession mounted, Henry Clay of Kentucky introduced into the Senate a series of compromise resolutions, proposing the admission of California as a free state, a stricter Fugitive Slave Law, the continuation of slavery in the District of Columbia, the extension of slavery in the new territory, and a prohibition of congressional interference with the interstate slave trade. The willingness of the Democratic party and the North to compromise on the issue of slavery sent Whitman literally raging into verse.

On March 2, 1850, William Cullen Bryant's *New York Evening Post* published Whitman's "Song for Certain Congressmen." Describing "dough" in an epigraph—"Like dough; soft; yielding to pressure; pale"— the poet speaks from the point of view of the Northern congressmen who had become like dough in the hands of Southern slaveholders:

> We are all docile dough-faces,
>     They knead us with the fist,
> They, the dashing southern lords,
>     We labor as they list;
> For them we speak—or hold our tongues,
> —For them we turn and twist.
>
> We join them in their howl against
>     Free soil and "abolition,"

> That firebrand—that assassin knife—
>     Which risk our land's condition,
> And leave no peace of life to any
>     Dough-faced politician.
>
>                              (*EPF*, p. 44)

Through the compromised voice and vision of these dough-faces, Whitman draws attention to the irony of seeking to save the Union by preserving the movement and barter of "nigger slaves":

> Things have come to a pretty pass,
>     When a trifle small as this,
> Moving and bartering nigger slaves,
>     Can open an abyss,
> With jaws a-gape for "the two great parties;"
>     A pretty thought, I wis!
> Principle—freedom!—fiddlesticks!
>     We know not where they're found.
> Rights of the masses—progress!—bah!
>     Words that tickle and sound;
> But claiming to rule o'er "practical men"
>     Is very different ground.
>
> Beyond all such we know a term
>     Charming to ears and eyes,
> With it we'll stab young Freedom,
>     And do it in disguise;
> Speak soft, ye wily dough-faces—
>     That term is "compromise."
>
>                              (*EPF*, pp. 44–45)

In the dough-faced politics of Senators Daniel Webster, Robert J. Walker, Donald S. Dickinson, and James Cooper—who are named in the poem—Whitman sees the collapse of revolutionary principle. Signing his name PAUMANOK, the native American name for Long Island, Whitman voices a sense of personal and communal betrayal, as symbolized in the figure of "young Freedom" stabbed by the spirit of compromise. His ironic tone is atypical, but as in such later poems as "A Boston Ballad" (1854) and "Respondez" (1856), it is a tone that arises out of the experience of fracture and incongruence in the political sphere and that continually threatens to undermine the optative mood of *Leaves of Grass*.

When on March 7, 1850, Daniel Webster, speaking "not as a Massachusetts man, nor as a northern man, but as an American," delivered

famous speech to the Senate declaring his willingness to compromise on
the issue of slavery in order to preserve the Union, Whitman turned
once again to poetry as a form of political protest. The poem "Blood-
Money," which was published by the Abolitionist Horace Greeley in the
New York *Tribune Supplement* on March 22, 1850, is a further sign of
Whitman's own Abolitionist sentiment. Once again he assumes the guise
of Paumanok to criticize the politics of compromise. His method in
"Blood-Money," however, is neither political invective nor rhymed verse
but a sustained use of biblical allusion and line. The poem is organized
around the New Testament text of Matthew, in which Christ is betrayed
by Judas:

> Of olden time, when it came to pass
> That the beautiful god, Jesus, should finish his work on earth,
> Then went Judas, and sold the divine youth,
> And took pay for his body.

*(EPF,* p. 47)

Moving from the biblical past to the present in a manner that anticipates
the spatial and temporal leaps of *Leaves of Grass*, Whitman draws the
analogy betwen Judas's selling the body of Christ and the current political
situation in America:

> The cycles, with their long shadows, have stalk'd silently forward,
> Since those ancient days—many a pouch enwrapping meanwhile
> Its fee, like that paid for the son of Mary.
> And still goes one, saying,
> "What will ye give me, and I will deliver this man unto you?"
> And they make the covenant, and pay the pieces of silver.

*(EPF,* p. 48)

These lines have both a specific and a more general political reference.
On the one hand, they suggest the Judas-like betrayal of Daniel Webster
and other congressmen, who, for personal gain, are willing to perpetuate
the sale and hunting of slaves in America. On the other hand, the lines
suggest a more general situation in which both black and white persons
have become marketable commodities.

The events leading up to the passage of the Compromise of 1850 in
September also inspired "The House of Friends," which was published in
the *Tribune* on June 14, 1850. Drawing on another biblical text—"I was
wounded in the house of my friends" from Zachariah—here again under
the pressure of political events, Whitman breaks the bounds of traditional
verse:

> If thou art balked, O Freedom,
> The victory is not to thy manlier foes;
> From the house of thy friends comes the death stab.
>
> (*EPF*, p.36)

As a kind of sequel to "Song for Certain Congressmen," Whitman once again employs the figure of Freedom stabbed as an emblem of his own and the nation's betrayal. Here, however, the threat comes not from the South, where the paradox of slavery in the land of the free is endemic, but from the North, where the republican commitment to freedom is presumably realized in word and deed:

> Virginia, mother of greatness,
> Blush not for being also mother of slaves.
> You might have borne deeper slaves—
> Doughfaces, Crawlers, Lice of Humanity—
> Terrific screamers of Freedom,
> Who roar and bawl, and get hot i' the face,
> But, were they not incapable of august crime,
> Would quench the hopes of ages for a drink—
>
> (*EPF*, p. 37)

Like Daniel Webster, the North was willing to compromise on freedom for the sake of commerce and comfort and so has died as the site of republican hope:

> A dollar dearer to them than Christ's blessing;
> All loves, all hopes, less than the thought of gain;
> In life walking in that as in a shroud.
>
> (*EPF*, p. 37)

The image of the dollar-enshrouded Northerners is incisive, registering Whitman's uneasy recognition of their increasingly capitalist ethos, an ethos of acquisition and consumption that has made them not only "deeper slaves" but also a deeper threat to the republic.

In the poem's final lines, Whitman summons the North to republican regeneration, urging the faithful few to revive the "elder blood" of the revolutionary fathers:

> Arise, young North!
> Our elder blood flows in the veins of cowards—
>  . . . . . . . .
> Fight on, band braver than warriors,
> Faithful and few as Spartans;
> But fear not most the angriest, loudest malice—

> Fear most the still and forked fang
> That starts from the grass at your feet.
> 
> (*EPF*, p. 37)

Betrayed by Northern congressmen and the Democratic party, Whitman in 1850 considered himself as one of the faithful band of Freedom's warriors who would continue to wield the pen as weapon against the "forked fang" that sprang from the ground of the republic itself.

Whitman's impassioned involvement in the slavery controversy was fired by his sense that what was happening in America was part of a universal advance from enslavement to freedom: "Not only here, on our beloved soil, is this democratic feeling infusing itself, and becoming more and more powerful," he wrote in an 1846 editorial on progress. "The lover of his race—he whose good-will is not bounded by a shore or a division line—looks across the Atlantic, and exults to see on the shores of Europe, a restless dissatisfaction spreading wider and wider every day. Long enough have priestcraft and kingcraft stalked over those lands, clothed in robes of darkness and wielding instruments of subjection" (*GF*, I, 12–13, 15). As Whitman wielded the power of the pen against the "instruments of subjection" at home, he was inspired by the signs of revolutionary ferment he saw spreading in Europe. "In France, the smothered fires only wait the decay of the false one, the deceiver Louis Philippe, to burst forth in one great flame," he wrote in February 1847. "The mottled empire of Austria is filled with the seeds of rebellion—with thousands of free hearts, whose aspirations ever tend to the downfall of despotism; and the numerous petty German states, too, have caught the sacred ardor" (*GF*, I, 30).

Whitman saw the political events being enacted in America as central to both the future of the republic and the fate of revolution throughout the world. In an article on American union, he wrote: "The perpetuity of the sacred fire of freedom, which now burns upon a thousand hidden, but carefully tended, altars in the Old World, waits the fate of our American Union. O, sad would be the hour when that Union should be dissolved!" (*GF*, I, 229–30). For all their revolutionary fire, Whitman's words are not historically innocent. Joining the Puritan vision of America as the city on the hill with the revolutionary vision of America as the beacon of republican liberty, Whitman was himself perpetuating the potent but baneful intermingling of idealism and imperialism that had marked and would continue to mark American history.

Whitman's vision of the sacred flame of freedom advancing throughout the world is also instructive in demonstrating the way that Romantic writers mythologized the historically specific struggles of the nineteenth century and then read these struggles back into history as part of an inevitable and providentially ordained revolutionary process. Like other champions of democracy at home and abroad, Whitman found confirmation for his revolutionary reading of history when in 1848 Louis Philippe was dethroned, a second French republic was declared, and this revolution set off a series of uprisings in Austria, Hungary, Germany, and Italy. Although these revolutions were later defeated, Whitman maintained his belief in the ultimate triumph of liberty, which he celebrated in "Resurgemus," published in the New York *Tribune* on June 21, 1850. Inspired by the revolutions in Europe, "Resurgemus" is the first of Whitman's early political poems to be included among the twelve untitled poems of *Leaves of Grass*.

Like "Blood-Money" and "The House of Friends," "Resurgemus" breaks the pentameter and turns on the polarized images of slavery and freedom. The text of "Resurgemus," however, is not the Bible but history. The poem's tone is hopeful and prophetic rather than bitter and ironic:

> Suddenly, out of its stale and drowsy lair, the lair of slaves,
> Like lightning Europe le'pt forth,
> Sombre, superb and terrible,
> As Ahimoth, brother of Death.
>
> God, 'twas delicious!
> That brief, tight, glorious grip
> Upon the throats of kings.

(EPF, p. 38)

The first line, which swings with a loose-cadenced rhythm based on parallelism and repetition, was retained in the 1855 version; the other lines were deleted, reinvented, or fused with suspension points in the following manner:

> Suddenly out of its stale and drowsy lair, the lair of slaves,
> Like lightning Europe le'pt forth . . . . half startled at itself,
> Its feet upon the ashes and the rags . . . . Its hands tight to the
>      throats of kings.

(LG 1855, p. 133)

The run-on lines of the first version are transformed into the thought-rhythms of Whitman's more mature style, and in accordance with his desire to avoid conventional literary devices, he eliminated the allusion to Ahimoth in the poem's final version.

Whitman mythologized the revolt of the people against the king as part of a universal movement from enslavement to liberation as power is transferred from the centralized authority of the state to the sovereign power of the individual. "The People scorned the ferocity of kings" and treated their oppressors with mercy rather than vengeance, and thus the "frightened rulers come back":

> Each comes in state, with his train,
> Hangman, priest, and tax-gatherer,
> Soldier, lawyer, sycophant;
> An appalling procession of locusts,
> And the king struts grandly again.
>
> (*EPF*, p. 38)

Each figure in the king's train represents the system that Whitman defied in his journalistic protests against capital punishment, religious oppression, high tariffs, war and the military, the machinations of lawyers, and the servants of authority. Even though the "king struts grandly again" and "corpses lie in new-made graves," Whitman envisions the triumph of liberty as part of the regenerative law of the universe:

> Not a grave of those slaughtered ones,
> But is growing its seed of freedom,
> In its turn to bear seed,
> Which the winds shall carry afar and resow,
> And rain nourish.
>
> (*EPF*, p. 39)

In the 1855 version of the poem, these lines are joined and recombined to stress the linkage between the struggle for freedom and the fluid, eternal processes of nature:

> Not a grave of the murdered for freedom but grows seed for
>    freedom . . . . in its turn to bear seed,
> Which the winds carry afar and re-sow, and the rains and the
>    snows nourish.
>
> (*LG* 1855, p. 134)

This equation between the historical and the natural, between the revolutionary struggle for liberty and the physical laws of the universe, is further accentuated in the 1855 version when the poet exhorts his readers to "Turn back unto this day, and make yourselves afresh."

During the early 1850s, Whitman was, as he said in *Specimen Days*, "occupied in house building in Brooklyn," a part-time activity that freed him to engage his artistic powers. Although his active participation in party politics abated, he continued to support the cause of Free Soil. On August 14, 1852, he wrote to John P. Hale, a New Hampshire senator and an outspoken opponent of slavery. Presenting himself as "a stranger, a young man, and a true Democrat," Whitman urged Hale to accept the nomination of the Free-Soil party in the 1852 election: "Out of the Pittsburgh movement and 'platform' it may be that a real live Democratic party is destined to come forth, which, from small beginnings, ridicule, and odium, (just like Jeffersonian democracy fifty years ago,) will gradually win the hearts of the people, and crowd those who stand before it into the sea. Then we should see an American Democracy with thews and sinews worthy this sublime age" (*Corr.*, I, 39). Whitman's letter is more than a plea to the senator to join in the work of reviving American democracy; it is also a kind of minielection manual in which he instructs Hale on how to conduct his presidential campaign: "Look to the young men—appeal specially to them," he advised. "Take two or three occasions within the coming month to make personal addresses directly to the people, giving condensed embodiments of the principal ideas which distinguish our liberal faith from the drag-parties and their platforms. Boldly promulge these, with that temper of rounded and good-natured moderation which is peculiar to you; but abate not one jot of your fullest radicalism" (*Corr.*, I, 39–40).

Whitman wrote to Hale as a representative citizen, assuring him of the radical flame and divine urge for freedom that still burned in the hearts of the American people. "I know the people," he said.

> I know well, (for I am practically in New York), the real heart of this mighty city—the tens of thousands of young men, the mechanics, the writers, &c &c. In all these, under and behind the bosh of the regular politicians, there burns, almost with fierceness, the divine fire which more or less, during all ages, has only waited a chance to leap forth and confound the calculations of tyrants, hunkers, and all their tribe. At this moment, New York is the most radical city in America. It would be the most anti-slavery city, if that cause hadn't been made ridiculous by the freaks of the local leaders here. (*Corr.*, I, 40)

In his vision of an America split between the "bosh of the regular politicians" and the divine fire of the people, Whitman was already preparing, at least mentally, the political diatribe that he would write in *The Eighteenth Presidency!*

When Hale lost the election, winning only 150,000 votes, Whitman determined to make his own direct appeal to the American people. In his notebooks he was actively engaged in finding the most effective means of arousing what he called the "divine fire" of revolutionary sentiment in the mass of common people. Before he turned his main energies to writing poetry, he dreamed of "ranging up and down The States," delivering his message to the people in a series of lectures, which he referred to as "lessons."[10] Whitman's friend Richard Maurice Bucke added: "I believe he had formed this intention some years before such a book as *Leaves of Grass* was planned or even thought of. Nor did he drop the notion of lecturing as an integral part of his scheme of self presentation after he began to write the *Leaves,* but held to it certainly until the war."[11] Whitman's objective in these lectures was expressly political: to invigorate the physical and spiritual life of the American republic by educating the people in the religion of democracy. In an early note, under the heading "Lectures" or "Lessons," he wrote: "The idea of strong live addresses directly to the people, adm. 10c., North and South, East and West . . . promulging the grand ideas of American ensemble liberty, concentrativeness, individuality, spirituality &c &c" (*WWW*, p. 33).

The main subject and inspiration of these "lessons" was slavery, a topic on which Whitman may have lectured in the late 1840s and early 1850s.[12] In one note on a real or imagined event, Whitman described in great detail the orator's liberating power:

> As of the orator advancing
> As, for example, having been engaged to deliver one of the "Lessons" to an Anti Slavery Meeting—he does not go, smiling and shaking hands, waiting on the platform with the rest—but punctual to the hour, appears at the platform-steps with a friend, and ascends the platform, silent, rapid, stern, almost fierce—and delivers an oration of liberty—up-braiding, full of invective—with enthusiasm. (*WWW*, p. 74)

Like his political poems of the early 1850s, Whitman's orations of liberty were stimulated by the Compromise of 1850 and the subsequent tightening of the Fugitive Slave Law. Invoking the principles of the Declaration of Independence and the provisions of the Constitution, Whitman's orator envisions himself as part of a small band of freedom's warriors who seek, as did the biblical evangelists, to spread the new gospel of democracy by teaching "the inalienable right of every human being to his life, his liberty and his rational pursuit of happiness" (*WWW*, p. 75).

Whitman's "lessons" are full of revolutionary sentiment. Having embraced the premises of life and liberty as inalienable rights of all—"of whatever numbers, ages, hues, or language or belief"—he asserts the inalienable right of the individual to defend himself if these rights are taken away. Besides appearing to support the principle of slave revolt, Whitman declares his own willingness to assist individuals who flee their oppressors: "As to assisting such a person, it is not likely I shall ever have the privilege, but if I can do it, whether he be black or whether he be

white, whether he be an Irish fugitive or an Italian or German or Carolina
fugitive, whether he come over sea or over land, if he comes to me he gets
what I can do for him" (*WWW*, pp. 76–77). By his own refusal to compromise and his determination to draw out revolutionary principles to their
ideal limit, Whitman had arrived at a very radical lesson indeed. Like
Frederick Douglass in his *Narrative* (1845) and Henry David Thoreau in
"Resistance to Civil Government" (1849), Whitman turned the founding
ideology against the government itself, advocating a violation of American
civil and legal codes in the name of personal liberty and inalienable rights.

Although Whitman did not publish this lesson in any of his written
work, he did inscribe its essence in section 10 of "Song of Myself":

> The runaway slave came to my house and stopped outside,
> I heard his motions crackling the twigs of the woodpile,
> Through the swung half-door of the kitchen I saw him limpsey and
>     weak,
> And went where he sat on a log, and led him in and assured him
> And brought water and filled a tub for his sweated body and
>     bruised feet,
> And gave him a room that entered from my own, and gave him
>     some coarse clean clothes,
> And remember perfectly well his revolving eyes and his awkwardness,
> And remember putting plasters on the galls of his neck and ankles;
> He staid with me a week before he was recuperated and passed north,
> I had him sit next me at table . . . . my firelock leaned in the corner.
>
> (*LG* 1855, pp. 33–34)

Assuming the Christ-like role of regenerator, Whitman "re-members"
the runaway slave into new life and shares with him the meal that is a sign
of both human community and the feast of life. Not only does he assist the
slave in his flight north, but the firelock leaning in the corner suggests that
he will defend his flight to freedom at gunpoint if necessary.

The runaway-slave passage in "Song of Myself" also registers Whitman's defiance of the Compromise of 1850, which had brought federal
commissioners into the states to enforce the Fugitive Slave Law. In his
lecture notes, he protested the federal government's violation of personal
liberty and sovereignty, invoking the "iron law of rebellion" that the "true
American freeman holds in reserve." Characterizing the government action as a "violent intrusion from abroad" of a foreign and tyrannical
power, he points to its analogy with the American Revolution: "Is this a
small matter?  —The matter of tea and writing paper was smaller.  —
Why what was it—that little thing that made the rebellion of '76—a little
question of tea and writing paper only great because it involved a great

principle. But this is in every way a large question—because among other points it involves the large principle whether we or a power foreign to us shall be master of our own special and acknowledged ground" (*WWW*, p. 78). Here again, Whitman advanced the radical premise that popular insurrection rather than political compromise might be the only true means of preserving the Union.

It was this insurrectionary sentiment that inspired Whitman's poem "A Boston Ballad," which was written in response to the arrest, trial, and return of the fugitive slave Anthony Burns in 1854. In accordance with the new Fugitive Slave Law, Burns was arrested in Boston on May 24, 1854, on a warrant from a federal commissioner, judged by a federal court to be an escaped slave, and ordered returned to his southern master. When the incident provoked a riot in which one man was killed, federal and state troops were called in to guard the prisoner and escort him to the ship that was to return him to slavery. The irony of federal, state, and local authorities defending the institution of slavery in the very seat of American liberty was not lost on its opponents. "The boast of the slaveholder is that he will catch his slaves under the shadow of Bunker Hill," said Judge George H. Russell at a mass protest meeting in Boston, and at the same meeting, Theodore Parker referred to Burns as "a man stolen in the city of our fathers," by the graves of John Hancock and John Adams.[13] In response to the Burns incident, William Lloyd Garrison burned a copy of the U.S. Constitution in Boston on July 4, 1854.

"A Boston Ballad," which along with "Resurgemus" was one of the earliest poems to be included in *Leaves of Grass*, is Whitman's poetic burning of the Constitution. The poem turns on the irony of Burns's arrest and return in Boston. Inverting both the traditions of the ballad and the heritage of the Revolution, Whitman protests the Fugitive Slave Law and the politics of compromise, as well as the larger betrayal of republican values in an orderly, complacent, but ultimately dead America. "Have the ages so rolled backward, and humanity with them, that what they went to war to *stop*, seventy years ago, we shall now keep up a war to *advance*," Whitman had asked in an *Eagle* editorial in 1846. "A Boston Ballad" raises the same question, ironically juxtaposing the revolutionary past with the compromised politics of the present. Breaking the rhyme and meter of the traditional ballad with a jarring and cacophonous free form, Whitman's ballad records not the traditions of the folk but their undoing by political compromise and moral complacency.

"A Boston Ballad" moves in the manner of a procession similar to the one that accompanied Burns in delivering him back to slavery. Burns himself, however, is absent from the poem. This failure to name him as the subject of the poem adds an unintended dimension of irony, which reveals as it conceals the racial phobia of Whitman and his age. Like the nameless black narrator of Ralph Ellison's *Invisible Man*, Burns's absence is, in effect, a sign of the cultural invisibility to which the black person *as a person* would continue to be sentenced, by Northerners and Southerners alike, despite the rhetoric of liberty and equality that accompanied the antislavery movement in the nineteenth century and the Civil Rights movement in the twentieth century.

The poem begins and ends with an address to Jonathan, a figure of Yankee virtue in the revolutionary past, who represents the failure of republican traditions in the present:

> Clear the way there Jonathan!
> Way for the President's marshal! Way for the government cannon!
> Way for the federal foot and dragoons. . . . and the phantoms
>    afterward.
>
> (LG 1855, p. 135)

The procession, in which Jonathan as a figure of revolutionary defiance is thrust aside by the "federal foot" of the president's marshal and government cannon, dramatizes the conflict between individual and state, personal liberty and public law, revolutionary past and compromised present evident in the Burns case. To stress the betrayal of republican ideals, Whitman summons phantoms from the American Revolution to witness the procession:

> What troubles you, Yankee phantoms? What is all this chattering
>    of bare gums?
> Does the ague convulse your limbs? Do you mistake your crutches
>    for firelocks, and level them?
> If you blind your eyes with tears you will not see the President's
>    marshal,
> If you groan such groans you might balk the government cannon.
>
> (LG 1855, pp. 135–36)

The apparently senile but actually revolutionary gesture of mistaking their crutches for firelocks implies that had these figures been present in Boston as more than impotent phantoms, they would have defied the violation of personal liberty and popular sovereignty represented by the president's marshal and the government cannon; their sickness and malaise underscore the impotence of revolutionary principle in America. In the "well-dressed" and "orderly" America of the present, the revolutionary gesture is out of place:

> For shame old maniacs! . . . . Bring down those tossed arms, and let
>     your white hair be;
> Here gape your smart grandsons . . . . their wives gaze at them from
>     the windows,
> See how well-dressed . . . . see how orderly they conduct themselves.
>
> Worse and worse . . . . Can't you stand it? Are you retreating?
> Is this hour with the living dead too dead for you?
>
> Retreat then! Pell-mell! . . . . Back to the hills, old limpers!
> I do not think you belong here anyhow.
>
> (*LG* 1855, p. 136)

Lacking the revolutionary vitality of its ancestors, the America of the present is willing to sacrifice the principles of liberty, self-sovereignty, and inalienable rights for personal comfort and material wealth.

In their willingness to roll back the heritage of the Revolution, the American people are, in effect, ready for the return of King George, and it is his corpse that is summoned in the final section of the poem:

> Dig out King George's coffin . . . . unwrap his quick from the
>     graveclothes . . . . box up his bones for a journey:
> Find a swift Yankee clipper . . . . here is freight for you blackbellied
>     clipper,
> Up with your anchor! shake out your sails! . . . . steer straight toward
>     Boston bay.
>
> (*LG* 1855, p. 136)

The return of King George aboard a clipper, which also served as the "blackbellied" vehicle of slaves, suggests the association of the king with the institutions of slavery, black or white. As a sign of both technological progress and the traffic in slaves, the "swift" and "blackbellied" clipper also points to the paradox of American progress in the mid-nineteenth century: The advance of America had become bound to the advance of the institutions of slavery.

To Whitman when he was working on the first edition of *Leaves of Grass*, the American republic appeared "cadaverous as a corpse," having lost "its bloom of love and its freshness of faith." Depressed by the moral apathy and complacency bred by material prosperity, he came to see the American republic itself as a kind of plantation economy:

> Folks talk of some model plantations where collected families of niggers grow
> sleek and live easy with enough to eat, and no care only to obey a thriving
> owner, who makes a good thing out of them and they out of him   —By God
> I sometimes think this whole land is becoming one vast model plantation

thinking itself well off because it has wherewithal to wear and no bother about its pork. (*WWW*, pp. 82–83)

At a time when industrial entrepreneurs were making a "good thing" out of a newly emergent working class, Whitman's vision registers his knowledge, albeit intuitive, that the sources of American failure were economic. The slave system had become a trope for America itself, the sign of a culture of abundance propelled not by republican virtue but by the economics of market capitalism.

"Nations sink by stages, first one thing and then another," Whitman wrote in one of his antislavery lecture notes. By the summer of 1854, the capture and return of Anthony Burns in Boston was only one of a number of signs that America might be falling rather than rising in glory. Before the Burns case there had been the controversy over the Wilmot Proviso, the Compromise of 1850, and the provisions of the new Fugitive Slave Law. In 1850 in New York City, striking workers had for the first time been killed by police in a labor dispute.[14] There had been factionalism, sectional rivalry, and betrayal in the Democratic party; and in the presidential election of 1852, both Franklin Pierce, the Democratic candidate, and Winfield Scott, the Whig candidate, supported the politics of compromise. In May 1854, with the endorsement of President Pierce, Congress passed the Kansas–Nebraska Act, which annulled the Missouri Compromise and allowed for the extension of slavery in the territories. At the same time, after a brief period of prosperity, the nation was plunged once again into economic depression.

The individual's loss of economic power was, in Whitman's view, a further sign of the increasingly oppressive "foot" of government evident not only on the national and state level but also in the local government of Brooklyn. On October 16, 1854, Whitman addressed a letter to the common council and the mayor of Brooklyn, urging them to lift all Sunday restrictions. He reminded them of the limits of national, state, and local government in words that echo Tom Paine: "The office of Alderman or Mayor or Legislator is strictly the office of an agent. . . . He is not so continually to go meddling with the master's personal affairs or morals. Such is the American doctrine, and the doctrine of common sense" (*UPP*, I, 260). Appealing once again to revolutionary principle, Whitman advocated the expansion rather than the limitation of human freedom as the source of personal, local, and national good: "The citizen must have room. He must learn to be so muscular and self-possessed; to rely more on the restrictions of himself than any restrictions of statute books, or city ordinances, or police. This is the feeling that will make live men and superior women. This will make a great, athletic, spirited city, of noble and marked character . . ." (*UPP*, I, 261).

This dream of athletic and self-possessed citizens creating a spirited city is countered by the impotence of Jonathan and the revolutionary fathers and the corpselike body of America that Whitman had evoked in "A Boston Ballad" only a few months earlier. Like America itself, Whitman was torn between a boundless vision of republican regeneration, linked with the past and the traditions of the founding moment, and a deep-seated anxiety that as the last of the revolutionary fathers died, the republican body of America would itself die. As long as Whitman dreamed the dreams of the revolutionary founders, his imagination was fired by the vision of a regenerated America and a regenerated world, but when he turned to the *realpolitik* of his time, with its legacy of personal complacency and political compromise, his lyricism was undercut by irony. "It is not events of danger and threatening storms that I dread," he wrote in his antislavery notes. "Give us turbulence, give us excitement, give us the rage and disputes of hell, all this rather than this lethargy of death that spreads like a vapor of decaying corpses over our land" (*WWW*, p. 81).

This vision of the land as a corpse, a vision that became with T. S. Eliot's *The Waste Land* a key figure in the modern literary landscape, was at the very foundation of Whitman's regenerative "leaves." It was, at least in part, out of his desire to revive the dead body of republican America that he turned his main energies in 1854 to completing the poems of *Leaves of Grass*. The poems were not, as is commonly assumed, a product of Whitman's unbounded faith in the democratic dream of America; on the contrary, they were an impassioned response to the signs of the death of republican traditions he saw throughout the land and his growing fear that the ship of American liberty had run aground.

# 4

# Aesthetics and Politics

> I consider "Leaves of Grass" and its theory experimental—as,
> in the deepest sense, I consider our American republic itself to
> be, with its theory.
>
> —WHITMAN, *A Backward Glance o'er Travel'd Roads*

In his 1844 essay "The Poet," Ralph Waldo Emerson described an American poet who did not exist. "Our logrolling, our stumps and their politics,
our fisheries, our Negroes, and Indians, our boasts, and our repudiations,
the wrath of rogues, and the pusillanimity of honest men, the northern
trade, the southern planting, the western clearing, Oregon, and Texas, are
yet unsung."[1] Critics have been quick to point out that Whitman's poems
appear to be a direct response to Emerson's call for an American poet.[2]
Emerson himself responded to the 1855 *Leaves of Grass* in an act of
literary recognition that was as generous as it was prophetic. Only a few
weeks after Whitman sent him a copy of his poems, Emerson wrote: "I
greet you at the beginning of a great career, which yet must have had a
long foreground somewhere, for such a start. I rubbed my eyes a little, to
see if this sunbeam were no illusion; but the solid sense of the book is a
sober certainty. It has the best merits, namely, of fortifying and encouraging" (*LGC*, p. 732). Gratified by Emerson's recognition, Whitman published the letter along with his own letter addressing Emerson as "dear
Friend and Master" in the 1856 edition of *Leaves of Grass.*

The Emerson–Whitman exchange over the 1855 *Leaves* is usually
cited as evidence that Emerson was the principal catalyst of Whitman's
poems. But Whitman's 1856 letter is full of ambiguity. The irony of the
bard of American democracy addressing the New England sage as Master
was probably unintended, but the letter itself is less a tribute to Emerson
than it is a critique of the institutions of literature that he represented.
Whitman must surely have realized that the program he announced—his
intent to root American literature in "sex, womanhood, maternity, desires,

lusty animations, organs, acts"—was an assault on both the genteel literary establishment that Emerson represented and the Puritan pieties of New England itself.[3]

Although Emerson's influence on Whitman seems unmistakable, it was not totalizing. "I was simmering, simmering, simmering; Emerson brought me to a boil," Whitman told John Trowbridge during the Civil War. But it is important to bear in mind that these words, which have often been cited as proof of Emerson's definitive role in Whitman's poetic formation, were recorded secondhand years after Whitman's death.[4] By dwelling on Emerson in our study of Whitman, we have tended to dehistoricize him, privileging the transcendental value of the religiospiritual prophet at the expense of the historically specific poet with roots in working-class culture in Brooklyn and New York.

What is in fact most interesting about the relationship between Whitman and Emerson is not their likeness but their difference. If Whitman had a literary parentage, it was not like Emerson's among the Puritan fathers, but among the Sons of Liberty. Whitman was not a student of E. T. Channing at Harvard nor a member of the Transcendental Club in Concord; his roots were not so much in the religious as in the political battles of his times. In his essay "The Poet," Emerson exhorts the poet to leave the world of affairs: "Thou shalt leave the world, and know the muse only. Thou shalt not know any longer the times, customs, graces, politics, or opinions of men, but shall take all from the muse. . . . Thou shalt lie close hid with nature, and canst not be afforded to the Capitol or the Exchange" (*CW*, III, 41). While Emerson was advocating a flight from the world into nature and spiritual transcendence, Whitman was living and working in New York City, in the world of banks and tariffs, newspapers and caucus, stump and crowd; and it was out of this world rather than out of the study or pine woods of Emerson's scholar that he emerged as a poet.

Whitman's democratic poetics—his attempt to create a democratic language, form, content, and myth commensurate with the experimental politics of America, to embody in his poetic persona America's unique political identity, and to engage the reader as an active participant in the republican politics of his poem—may be best understood not in relation to Emerson and the Transcendentalists but in relation to the body of political and aesthetic thought that emerged from the American Revolution.

The poet Whitman describes in his 1855 preface to *Leaves of Grass* and celebrates in his twelve untitled poems bears the rhetorical traces of the revolutionary enlightenment. "The attitude of great poets is to cheer up slaves and horrify despots," Whitman declares. "The turn of their

necks, the sound of their feet, the motions of their wrists, are full of
hazard to the one and hope to the other. Come nigh them awhile and
though they neither speak or advise you shall learn the faithful American
lesson." The American lesson is the lesson of liberty and equality be-
queathed by the revolutionary fathers: "They are the voice and exposition
of liberty," he says of the American poet. "They out of ages are worthy the
grand idea . . . to them it is confided and they must sustain it" (*LG* 1855,
p. 15). The American poet is also an "equable man": "He is the equalizer
of his age and land," and his art reflects the enlightenment ideals of
equality and balance: "About the proper expression of beauty there is
precision and balance . . . one part does not need to be thrust above
another" (*LG* 1855, pp. 8, 12). Echoing the anticlerical rhetoric of Paine's
*The Age of Reason* and Volney's *The Ruins,* Whitman invests these so-
cioaesthetic and politically contingent ideals of liberty, equality, and balance with the inevitability of natural law: "The whole theory of the special
and supernatural and all that was twined with it or educed out of it departs
as a dream. What has ever happened . . . what happens and whatever may
or shall happen, the vital laws enclose all . . . they are sufficient for any
case and for all cases" (*LG* 1855, p. 14).

Unlike Poe, who regarded science as a vulture of dull realities prey-
ing on the poet's heart, Whitman embraced scientific knowledge as a
source of poetic possibility. "Exact science and its practical movements
are no checks on the greatest poet but always his encouragement and
support. . . . The sailor and traveler . . . the anatomist chemist astrono-
mer geologist phrenologist spiritualist mathematician historian and lexicographer are not poets, but they are the lawgivers of poets and their
construction underlies the structure of every perfect poem" (*LG* 1855, p.
14). Like the writers of the revolutionary enlightenment, Whitman regarded advances in science and technology as part of the progressive
amelioration of the world: "For the eternal tendencies of all toward
happiness," he says, "make the only point of sane philosophy" (*LG*
1855, p. 15).

Even the expressive power of the English language originated, in
Whitman's view, in the revolutionary heritage of "a race who through all
change in circumstances was never without the idea of political liberty,
which is the animus of all liberty. . . . It is the powerful language of
resistance . . . it is the dialect of common sense. . . . It is the chosen
tongue to express growth faith self-esteem freedom justice equality friend-
liness amplitude prudence decision and courage. It is the medium that
shall well nigh express the inexpressible" (*LG* 1855, pp. 22–23).

As an experiment in poetic form, *Leaves of Grass* is marked by the revolutionary politics that Whitman announced in the 1855 preface. If the book bears the impress of Emerson's "The Poet," it also responds to the call for an indigenous American art that had roots in the revolutionary period. Wrestling with the same problems that Philip Freneau and Joel Barlow had in trying to create a distinctively American literature, Whitman seeks to reconcile politics and poetry, activism and art, revolutionary ideology and a revolutionary creation.

When in *A Grammatical Institute* (1783), Noah Webster declared that America must "be as distinguished by the superiority of her literary improvements, as she is already by the liberality of her civil and ecclesiastical constitutions," he was the first to recognize and insist on the relationship between the political republic and the republic of letters in America.[5] Simultaneously with the move to create a political Constitution, men of letters set about the work of constituting—and in effect controlling—America culturally by writing a national epic. In 1785, Timothy Dwight published *The Conquest of Canaan*, America's first epic, and in 1787, Joel Barlow published *The Vision of Columbus*. As epics of revolutionary America, both texts are governed by neoclassical aesthetic codes, Puritan religious structures, and the conservative interests of the ruling social class.

In *The Columbiad* (1807), a revision of *The Vision of Columbus* that Barlow wrote during his own ideological engagement with the French revolutionary struggle, he became increasingly conscious of the potential conflict between conservative form and radical content. Calling attention to the conflict between the more radical premises of republican ideology and the essentially conservative standards of neoclassical taste in the preface to *The Columbiad*, he attempts to mend this gap by distinguishing between poetic means and moral end: "There are two distinct objects to be kept in view in the conduct of a narrative poem: the *poetical* object and the *moral* object. The poetical is the fictitious design of the action; the moral is the real design of the poem." Barlow's formulation is in fact the reverse of T. S. Eliot's Modernist dictum that a poem's content should divert the reader while the poem does its work. Although Barlow is intrigued by the fictitious design of Homer and Virgil, he is troubled by their political design: Each encouraged militarism and the subjugation of the people to a princely power. The "real design" of Barlow's epic is the construction of the American republic: "My object is altogether of a moral and political nature. I wish to encourage and strengthen, in the rising generation, a sense of the importance of republican institutions; as being the great foundation of public and private happiness, the necessary aliment of future and permanent meliorations in the condition of human nature."6 Despite his effort at reconciliation, however, Barlow remains divided between the poetical and the political design of his poem. Harking back to the Horatian concept of teaching and delighting, he seeks to achieve his avowedly political designs on his readers by delighting them with "fictitious" neoclassical structures. Although he experiments with language and orthography in *The Columbiad*, he never fully challenges the political bounds of neoclassical aesthetics by reinventing both the form and the content of the American epic.

During the early national period, literature became a rhetorical battle-field in which writers carried on the political struggle between Federalists and anti-Federalists over who would name and preside over the constitution of the American republic. After the French Revolution and the Reign of Terror, while Barlow was revising his epic to reflect the politics of radical republicanism, there was a significant move among cultural critics away from the democratic implications of revolutionary ideology. In 1807, the same year that Barlow published *The Columbiad*, the Reverend Theodore Dehon delivered a "Discourse upon the Importance of Literature to Our Country" to the Phi Beta Kappa Society at Harvard, warning against the danger of democratizing the republic of letters. "Shall we be pardoned the expression, if we further observe, that through the innovating spirit of the times the *republick* of letters may have its dignity and prosperity endangered by sliding inadvertently into a *democracy?*"7 Dehon's lecture articulates three essential features of American literary "discourse" as it developed in the early nineteenth century: the idea of the "importance" of literature as a vehicle of national acculturation and control, the idea that the "*republick* of letters" should be governed in the interests of the existing social order, and the idea of literary democracy as a threat to traditional hierarchies of authority and power.

As relations of power in the American republic consolidated after the defeat of England in the War of 1812, cultural nationalism rose, but the call for an American art was curiously devoid of any concept of art as a democratic structure. Even in the prestigious *North American Review*, which was established in 1815 to help create and disseminate original American writing, the critical stress was on the artist as a kind of national keeper of public morals who speaks to and for rather than of and from the people. Despite the rhetoric of cultural nationalism, the critical emphasis on literature as a moral instrument tended to de-politicize the role of the writer and relocate literature in the private rather than in the public sphere.8

This move away from the revolutionary emphasis on the pen as an expressly political power toward the Romantic concept of art as a moral and religious force may have been in part a response to the increasing commercialization of art and artist in the literary marketplace. As the constituted authority of the republic of letters became subject to the tastes of a mass audience and the literary economics of supply and demand, the writer retreated from the public sphere in a paradoxical attempt to recuperate lost social power by claiming absolute moral power. In "Remarks on National Literature," which appeared in *The Christian Examiner* in 1830, William Ellery Channing stressed the paradoxically apolitical nature of American national literature. "We love our country much," he said, "but mankind more."⁹ As defined by Channing, literature was an expression of neither the democratic people nor the republican principle; rather, it was an expression by superior minds of religious and moral truths. Placing the writer not among but above the people, and apart from rather than within the political sphere, Channing argued that literature would eventually replace politics as the sphere of the best minds and the widest influence.

Emerson proposed a similar separation of the poet from the political life of his times in his famous 1846 "Ode" inscribed to the politically active W. H. Channing:

> Though loath to grieve
> The evil time's sole patriot,
> I cannot leave
> My honied thought
> For the priest's cant,
> Or the statesman's rant.
>
> If I refuse
> My study for their politique,
> Which at the best is trick,
> The angry Muse
> Puts confusion in my brain.
> (*CW*, IX, 76)

Presenting the poet as a speaker of absolute truths untarnished by the "politique" of the "priest's cant" and the "statesman's rant," Emerson's "Ode" inscribes the Romantic split between poetry and politics that would continue to govern Modernist and New Critical notions of literary purity and aesthetic value.

Whereas Emerson insisted on the necessary isolation of the artist from the life of his times, the young socialist Orestes Brownson anticipated Whitman in predicting that a truly indigenous American literature would come not from the study of the scholar but from the struggles of the democratic masses, particularly the struggle between labor and capital. Critical of the separation of American writers from the people, Brownson observed: "The literary men of our country have not sympathized with the people [and] have had in their minds no clear conceptions of the great doctrine of equal rights, and social equality, to which this nation stands pledged . . . they have not been true democrats in their hearts." To create a national literature, he commented, in words that foreshadow Whitman's democratic persona, American writers must be the "impersonations" of the people's "wishes, hopes, fears, sentiments."[10] Extending the egalitarian rhetoric of Jefferson and Jackson to the republic of letters, Brownson, along with the Young America group at the *Democratic Review*, reacted against the concept of literature as a privileged order existing apart from the energies and interests of the people. "The favorite topics of the Poet and the People," wrote William A. Jones in the *Democratic Review*, are the "necessity and dignity of labor . . . the native nobility of an honest and brave heart; the futility of all distinctions of rank and wealth . . . the brotherhood and equality of men."[11] The article, which Whitman had among his files, is a virtual checklist of the republican ideals he himself invoked first as a journalist and later as the poet and critic of democracy.

In his own journalistic agitation for a national art, Whitman was closer to the political rhetoric of Brownson and the Young America group than he was to the transcendental ideals of Channing and Emerson. In articles on literature, music, drama, and painting, he admonished American artists to overthrow the tyranny of old-world structures. Like Brownson, he was particularly critical of the antidemocratic influence of the British in American art: "It must not be forgotten, that many of the most literary men of England are the advocates of doctrines that in such a land as ours are the rankest and foulest poison.   —Cowper teaches blind loyalty to the 'divine right of kings,'—Johnson was a burly aristocrat—and many more of that age were the scorners of the common people, and pour adulation on the shrine of 'toryism.' " These writers and several others, Whitman said, "exercise an evil influence through their books . . . for they laugh to scorn the idea of republican freedom and virtue" (*UPP*, I, 122).

To counteract the antirepublican influence of old-world culture, Whitman urged American artists to create an art commensurate with new-world democracy. In an article entitled "Miserable State of the Stage," he called upon "some *American*" to "revolutionize the drama," to create plays "fitted to American opinions and institutions" (*GF*, II, 313–14). The art he imagined was an art that would both express and stimulate the republican spirit of the nation. Critical of the slavish dependence of American drama on English managers, actors, and plays, Whitman said: "The drama of this country *can* be the mouth-piece of freedom, refinement, liberal philanthropy, beautiful love for all our brethren, polished manners and an elevated good taste. It can wield potent sway to destroy any attempt at despotism" (*GF*, II, 314–15).

To promote republican habits of mind and character among his readers, Whitman introduced a book review section in the *Eagle*. In reviews of over a hundred books ranging in subject from native and foreign literature to ancient and modern history, religion, astronomy, phrenology, marriage, and women's diseases, he applied a similarly political standard of measure: How would the book affect the physical, mental, and spiritual growth of the American republic? William Hazlitt's *Napoleon* was a healthy influence on the budding republic: "A noble, grand work!, a democratic work! Let every lover of the *race* before classes . . . every encourager of the great rights of man—help promulge this book. It is a wholesome book for the young fresh life of our Republic" (*GF*, II, 285). Walter Scott and William Shakespeare, however, were less admirable; although they were two of Whitman's own guilty pleasures, he feared the antidemocratic structures of their works. "Scott was a Tory and a High Church and State man. The impression after reading any of his fictions where monarchs or nobles compare with patriots and peasants, is dangerous to the latter and favorable to the former. . . . In him as in Shakespere [*sic*], (though in a totally different method) 'there's such divinity does hedge a king,' as makes them something more than mortal—and though this way of description may be good for poets or loyalists, it is poisonous for freemen" (*GF*, II, 264–65).

Whitman was Brownson's true democrat of the heart. Just as during the revolutionary era, verse was used in newspapers, broadsides, and pamphlets as a popular form of political debate, so Whitman's poems were born out of a tradition in which poetry was not separate from but actively engaged in the political struggles of the time. To Whitman the "politique" of America was a found poem. In one of his early notebooks he referred to the "advent of America" and the events since the American Revolution as the most "inspiring" occurrence to be given to a poet in all times (*CW*, IX, 106). Defying the endless plaints about the inadequacy of America as an artistic subject, he was the first writer to respond to the challenge of the new poetic vistas opened by democracy. "The United States themselves are essentially the greatest poem," he declared in the 1855 preface. "Here are the roughs and beards and space and ruggedness and nonchalance that the soul loves" (*LG* 1855, p. 5).

Reversing the archetype of the poet as alien to the material culture of America—an archetype that was anticipated by Freneau and received its fullest formulation in the work of Poe—Whitman installed his muse amid the gadgetry and bric-a-brac of the modern world. "Imagination and actuality must be united," he asserted, as he sought to "give ultimate vivification to facts, to science, and to common lives, endowing them with the glows and glories and final illustriousness which belong to every real thing, and to real things only" (*CW,* IX, 127; *LGC,* p. 564). Whitman's desire to "give ultimate vivification to facts, to science, and to common lives" might be read as a response to the essentially unpredictable workings of a new market economy in which "real things" were being emptied of place and significance.

Seeking to endow the facts of modern industry and technology with "the glows and glories and final illustriousness" of an imaginative and ultimately spiritual economy, Whitman sang of the city crowd, of butchers, mechanics, and day laborers; of factories, foundries, workshops, and the "blab of the pave"; he celebrated steam power and express lines, the Hoe press and the cotton gin, gas and petroleum, the Suez Canal and the Atlantic cable. But in his urge to make things mean, to give the potentially alienating modern landscape "ultimate" signification, he had to glide over or silence the more sordid conditions of industrial and technological transformation, making these conditions seem right, natural, and divinely ordained. His attempt to unite imagination and actuality, an essentially visionary political community with the fact of a modern industrial economy, could also work paradoxically to justify the very conditions of impoverishment and dehumanization he set out to challenge.

While Whitman's Romantic contemporaries in America were trying to remove art from the political sphere by advocating, like Emerson, a poetics of transcendence or, like Poe, an aesthetics of pure poetry, Whitman openly proclaimed the political contingency of art and artist. Just as the political creation of America had required the overthrow of the Old World's political system, so the poetic creation of America would require the overthrow of the Old World's artistic structures: "Has not the time arrived when, (if it must be plainly said, for democratic America's sake, if for no other) there must imperatively come a readjustment of the whole theory and nature of Poetry?" (*LGC,* pp. 566–67).

Whitman's attempt to readjust the theory and nature of poetry centered on the issue of class: "The tendency permitted to literature has always been and now is to magnify and intensify its own technism, to isolate itself from general and vulgar life, and to make a caste or order" (*CW,* IX, 197). With its focus on the past, on elevated character and action, on aestheticism and closed form, and on an economy of hierarchy, order, and restraint, old-world literature was a class act, reflecting the aristocratic values and viewpoint of an exclusive group. Whitman was conscious of the political inscription of poetry, even among putatively apolitical Romantic writers: "Of the leading British poets many who began with the rights of man abjured their beginning and came out for kingcraft, obedience and so forth. Southey, Coleridge, and Wordsworth did so" (*CW*, IX, 116). Whereas his American contemporaries found antecedents in English and German Romanticism, Whitman found antecedents more relevant to American democratic experience among French writers—in the works of Volney and Rousseau, Michelet and Sand, Béranger and Hugo—whose works were shaped by the egalitarian and libertarian ideals of the revolutionary era.[12]

In creating the content and form of *Leaves of Grass*, Whitman sought to overturn the poetic conventions he associated with the literature of caste and privilege:

> For grounds for "Leaves of Grass," as a poem, I abandon'd the conventional themes, which do not appear in it: none of the stock ornamentation, or choice plots of love or war, or high, exceptional personages of Old-World song; nothing, as I may say, for beauty's sake—no legend, or myth, or romance, nor euphemism, nor rhyme—But the broadest average of humanity and its identities in the now ripening Nineteenth Century, and especially in each of their countless examples and practical occupations in the United States to-day. (*LGC*, p. 564)

Whitman's Bloomian "swerve" from his poetic fathers was not always successful. Although he broke with the conventions of linear narrative, he continued to use the rituals of myth and romance. Themes of love and war were at the very center of his verse; the "high exceptional" personage of President Abraham Lincoln was the subject—or at least the occasion—of his famous Civil War elegy "When Lilacs Last in the Dooryard Bloom'd"; and in the postwar years, he often returned to the conventions of narrative, public occasion, and even rhyme and meter. In fact, insofar as the experimental poetics of *Leaves of Grass* was, as Whitman suggested, a response to the experimental politics of the American republic, his return to more traditional forms during and after the Civil War might be read as a sign of his increasing disillusion with the entire democratic enterprise.

Nevertheless, in his early phase at least, Whitman did make a revolutionary move against the past. As a model democrat, Whitman's poet is a breaker of bounds: He is female and male, farmer and factory worker, prostitute and president, citizen of America and citizen of the world;

shuttling between past, present, and future, he is an "acme of things accomplished" and an "encloser of things to be" (*LG* 1855, p. 77). His songs are songs not only of occupations but also of sex and the body. Celebrating the body as the luxuriant growth of nature and sexual energy as the regenerative law of the universe, Whitman sang of masturbation, the sexual organs, and the sexual act; he was one of the first poets to write of the "body electric," female eroticism, homosexual love, and the anguish of repressed desire. By equating democracy with sexual liberation, Whitman was also the first poet to provoke among his unsympathetic readers what was (and perhaps still is) the deepest fear of democracy in America, namely, that in its purest form democracy would lead to a blurring of sexual bounds and thus the breakdown of a social and bourgeois economy based on the management of the body and the polarization of male and female spheres.

Whitman's move against tradition placed him in a potentially oppositional relation to the mass audience he wanted to reach and to teach. There is a paradoxical tension in his work between his revolutionary commitment to an experimental poetics and his desire to write as accessibly as possible. Moreover, in his attempt to fuse the vitality of the artisan culture from which he arose with the aesthetic codes of European and American Romanticism, Whitman further distanced himself from a more general readership. Despite, or perhaps because of, his revolutionary poetics, Whitman was himself caught in the "tendency" of literature "to magnify its own technism, to isolate itself from the general and vulgar life," and to make a separate class.

In asking "how best can I express my own distinctive era and surroundings, America, Democracy?" Whitman decided that "the trunk and centre whence the answer was to radiate . . . must be an identical body and soul, a personality—which personality, after many considerations and ponderings I deliberately settled should be myself" (*LGC*, p. 569). Just as the American Revolution had led to a relocation of authority inside rather than outside the individual, so Whitman's myth of origins focused not on the exploits of an historical or mythical figure of the past but on the heroism of a self who was, like the nation, in the process of creation. By concentrating on the heroism not of a completed action but of a self in the process of formation, Whitman created a self and a book that could and would grow, evolve, and change in response to corresponding changes in his own and the nation's development.[13]

The self of Whitman's songs is typical, not exceptional. Unlike Barlow's *Columbiad*, which still reflects the epic convention of a hero elevated above the people, Whitman invests the poet hero of *Leaves of Grass* with the energies of the people: "All others have adhered to the principle [that] the poet and savan form classes by themselves, above the people, and more refined than the people; I show that they are just as great when of the people, partaking of the common idioms, manners, the earth, the rude visage of animals and trees, and what is vulgar" (*CW*, IX, 36–37). Intimately acquainted with the hawkers, roughs, and gangs of the New York Bowery, Whitman sought to practice a streetwise poetics that collapsed the traditional hierarchies separating hero and people, poet and audience.

For Whitman, as for Barlow and Webster, the question of democratic creation—personal, political, and artistic—came to center on the question of language. "No country can have its poems without it have its own names," Whitman wrote in *The Primer of Words*, a series of notes on language he composed in the 1850s.[14] This was the period of Whitman's most sustained thought about the connection between language and politics in America. In addition to the *Primer*, he wrote and collected extensive materials on language. In 1856 he published in *Life Illustrated* an article on the English language entitled "America's Mightiest Inheritance"; in the "Appendant for Working-People, Young Men and Women, and for Boys and Girls," he included "A Few Foreign Words, Mostly French, Put Down Suggestively." According to C. Carroll Hollis, in 1859 Whitman also wrote with William Swinton a book on language entitled *Rambles Among Words*.[15]

In the *Primer*, which was originally intended as a lecture "For American Young, Men, and Women, for Literats, Orators, Teachers, Musicians, Judges, Presidents, &c.," Whitman called upon Americans to "throw off" the rule of standard English in order to create a language commensurate with the "new occasions, new facts, new politics, new combinations" of America (*DN*, III, 734). "I have heard it said that when the spirit arises that does not brook submission and imitation, it will throw off the ultramarine names. —That spirit already walks the streets of the cities of These States—I, and others, illustrate it. I say America too shall be commemorated—shall stand rooted in the ground in names—" (*DN*, III, 755).

Like the political revolution, the linguistic revolution that Whitman envisioned meant relocating power in the hands of the people and shifting authority from the rules of dictionaries and grammars to the "real grammar" of the spoken language. "The words continually used among the people are, in numberless cases, not the words used in writing, or recorded in the dictionaries by authority. —There are many words in daily use, not inscribed in the dictionary, and seldom or never in any print. —

Alas, the forms of grammar are never persistently obeyed, and cannot be. —The Real Dictionary will give all words that exist in use, the bad words as well as any" (*DN*, III, 734–35).

The issue of language was once again a class issue. By opening the privileged domain of language to the idioms of the people, Whitman attempted to break down the hierarchies of a class system based on linguistic usage.

> Do you suppose the liberties and brawn of These States have to do only with delicate lady-words? with gloved gentleman-words? Bad Presidents, bad judges, bad clients, bad editors, owners of slaves, and the long ranks of Northern political suckers, monopolists, infidels (robbers, traitors, suborned,) castrated persons, impotent persons, shaved persons, supplejacks, ecclesiastics, men not fond of women, women not fond of men, cry down the use of strong, cutting, beautiful rude words—To the manly instincts of the People they will forever be welcome. (*DN*, III, 746)

Recognizing the interrelation between linguistic, class, and political systems, Whitman represents language as a site of social struggle, a relationship of power in which the "rude words" of the people struggle against the lady and gentleman words of the dominant class.

Whitman experimented with language as both product and agent of cultural creation. If language could work as a form of social control, it could also be reworked as a medium of social transformation. Insisting on the relationship between linguistic change and political change, he filled his *Primer* with extensive lists of words and categories of words that he wanted to introduce into spoken and written English. One note reads:

> Words of Men and Women—the hundreds of different nations, tribes,
>     colors, and other distinctions,
> Words of the Sea
> Words of Modern Leading Ideas,
> Words of Modern Inventions, Discoveries, engrossing Themes, Pursuits,
> Words of These States—the Revolutionary Year 1, Washington, the Primal
>     Compact, the Second Compact, (namely the Constitution)—trades,
>     farms, wild lands, iron, steam, slavery, elections, California, and
>     so forth,
> Words of the Body, Senses, Limbs, Surface, Interior.
>
>                                                     (*DN*, III, 750)

Seeking to enlarge the range of expression with the words of prostitutes, fighting men, mechanics, farmers, blacks, factories, manufacturers, technology, science, politics, the body, emotions, and sex, Whitman's *Primer* is itself a kind of preliminary sketch toward what he called the "Real Dictionary" of the American people.

In fact, Whitman's extensive manuscript notes on language suggest that he intended not only to deliver a series of lectures on language but also to compose some kind of dictionary of American democracy. His notes include long lists of foreign and American words and phrases, a heavily corrected manuscript "Our Language and Literature," and a homemade volume entitled *Words*, which contains lists of slang, common idioms, and foreign terms; clippings on language; and entries on etymology, orthography, philology, and American names. In these notes, Whitman, like Barlow in the *Columbiad*, experiments with orthography and dreams of a world united in peace and freedom by a global language.

"Talk to everybody everywhere," he wrote in a directive to himself. "Keep it up—*real* talk—no airs—real questions—no one will be offended" (*DN*, III, 675). In a note on "Language and Literature," he observes: "I love to go away from books, and walk amidst the strong coarse talk of men as they give muscle and bone to every word they speak. —I say The great grammar, and the great Dictionary of the future must . . . embody [those]" (*DN*, III, 811). The notebooks he kept as he went about talking to "everybody everywhere" include detailed notes on regional accents, slang phrases, and popular idioms. One note reads "the New York Bowery boy—Sa-a-a-y! What—a—t?" (*DN*, III, 669), and another reads " 'cut loose' (railroad men's term Fred's explanation)" (*DN*, III, 674). In one entry on American idioms, which is probably the basis of his 1860 poem "So Long!" Whitman writes: " 'So long'—(a delicious American—New York—idiomatic phrase at parting equivalent to 'good bye' 'adieu' " (*DN*, III, 669).

Whitman's intent was both to record and to reinvent American English as a democratic medium. "Whigs," he wrote in a list of political terms, "what a ridiculous name for an American party" (*DN*, III, 683), and among "Words Wanted" he lists "A word which *happily* expresses the idea of An Equal Friend of All These States" (*DN*, III, 709). Recognizing the symbolic and creative dimension of language, he proposed renaming the months of the year, as well as many counties, islands, rivers, and cities in America. "What is the name of Kings' County? or of Queens' County to us?" he asks, "or St. Lawrence County? Get rid as soon as convenient of all the bad names—not only of counties, rivers, towns,—but of persons, men and women—" (*DN*, III, 701). In his attempt to democratize relations between men and women by reinventing the American language, he experimented with agent-nouns ending in *ess:* revolter, revoltress; tailor,

tailoress; orator, oratress. He also anticipated twentieth-century transformations in the conventions of naming: "The woman should preserve her
own name, just as much after marriage as before[.] Also all titles must be
dropped—no Mr. or Mrs. or Miss any more" (*DN*, III, 712).

Although Whitman never composed his "real grammar" or dictionary
of the American people, his *Primer*, his *Words* book, and his extensive
notes on language served as a kind of laboratory for the linguistic experiments of *Leaves of Grass*. "This subject of language interests me—
interests me," he told Traubel. "I never quite get it out of my mind. I
sometimes think the Leaves is only a language experiment—that it is an
attempt to give the spirit, the body, the man, new words, new potentialities
of speech—an American, a cosmopolitan . . . range of self-expression.
The new world, the new times, the new peoples, the new vista, need a
tongue according—yes, what is more, will have such a tongue—will not be
satisfied until it is evolved" (*DN*, III, 729 n).

The range of Whitman's language experiment in *Leaves*, which is
particularly evident in the 1855, 1856, and 1860 editions, is indicated by
the almost fourteen thousand words he used in his poems, of which over
half were used only once.[16] "In most instances," he wrote in the *Primer*,
"A characteristic word once used in a poem, speech, or what not, is then
exhausted" (*DN*, III, 750). By using strong words only once and by drawing on slang phrases and common idioms, Whitman sought to open the
"potentialities of speech" as a means of extending the possibilities of
democracy. His poems are alive with street talk, with words such as *rowdy*,
*swap off*, *top-knot*, and *duds;* and colloquial speech rhythms such as
"washes and razors for foofoos . . . . for me freckles and a bristling beard"
or "You there, impotent, loose in the knees, open your scarfed chops till I
blow grit within you" (*LG* 1855, pp. 46, 70).

Much of Whitman's linguistic experimentation in *Leaves* emerged
from his notebooks on popular usage. His notebook on *Words* contains the
following entry: "plentiful crops of words, or new applications of words
arising out of the general establishment and use of new inventions, such
as the words from the steam-engine, and its various moving and stationary
structures, on land and water—words from the electric telegraph, the
sewing-machine, the daguerreotype, the modern newspaper press. Many
of the above are words of *Personnel*—of the names applied to the men and
women who have to do with the new inventions" (*DN*, III, 710). The note
anticipates "A Song for Occupations" (1855), which is a kind of poetic
primer of words that introduces the "personnel" and "plentiful crops of
words" arising out of new inventions:

The pump, the piledriver, the great derrick . . the coalkiln and
  brickkiln,
Ironworks or whiteleadworks . . the sugarhouse . . steam-saws,
  and the great mills and factories;
The cottonbale . . the stevedore's hook . . the saw and buck of the
  sawyer . . the screen of the coalscreener . . the mould of the
  moulder . . the workingknife of the butcher;
The cylinder press . . the handpress . . the frisket and tympan . . the
  compositor's stick and rule,
The implements for daguerreotyping . . . . the tools of the rigger or
  grappler or sailmaker or blockmaker.

(LG 1855, pp. 94–95)

In another note, Whitman lists Western nicknames for people of different
states: Badgers for Wisconsin people, Buckeyes for Ohioans, Hoosiers for
Indianians, Tuckahoes for Virginians, and Kanucks for Canadians (*DN*,
III, 813–14). These nicknames became part of his corporate identity in
"Song of Myself": "Kanuck, Tuckahoe, Congressman, Cuff, I give them
the same, I receive them the same" (*LG* 1855, p. 29).

Whitman's language experiments did not go unnoticed by the genteel
literary establishment. Whitman, said Oliver Wendell Holmes, "takes into
his hospitable vocabulary words which no English dictionary recognizes as
belonging to the language. . . ."[17] His linguistic innovations were in fact
part of a national debate about the relationship between language and
culture in America that reaches back at least as far as Captain John Smith
who, in his *Map of Virginia* (1612), invented several neologisms to describe
the unfamiliar flora and fauna of America. This debate, which came to a
head during the revolutionary period, centered on whether America
should have a national language policy and on whether America should
write and speak in "pure" or American English.

From the first, those favoring an American standard of English insisted on the connection between the American language and the American political system. In a letter to the Continental Congress in 1780
proposing the establishment of a national language academy, John Adams
stated: "It is not to be disputed that the form of government has an
influence upon language, and language in its turn influences not only the
form of government, but the temper, the sentiments, and the manners of
the people."[18] Arguing a similar relation between language and politics,
Noah Webster wrote what was in effect America's declaration of linguistic
independence: "As an independent nation," he said in *Dissertations on the
English Language* (1789), "our honor requires us to have a system of our

own, in language as well as government. Great Britain, whose children we are, and whose language we speak, should no longer be *our* standard; for the taste of her writers is already corrupted, and her language on the decline."[19]

Webster's views, though put into practice by Barlow in *The Columbiad,* were by no means universal. The concept of a new American language was vigorously contested not only in the British press but also by several prominent Americans, including Benjamin Franklin. The case for the opposition was summed up in 1816 by the distinguished nineteenth-century linguist, John Pickering:

> The language of the United States has perhaps changed less than might have been expected, when we consider how many years have elapsed since our ancestors brought it from England; yet it has in so many instances departed from the English standard, that our scholars should lose no time in endeavouring to restore it to its purity, and to prevent future corruption. . . . As a general rule also, we should avoid all those words which are noticed by English authors of reputation, as expressions with which *they are unacquainted;* for although we might produce some English authority for such words, yet the very circumstance of their being thus noticed by well educated *Englishmen,* is proof that they are not in use at this day in England, and, of course, ought not to be used elsewhere by those who would speak *correct English.*[20]

Like Adams and Webster, Whitman insisted on the connection between American language and American polity. In seeking to make the American language an index and instrument of national creation, he refused to fasten language to any absolute model: "It is not a polished fossil language," he remarked, "but the true broad fluid language of democracy. Then have we upon it great improvements to make—very great ones. — It has yet to be acclimated here, and adapted still more to us and our future—many new words are to be formed—many of the old ones conformed to our uses."[21]

In his attempt to keep language open, flexible, and responsive to the changing contours of American democratic experience, Whitman was at odds with other poets of his time. Poe's pursuit of pure poetry engaged him in a perpetual struggle to strip language of its worldly dross; his experiments in the sound, meter, and imagery of verse were part of an effort to push language "out of space—out of time." Emerson, too, was fundamentally ahistorical in his view of language. His oft-quoted dictum "Words are signs of natural facts" expresses a fixed and purist concept of language; in pursuit of some primitive state in which words are in absolute accord with things, the Emersonian poet must excavate his language—or fossil poetry—from under the layers of culture in which it is embedded. For all his belief that the "experience of each new age requires a new confession," Emerson used a language that was—like the language of Bryant, Longfellow, Irving, and Cooper—indistinguishable from the language of his British contemporaries. "Whatever differences there may be," observed British lexicographer Sir William Craigie, "between the language of Longfellow and Tennyson, of Emerson and Ruskin, they are differences due to style and subject, to personal choice or command of words, and not to any real divergence in the means of expression."[22] In his later years, Whitman commented rather uncharitably on "Emerson's Books" as a "class" act, "superrefined" and seemingly "demarcated" from the lives of the American masses. "Suppose," he said, "these books becoming absorb'd, the permanent chyle of American general and particular character—what a well-washed and grammatical, but bloodless and helpless, race we should turn out! No, no, dear friend; though the States want scholars, undoubtedly . . . they don't want scholars, or ladies and gentlemen, at the expense of all the rest" (*PW*, II, 516).

If for Emerson the sources of language were in nature, for Whitman the sources of language were in democratic culture, which included—but was not limited to—natural facts. Language, Whitman argued, must express the multiplicity of habits, heritages, and races that make up the American nationality: "The immense diversity of race, temperament, character—the copious streams of humanity constantly flowing hither—must reappear in free, rich growth of speech. From no one ethnic source is America sprung: the electric reciprocations of many stocks conspired and conspire. This opulence of race-elements is in the theory of America."[23] Into the purity of New England English, Whitman introduced the ethnic and idiomatic color of American speech. His desire to keep language and literature open and responsive to the multiethnic sources of American nationality corresponded to his political desire to keep the country open to the immigrants who, after 1850, began coming to America in ever-increasing numbers.

In search of a language demonstrating the "opulence of race elements in the theory of America," Whitman was particularly interested in the Negro dialect. One notebook entry reads: "In the South, words that have sprouted up from the dialect and peculiarities of the slaves—the Negroes— The south is full of negro-words. —Their idioms and pronunciation are heard every where" (*DN*, III, 695). Fascinated by the expressive possibilities of black speech, he once speculated that the sources of a truly American music might be found not in the transplanted accents of New England English but in Negro dialect. This dialect, he pointed out in the *Primer*, "furnishes hundreds of outre words, many of them adopted into the common speech of the mass of the people.—" In black speech, Whitman finds "hints of the future theory of the modification of all the words of the English language, for musical purposes, for a native grand opera in America" (*DN*, III, 730). Whitman's words prophesy the development of black blues and jazz, which are in effect "a native grand opera" and one of America's most important contributions to world culture.

Although he did not draw on the musical possibilities of black dialect in composing *Leaves of Grass*, Whitman did introduce several foreign terms—including Spanish, Italian, and French—as part of his effort to establish a native American idiom. These foreign borrowings are neither merely ignorant nor merely arrogant, as is commonly assumed; rather, they are part of Whitman's effort to create a racially and ethnically mixed language to match America's democratic pluralism.[24] Whitman was particularly fond of borrowings from the French, which he used to reflect the French contribution to American nationality and to connect America's experience to France's enlightened, republican, and revolutionary heritage. The French language was a consistent feature of Whitman's poems of international embrace and cosmopolitan philosophy, and through all editions of *Leaves of Grass*, his favorite French borrowings were words of unity, bonding, and affectionate address: *rapport, ensemble, en masse, rondure, mélange, résumé, mon cher, mon enfant, ma femme, compagnon,* and *ami.*

Whitman also experimented with the sound and sense of native American terms. "What is the fitness—What the strange charm of aboriginal names? —Monongahela—it rolls with venison richness upon the palate" (*DN*, III, 752). In his verse he introduced several native American words, including *moccasin, squaw, quhaug, wigwam, powow, sachem,* and *titi;* and in accordance with his desire to rename American places, Whitman commonly referred to New York and Long Island by their aboriginal names: Mannahatta and Paumanok.

In pursuit of a revolutionary formation, Whitman sought to reinvent both the language and substance of verse and the genre of poetry itself. "Poetry Fetter'd Fetters the Human Race," wrote William Blake in his *Prophetic Books.*[25] Whitman's free verse originated from a similar desire to release humanity from the fetters of external form, political or artistic. As we have seen, it was in his early political poems defending liberty that he first broke away from traditional rhyme and meter. This refusal to be bound by the external rules of poetic form is a key element in the theory of free verse and organic form that Whitman set forth in the 1855 preface: "The rhyme and uniformity of perfect poems show the free growth of metrical laws and bud from them as unerringly and loosely as lilacs, or roses on a bush, and take shapes as compact as the shapes of chestnuts and oranges and melons and pears, and shed the perfume impalpable to form" (*LG* 1855, p. 10). Whitman's defense of poetic liberty is rooted in a revolutionary appeal to the self-evident truths of nature. Like Jefferson's theory of the American republic, Whitman's theory of organic form is grounded in the idea that law, whether political or poetic, is embedded in the processes of natural creation.

In exploring the expressive possibilities of poetry as a democratic form, Whitman not only liberated verse from the traditional rules of rhyme, meter, and stanza division; he also urged a blurring of the generic distinction between poetry and prose.

The time has arrived to essentially break down the barriers of form between prose and poetry. I say the latter is henceforth to win and maintain its character regardless of rhyme, and the measurement-rules of iambic, spondee, dactyl, &c. . . . the truest and greatest *Poetry*, (while subtly and necessarily always rhythmic, and distinguishable easily enough,) can never again, in the English language, be express'd in arbitrary and rhyming meter, any more than the greatest eloquence, or the truest power and passion. (*PW*, II, 519)

Having discarded the "arbitrary" rules of rhyme, meter, and stanza division, Whitman returned to a freer and more ancient prosodic method based on periodic stress, rhythmic recurrence, parallelism, repetition, alliteration, and assonance.[26]

At the base of Whitman's free-verse poetics is the catalogue. Among his notes preparatory to writing *Leaves of Grass* is the following idea for a poem: "Poem of Pictures. Each verse presenting a picture of some characteristic scene, event, group, or personage—old or new, other countries or our own country" (*CW*, X, 32). The catalogue, with each verse containing a picture, became an instrument of Whitman's poetic democracy, a means of embodying the universe in a poem. He tried out the technique in the pre-1855 poem "Pictures," which inventories the pictures lodged in the gallery of his mind. "In a little house I keep many pictures," the poem begins. The house contains pictures drawn from Whitman's own experience "going up and down Manahatta," portraits of his family, scenes of Athens based on his reading of Wright's *A Few Days in Athens,* and sketches of the past and the present, the United States and the world:

—Here! do you know this? This is cicerone himself;
And here, see you, my own States—and here the world itself,
    bowling
                through the air
    rolling
And there, on the walls hanging, portraits of women and men, carefully
    kept,
This is the portrait of my dear mother—and this of my father—and these
    of my brothers and sisters;
This (I name every thing as it comes,) This is a beautiful statue, long
    lost, dark buried, but never destroyed—now found by me, and
    restored to the light;
There five men, a group of sworn friends, stalwart, bearded, determined,
    work their way together through all the troubles and impediments
    of the world.

                                                        (*LGC*, p. 642)

As a round house—"but a few inches from one side of it to the other
side"—Whitman's picture gallery is an early version of his composite
democratic persona, and his catalogue of images is an embryonic version
of the thought rhythms—the end-stopped lines linked by parallelism,
repetition, and periodic stress—that are the fundamental unit of his verse.

In pursuit of a new measure and a new way of measuring expressive of
the modern democratic world, Whitman tried to avoid simile, metaphor,
and the highly allusive structure of traditional verse. Rather, his catalogues work by juxtaposition, image association, and metonymy to suggest
the interrelationship and identity of all things. By basing his verse on the
single, end-stopped line at the same time that he fuses this line through
various linking devices with the larger structure of the whole, Whitman
weaves an overall pattern of unity in diversity.

What the New Critics regarded as formlessness and failure to discrimi-
nate was in fact part of Whitman's attempt to invent an egalitarian poetics.[27] His verse form, like the catalogue technique in which it is rooted, is
a poetic analogue of democracy, inscribing a pattern of many in one. The
following catalogue from "Song of Myself" undoes traditional hierarchies
by presenting each person as part of a seemingly indiscriminate mass:

The bride unrumples her white dress, the minutehand of the clock
    moves slowly,
The opium eater reclines with rigid head and just-opened lips,
The prostitute drags her shawl, her bonnet bobs on her tipsy and
    pimpled neck,
The crowd laugh at her blackguard oaths, the men jeer and wink
    to each other,

> (Miserable! I do not laugh at your oaths nor jeer at you,)
> The President holds a cabinet council, he is surrounded by the
>     great secretaries,
> On the piazza walk five friendly matrons with twined arms.
>
> (*LG* 1855, pp. 38–39)

Presented in a sequence of separate and end-stopped images, these figures are independent and yet related through the parallel structure of the lines, the patterns of rhythmic stress, and the summary statement of the poet at the end of the catalogue:

> And these one and all tend inward to me, and I tend outward to
>     them,
> And such as it is to be of these more or less I am.
>
> (*LG* 1855, p. 40)

The total effect of the passage is to equalize and fuse in one chain brides and opium eaters, prostitutes and presidents, men and women, by presenting them paratactically on a horizontal plane. In a further display of poetic democracy, the passage turns on the figure of the prostitute by swelling to an eight-stress pattern in the lines that describe her, by presenting her before and next to the president, by dwelling on her for three lines—as opposed to the single line allotted to the other figures in the catalogue—and by expressing sympathy for her in a first-person parenthetical aside that changes the rhythm and tone of the sequence.

Whitman achieves a similarly equalizing and unifying effect through his use of parallelism and reiteration and his extensive use of coordinate, conjunctive, and prepositional constructions:

> Through me many dumb voices,
> Voices of the interminable generations of slaves,
> Voices of prostitutes and of deformed persons,
> Voices of the diseased and despairing, and of thieves and dwarfs,
> Voices of cycles of preparation and accretion,
> And of the threads that connect the stars—and of wombs, and of the
>     fatherstuff,
> And of the rights of them the others are down upon,
> Of the trivial and flat and foolish and despised,
> Of fog in the air and beetles rolling balls of dung.
>
> Through me forbidden voices,
> Voices of sexes and lusts . . . . voices veiled, and I remove the veil,
> Voices indecent by me clarified and transfigured.
>
> (*LG* 1855, p. 48)

Parallelism and repetition, alliteration and assonance, conjunctions and prepositions are interwoven to connect the "long dumb voices" of the universe on a single spatial and temporal plane. These democratizing strategies are particularly evident in the first edition of *Leaves,* in which Whitman's occasional elimination of commas to separate items in a series has the effect of combining objects in a single mass. And as in this passage, Whitman's use of ellipsis points for pause and emphasis serves both to particularize and to equalize objects and prepositional phrases on a horizontal plane.

As part of his democratic strategy, Whitman also experimented with open rather than closed forms. "The expression of the American poet is to be transcendent and new," he said in his 1855 preface. "It is to be indirect and not direct or descriptive or epic. . . . Let the age and wars of other nations be chanted and their eras and characters be illustrated and that finish the verse. Not so the great psalm of the republic. Here the theme is creative and has vista" (*LG* 1855, p. 8). Whitman conceived of his democratic art not as a finished object but as part of an open-ended process of personal and national creation. Emphasizing art as process rather than product, he turned away from the direct narrative presentation of the traditional epic toward an indirect, nonlinear poetic presentation, which moves by juxtaposition and image association rather than by cause-and-effect progression and which focuses not on the completion of a past action but on the act of creating present and future. Making extensive use of the present tense, the participial form, and other grammatical constructions that link his words in a fluid, temporal process, Whitman created poems that exist in a kind of perpetual present, embodying the past and growing toward the future.

This "indirect" presentation is at the hub of Whitman's democratic poetics.[28] It was a means of making a point symbolistically through the suggestiveness of words and images; of embodying poetically the incompletion and vista of the American republic; and of implying the religiospiritual dimension of the universe, what Whitman called the "well nigh . . . inexpressible." But most importantly, his indirect method was a means of engaging the reader as an active participant in the republican politics of his poetry.

In "Song of Myself" Whitman addresses the reader thus:

> Stop this day and night with me and you shall possess the origin of
>     all poems,
> You shall possess the good of the earth and sun . . . . there are
>     millions of suns left,
> You shall no longer take things at second or third hand . . . . nor look

> through the eyes of the dead . . . . nor feed on the spectres in
> books,
> You shall not look through my eyes either, nor take things from me,
> You shall listen to all sides and filter them from yourself.
>
> (*LG* 1855, p. 26)

The repetition of *you shall* as both command and prophecy drives the force of the passage toward the reader and the future. The origin of all poems to which Whitman refers is not nature, as is commonly assumed, but the self. By urging the reader to filter all sides from herself or himself, he beckons to the reader to participate in the process of democratic creation.

The active role of the reader required by Whitman's indirect method was, in his view, one of the "point characteristics" of *Leaves of Grass:* "The word I myself put primarily for the description of them as they stand at last, is the word Suggestiveness," he wrote in "A Backward Glance o'er Travel'd Roads" (1888). "I round and finish little, if anything; and could not, consistently with my scheme. The reader will always have his or her part to do, just as much as I have had mine. I seek less to state or display any theme or thought, and more to bring you, reader, into the atmosphere of the theme or thought—there to pursue your own flight" (*LGC,* p. 570).

The *you* of the reader, whom Whitman addresses as both singular and plural, present and future, female and male, is the ultimate hero of *Leaves of Grass.*[29] Here again, Whitman stood apart from his literary contemporaries. He was the first major poet to conceive of himself addressing a female as well as a male reader. Exhorting, questioning, caressing, and challenging his female and male readers, he also practiced poetry as a form of political action—as agency, speech act, and social event—at the very moment when, under the pressure of Romanticism, writers were beginning to aestheticize and privatize literature as a world apart.

Whitman's insistence on the reader's creative role was part of his revolutionary strategy, his attempt to collapse the traditionally authoritarian relation between poet and audience, text and reader by transferring the ultimate power of creation to the reader. "A great poem is no finish to a man or woman but rather a beginning," he said in the 1855 preface. "Has any one fancied he could sit at last under some due authority and rest satisfied with explanations and realize and be content and full? To no such terminus does the greatest poet bring . . . he brings neither cessation or sheltered fatness and ease. The touch of him tells in action" (*LG* 1855, p. 22). Whitman refused the traditional authority and closure of art; his revolutionary aesthetics is an activist aesthetics that incites the reader to the final act of creation—of self and poem, nation and world.

# 5

# Leaves of Grass *and the Body Politic*

> On women fit for conception I start bigger and nimbler babes,
> This day I am jetting the stuff of far more arrogant republics.
>
> —WHITMAN, "Song of Myself"

The publication of *Leaves of Grass* in 1855 was not an escape from politics but a continuation of politics by other means. The poems were, in Whitman's words, an attempt to provide "some worthy record of that entire faith and acceptance ('to justify the ways of God to man' is Milton's well-known and ambitious phrase) which is the foundation of moral America" (*PW*, II, 729). Whitman's reference to John Milton's *Paradise Lost* is instructive, for like his epic forebear, he attempts to mythologize the historically specific political struggles of his time as universal and divinely ordained. As a work of visionary democracy, *Leaves of Grass* is split between Whitman's desire to speak for and against his age. While the self-sovereign individual and utopian social community of his poems represent an implicit critique of the actual conditions of industrial and commercial transformation, Whitman's ideological investment in the political system led him to "justify" when he did not actually silence the more disagreeable workings of free-enterprise democracy in nineteenth-century America.

Although Whitman was cognizant of the ways that past literature participated in and justified the domination, power, and interests of an aristocratic class system, he never fully acknowledged the extent of his own ideological complicity as the celebrator of American democracy. The fervent Tom Paine democrat of *Leaves of Grass* is (mis)represented as a nonpartisan citizen of the world, a figure who stands simultaneously inside and outside social time and above the squalor of political contest:

> Backward I see in my own days where I sweated through fog with
> linguists and contenders,
> I have no mockings or arguments . . . . I witness and wait.
>
> *(LG 1855,* p. 28)

Masking the fact of his own ideological involvement in a particular system
of power, Whitman eternalized the democracy of his poems as part of
what he called the "politics of Nature." "All you are doing and saying is to
America dangled mirages,/You have not learn'd of Nature—of the politics of Nature you have not learn'd the great amplitude, rectitude, impar-
tiality," he wrote in a poetic address to President James Buchanan *(LGC,*
p. 272). Naturalizing his partisan voice as the voice of nature, the poet
asks disingenuously in "Song of Myself":

> Do you guess I have some intricate purpose?
> Well I have . . . . for the April rain has, and the mica on the side of
> a rock has.
>
> *(LG 1855,* p. 42)

Critics in their turn have tended to subscribe to Whitman's own self-
representation, reading *Leaves of Grass* as a work of mystical and ideological transcendence rather than as a historically marked text fully engaged
in and with the "mockings and arguments" of its time.

Whitman mythologized what he called the "entire faith and acceptance" of the American republic in a poetic persona who is at once a
model of democratic character and a figure of democratic union. Speaking of the analogy between the individual and the body politic, he said:
"What is any Nation, after all—and what is a human being—but a struggle
between conflicting, paradoxical, opposing elements—and they themselves and their most violent contests, important parts of that One Identity, and of its development?"[1] Through the invention of an organic self
who is like the Union, many in one, Whitman seeks to manage and resolve
poetically the conflicting and paradoxical energies of the nation.

"We are not one people. We are two peoples," the New York *Tribune*
noted in 1855. "We are a people for Freedom and a people for Slavery.
Between the two, conflict is inevitable."[2] Just as during the revolutionary
period, phrase the *E Pluribus Unum* became the seal of a people inventing
a united nation out of a plurality of conflicting interests, so during the
1850s—as factionalism and sectional rivalry threatened to fracture the
Union—Whitman attempted to seal the Union imaginatively by placing
the paradox of many and one at the thematic and structural center of
*Leaves of Grass.* The problem of reconciling private or factional interests
within the single identity of the Union was at the root of the political crisis

of the 1850s as it was at the foundation of the American republic. Addressing the problem of reconciling factional interest and public good within
the framework of popular government, James Madison wrote in *Federalist
Paper* No. 10: "To secure the public good and private rights against the
danger of such a faction, and at the same time to preserve the spirit and
the form of popular government, is then the great object to which our
inquiries are directed."[3]

During the 1850s, when the Constitution of the United States, as both
a body politic and a set of laws, was being contested, Whitman, too,
grappled with the problem of reconciling personal freedom with public
good and factional interest with democratic union. Just as during the
postrevolutionary period, the Federalists supported the idea of a strong
national government in order to weld the diversity of personal, local, and
state interests into a Union, so in the antebellum period, as the tensions
among Northern and Southern states began to expose the essentially
fictive nature of the Union, the federal government assumed an increasingly central role, particularly in enforcing the Fugitive Slave Law and
legislating on the issue of slavery in the territories.

As a Jeffersonian democrat, Whitman resisted the aggressive role
assumed by the central government to hold the Union together. He celebrated the ideals of prudence and self-regulation, with the individual
balanced between personal power and social love, as a kind of nineteenth-
century poetic equivalent of the republican ideals of personal sacrifice and
public virtue. The poet he imagines in the 1855 preface is, like his ideal
republic, balanced between self and other: "The soul has that measureless pride which consists in never acknowledging any lessons but its own.
But it has sympathy as measureless as its pride and the one balances the
other and neither can stretch too far while it stretches in company with the
other. The inmost secrets of art sleep with the twain. The greatest poet
has lain close betwixt both and they are vital to his style and thoughts" (*LG*
1855, p. 12).

This vision of a poet stretching within a universe bounded by pride and
sympathy had as its political analogue the paradox of an American republic
poised between self-interest and public virtue, liberty and union, the interests of the many and the good of the one. The secret of Whitman's art and
the American Union, the paradox of many in one, eventually became the
opening inscription and balancing frame of *Leaves of Grass:*

> One's-Self I sing, a simple separate person,
> Yet utter the word Democratic, the word En-Masse.
>                                                        (*LGC*, p. 1)

Balanced between the separate person and the en masse, the politics of *Leaves of Grass* is neither liberal nor bourgeois in the classical sense of the terms; rather, the poems represent the republican ideals of early-nineteenth-century artisan radicalism, emphasizing the interlinked values of independence and community, personal wealth and commonwealth.

The drama of identity in the initially untitled "Song of Myself," the first and longest poem in the 1855 *Leaves*, is rooted in the political drama of a nation in crisis—a nation, as Lincoln observed at the time, living in the midst of alarms and anxiety in which "we expect some new disaster with each newspaper we read."⁴ The poet's conflict between the separate person and the en masse, pride and sympathy, individualism and equality, nature and the city, the body and the soul symbolically enacts the larger political conflicts in the nation, which grew out of the controversy over industrialization, wage labor, women's rights, finance, immigration, slavery, territorial expansion, technological progress, and the question of the relation of individual and state, state and nation. The self that emerges in "Song of Myself" is united by the same constitutional system of checks and balances—between the one and the many, self and other, liberty and union, urban and agrarian, material and spiritual—that Whitman envisioned for the American republic.

Reversing the traditions of the epic, "Song of Myself" opens with words that both challenge and command the reader:

> I celebrate myself,
> And what I assume you shall assume,
> For every atom belonging to me as good belongs to you.
>
> *(LG* 1855, p. 25)

The poet's subject is not arms and the man or man's first disobedience but the self that is at the very center of the American myth of origins. If the first line of the poem—"I celebrate myself"—isolates the separate person of the poet as the hero of the poem, the following two conjunctive and parallel phrases link the *I* of the poet with the *you* of the reader. These opening lines immediately engage the reader as a participant in the action of the poem, which is simultaneously about the creation of a democratic poem and the creation of a democratic self/nation/world. Beginning with the *I* of the poet and ending with the *you* of the reader, the opening lines mark the poles between which the poem swings. The opening *I* and the closing *you* of the poem are the bounds of an agonistic arena in which the poet commands, questions, mocks, challenges, wrestles, fondles, and instructs his reader, finally sending him or her back into the world bearing the seeds of democratic potency.

From the first, critics have tried to transform the indirection of "Song of Myself" into a traditional narrative by discussing the poem in relation to a pattern or story external to the poem itself.[5] But while the poem, like the final edition of *Leaves of Grass,* inscribes an arc of development from life to death, body to spirit, summer to autumn, dawn to dusk, self to other, poet to reader, it has no traditional beginning, middle, and end. The poem moves in the form of a circle, not by narrative line, but by association and recurrence. At the beginning of the poem the poet contemplates a spear of summer grass:

> I loafe and invite my soul,
> I lean and loafe at my ease . . . . observing a spear of summer grass.
>
> > (*LG* 1855, p. 25)

The poem ends in autumn and dusk, with the poet dissolving into the elements, becoming himself a spear of grass:

> I bequeath myself to the dirt to grow from the grass I love,
> If you want me again look for me under your bootsoles.
>
> > (*LG* 1855, p. 86)

Between the poet's observing the spear of summer grass and the reader's observing the poet in the grass, Whitman engages in a sustained meditation on the linked concepts of individual power and cosmic union, concepts that he seeks to pass on to the reader through a pattern of statement, illustration, and reiteration. Just as he seeks on the thematic level to strike a balance between pride and sympathy, the many and the one, so on a structural level, the fifty-two sections of the poem differ in tone, mood, strategy, and substance at the same time that they are linked by an underlying pattern of unity in diversity.[6]

At the outset of "Song of Myself," the poet is balanced between his separate self, identified with a particular place and time, and a unitary self, standing apart from the pulling and hauling of the actual world:

> My dinner, dress, associates, looks, business, compliments, dues,
> The real or fancied indifference of some man or woman I love,
> The sickness of one of my folks—or of myself . . . . or ill-
>      doing . . . . or loss or lack of money . . . . or depressions or
>      exaltations,
> They come to me days and nights and go from me again,
> But they are not the Me myself.
>
> Apart from the pulling and hauling stands what I am,
> Stands amused, complacent, compassionating, idle, unitary,

> Looks down, is erect, bends an arm on an impalpable certain rest,
> Looks with sidecurved head curious what will come next,
> Both in and out of the game, and watching and wondering at it.
>
> (*LG* 1855, p. 28)

The "Me myself" of this passage is not, like Emerson's poet, an isolated introvert who defines himself apart from other selves. "Compassionating" and unitary, Whitman's "Me myself" defines himself as part of a commonality. With bent arm and side-curved head, his posture is in fact that of the common man who appears in the frontispiece of the 1855 *Leaves*. It is this unitary figure who bears the poem's democratic scripture. "Both in and out of the game," he speaks a divine grammar of reconciliation that smiles beyond the fact of indifference, sickness, ill-doing, loss, lack, and depression in the social world.

This divine grammar comes from inside the poet, from the erotic union of body and soul. In section 5, the poet invites his soul:

> Loafe with me on the grass . . . . loose the stop from your throat,
> Not words, not music or rhyme I want . . . . not custom or lecture,
>    not even the best,
> Only the lull I like, the hum of your valved voice.
>
> I mind how we lay in June, such a transparent summer morning;
> You settled your head athwart my hips and gently turned over
>    upon me,
> And parted the shirt from my bosom-bone, and plunged your tongue
>    to my barestript heart,
> And reached till you felt my beard, and reached till you held my feet.
>
> (*LG* 1855, pp. 28–29)[7]

The passage sensualizes the soul and spiritualizes the body, revising the traditional debate between body and soul by integrating them on an egalitarian plane. Challenging the Puritan distrust of the senses by associating spiritual energy with sexual force, Whitman transforms the traditional mystical experience into a validation of democracy.

The union of body and soul leads the poet not to forget himself but to remember the family of humanity:

> Swiftly arose and spread around me the peace and joy and knowl-
>    edge that pass all the art and argument of the earth;
> And I know that the hand of God is the elderhand of my own,
> And I know that the spirit of God is the eldest brother of my own,
> And that all the men ever born are also my brothers . . . . and the
>    women my sisters and lovers,

> And that a kelson of the creation is love;
> And limitless are leaves stiff or drooping in the fields,
> And brown ants in the little wells beneath them,
> And mossy scabs of the wormfence, and heaped stones, and elder
>     and mullen and pokeweed.
>
> (*LG* 1855, p. 29)

The revelation that the poet receives is the democratic knowledge of a universe bathed in an erotic force that links men, women, God, and the natural world. The use of the conjunctive *and* in a passage that runs up and down the evolutionary scale from universal to particular, spiritual to material, singular to plural—from God to man to brown ants and heaped stones—serves both to separate and link persons and objects in a single nonhierarchical plane. The use and repetition of *and* is in effect a syntactic enactment of the principle of many and one that underlies the passage. This egalitarian vision of a world linked as one loving family is the essence of Whitman's democratic faith, the source of his poetic utterance, and the underlying myth of his poem. The biblical language underscores the self-consciously prophetic role that the poet assumes as he sets out to spread the entire faith and acceptance that is the moral foundation of the American republic and the base of a new religion of humanity.

Whereas in section 5 the grass is the site of democratic revelation, in section 6 the grass becomes itself a sign, simultaneously symbolizing and naturalizing democracy, not as a particular political system, but as a fundamental law of the universe:

> A child said, What is the grass? fetching it to me with full hands;
> How could I answer the child? . . . . I do not know what it is any
>     more than he.
>
> I guess it must be the flag of my disposition, out of hopeful green stuff
>     woven.
>
> Or I guess it is the handkerchief of the Lord,
> A scented gift and remembrancer designedly dropped,
> Bearing the owner's name someway in the corners, that we may see
>     and remark, and say Whose?
>
> Or I guess the grass is itself a child . . . . the produced babe of the
>     vegetation.
>
> Or I guess it is a uniform hieroglyphic,
> And it means, Sprouting alike in broad zones and narrow zones,
> Growing among black folks as among white,
> Kanuck, Tuckahoe, Congressman, Cuff, I give them the same, I
>     receive them the same.
>
> (*LG* 1855, p. 29)

Linking the egalitarian ideals of the Revolution with the regenerative potency of the earth, the poet reads in the hieroglyphic of the grass the politics of democracy. As the overarching figure of *Leaves of Grass* and the central image of "Song of Myself," the grass signifies many in one. Sprouting among blacks and whites, Northerners and Southerners, Congressman and Cuff, the grass is a "uniform hieroglyphic" that fuses the earth in a single democratic growth. As the "produced babe of the vegetation" and the "beautiful uncut hair of graves," connecting life and death, and past, present, and future, the grass also signifies the process of regeneration—personal, political, and cosmic—that is the democratic and revolutionary lesson of the poem.

Having set forth the communal and egalitarian vision that is the democratic substance of his leaves, Whitman illustrates it by demonstrating his own sense of connectedness with the en masse. "Who need be afraid of the merge?" he asks in section 7, as if in response to the political skepticism of John Adams, Alexis de Tocqueville, James Fenimore Cooper, and others who feared the leveling tendency of democracy. Speaking the divine grammar of the poet, he presents himself as a kind of American hieroglyph who is, like the grass, both uniform and multiple. But while he seeks to unite and balance individual and en masse, city and country, East and West within the republic of the self, his descriptive passages bear the impress of the dis-ease and imbalance of his time.

For all Whitman's reputation as the celebrator of the modern city, the city he describes is the site of suicide, sickness, mobs, brawling, crime, and starvation:

> What groans of overfed or half-starved who fall on the flags
> sunstruck or in fits,
> What exclamations of women taken suddenly, who hurry home and
> give birth to babes,
> What living and buried speech is always vibrating here . . . . what
> howls restrained by decorum,
> Arrests of criminals, slights, adulterous offers made, acceptances,
> rejections with convex lips,
> I mind them or the resonance of them . . . . I come again and again.
>
> (*LG* 1855, p. 32)

The country, on the other hand, is the site of abundance and joy, as the poet comes stretched atop a harvest of hay:

> I felt its soft jolts . . . . one leg reclined on the other,
> I jump from the crossbeams, and seize the clover and timothy,
> And roll head over heels, and tangle my hair full of wisps.
>
> (*LG* 1855, p.32)

Placed side by side in the poem, the sections on country and city bear witness to a poet and a nation split between nostalgia for a rural life that was quickly passing away and the promise of a new urban world that was the site of both desire and desolation.

The poet's grammar of union contains marginal and potentially dangerous figures in the landscape of America: the trapper marrying a red girl in the "open air in the far-west," the "sweated" and "bruised" body of the fugitive slave, and the highborn woman observing a group of nude men bathing in the sea. This last episode, which Frederik Schyberg described as "entirely outside the design or framework of the poem,"[8] is in fact central to the poem's democratic text. The figure of the desiring woman represents the force of eros as the ultimate democratizer in the political economy of *Leaves of Grass:*

> Twenty-eight young men bathe by the shore,
> Twenty-eight young men, and all so friendly,
> Twenty-eight years of womanly life, and all so lonesome.
>
> She owns the fine house by the rise of the bank,
> She hides handsome and richly drest aft the blinds of the window.
>
> Which of the young men does she like the best?
> Ah the homeliest of them is beautiful to her.
>
> (*LG* 1855, p. 34)

The house and clothes, wealth and social status are inhibiting structures that separate the individual from the mass of humanity and the abundance of the physical universe, here symbolized by the twenty-eight young men engaged in a recreation that was common in mid-nineteenth-century workingmen's culture. The sexual urge that draws the woman to the young man is the kelson of creation—the unifying and equalizing energy of eros—that cuts across sexual, social, and class lines and joins her in imagination with the democratic crowd.

As the "twenty-ninth bather" dances imaginatively from house to sea, the poet himself merges with the consciousness of the female. "For I see you," he says earlier in the sequence, divided between the narrative *I* of the poet and the female *you* of his subject. In the final lines of the sequence, however, as the narrative point of view shifts from the woman in the house to the men in the water, the distinction between I and you, poet and woman breaks down:

> The beards of the young men glistened with wet, it ran from their
>     long hair,
> Little streams passed all over their bodies.

An unseen hand also passed over their bodies,
It descended tremblingly from their temples and ribs.

The young men floated on their backs, their white bellies swell to the
    sun . . . . they do not ask who seizes fast to them,
They do not know who puffs and declines with pendant and
    bending arch,
They do not think whom they souse with spray.

                                                    (*LG* 1855, p. 34)

On one level, of course, the woman is a mask for Whitman himself
anonymously fondling the bodies of the young men in solitary fancy. But
by displacing a homoerotic fantasy onto a daydreaming woman, the passage becomes doubly subversive in blurring traditional gender bounds.
The unseen and tremblingly eager hand of the handsome and rich
woman—who would become a model for the desiring heroine of Kate
Chopin's *The Awakening* (1899)—overturns nineteenth-century notions of
true womanhood, by suggesting the erotic energies of the female and by
presenting her as an active rather than a passive partner in sexual union.
Even more subversively, the image of one woman with twenty-eight young
men becomes a kind of tall tale of female eroticism revealing female
sexual energies that not only break bounds but that know no bounds.

In refusing to make distinctions ("Ah the homeliest of them is beautiful
to her") and in her uniform embrace of all the young men, the woman
becomes identified with the unifying and democratizing energies of the
universe. This identification is underscored by parallel passages that associate the woman's unseen hands with the streams of the sea passing over the
bodies of the young men. Like the sea, the female's unseen hand links the
twenty-eight young men, just as Whitman hoped to see the states and
eventually the world linked as brothers and sisters in a single loving embrace. The number twenty-eight, which suggests both the lunar cycle and
the cycle of female fertility, further associates the female and also the entire
episode with the regenerative processes of nature and the democratic pattern of the poem, whereby the diversity and multiplicity of the universe are
continually balanced and reconciled as part of a single unitary being.[9]

The poet's attempt to unite "every hue and trade and rank" reaches its
apogee in section 15, in which he absorbs and fuses within himself the
wagon driver and the half-breed, the woolly pate and the gentleman, the
squaw and the connoisseur, the bride and the opium eater, and the prostitute and the president. But although Whitman seeks to collapse the distinctions of race, class, and gender in a dazzling poetic inventory of the
modern world, his democratic catalogues bear the traces of an oppressive,
hierarchic order:

The lunatic is carried at last to the asylum a confirmed case,
. . . . . . . .
The quadroon girl is sold at the stand . . . . the drunkard nods by the
    barroom stove,
. . . . . . . .
The woollypates hoe in the sugarfield, the overseer views them from
    his saddle;
. . . . . . . .
The company returns from its excursion, the darkey brings up the
    rear and bears the well-riddled target,
The squaw wrapt in her yellow-hemmed cloth is offering moccasins
    and beadbags for sale.

(*LG* 1855, pp. 37–38)

For all their poetic democracy, Whitman's catalogues could operate paradoxically as a kind of formal tyranny, muting the fact of inequality, race
conflict, and radical difference within a rhetorical economy of many and
one. Whitman's captivatingly "large" persona tries to "contain" these
signs of historic contradiction: "Do I contradict myself?/ Very well
then . . . . I contradict myself;/I am large . . . . I contain multitudes" (*LG*
1855, p. 85). But over and over the poet's "multitudes" of contradiction
disrupt the harmonious democratic order that he enunciates and seeks to
achieve stylistically by assonance and consonance, alliteration and internal
rhyme, and parallelism and repetition.

The poet's act of self-naming in section 24 inscribes the figure of
many and one that is the founding myth of both poem and nation:

Walt Whitman, an American, one of the roughs, a kosmos,
Disorderly fleshy and sensual . . . . eating drinking and breeding,
No sentimentalist . . . . no stander above men and women or apart
    from them . . . . no more modest than immodest.

(*LG* 1855, p. 48)

Moving from self to nation to world, the lines mark the poles of Whitman's poetic identity: He is both single and multiple, common man and
en masse, American and kosmos. As "one of the roughs," he also identifies with an unruly segment of the American populace: "A more despica-
ble, dangerous, and detestable character than the New York rough does
not exist," wrote a correspondent for one of the journals of Whitman's
time.[10] By naming himself as one of the roughs and by asserting his
disorderly and sensual nature in an unpunctuated sequence that mirrors
the unruly flow of the senses, Whitman refuses decorum, hierarchy, and
stock sentiment as both life-style and literary style. He identifies himself

with those very qualities of disorder and animal passion most feared by the critics of democracy.

Whitman's fleshy persona, who became in later editions "Turbulent, fleshy, sensual, eating, drinking and breeding," represents in the body of the poet the turbulence of the democratic masses in the body of the republic. The turbulence of the masses, which was from the first a cause for alarm among the critics of democracy, was in Jefferson's view the necessary safeguard of republican government. Under a republican government, Jefferson wrote to James Madison in 1787, "the mass of mankind . . . enjoys a precious degree of liberty and happiness. It has its evils too; the principal of which is the turbulence to which it is subject. . . . Even this evil is productive of good. . . . I hold it that a little rebellion now and then is a good thing, and as necessary in the political world as storms in the physical."[11]

In his political writings, Whitman, too, consistently supported the idea that the turbulence of the people was a source of regenerative vitality, like storms in the physical sphere, or, as in "Song of Myself," like passions in the body. "To attack the turbulence and destructiveness of the Democratic spirit, is an old story," he wrote in the *Eagle*. "Why, all that is good and grand in any political organization in the world, is the result of this turbulence and destructiveness; and controlled by the intelligence and common sense of such a people as the Americans, it never has brought harm, and never can. A quiet contented race sooner or later becomes a race of slaves—. . . . But with the noble Democratic spirit—even accompanied by its freaks and its excesses—no people can ever become enslaved" (*GF*, I, 3–4). "Swing Open the Doors!" Whitman had declared in another *Eagle* editorial, "We must be constantly pressing onward—every year throwing the doors wider and wider—and carrying our experiment of democratic freedom to the very verge of the limit" (*GF*, I, 10). If in the political sphere the spirit of compromise had led to a retreat from the more radical premises of democracy, in his poems Whitman carried on his effort to push the democratic experiment to "the very verge of the limit."

Like the American republic, "Song of Myself" is an experiment in self-governance that both tests and illustrates the capacity of a muscular and self-possessed individual for regulation from within. The poem is a democratic performance in which the poet dances on "the very verge of the limit" of sexual appetite and hellish despair but is continually restored to an inward economy of equity and balance. "I speak the password primeval. . . . I give the sign of democracy," Whitman declares as he absorbs into himself and his poem the turbulence of the democratic spirit with all "its freaks and its excesses."

It is on the sexual plane, through a release of libidinous energies, that Whitman's democratic persona undergoes his first major trial of self-mastery:

> Something I cannot see puts upward libidinous prongs,
> Seas of bright juice suffuse heaven.
>
> The earth by the sky staid with . . . . the daily close of their junction,
> The heaved challenge from the east that moment over my head,
> The mocking taunt, See then whether you shall be master!
>
> > (*LG* 1855, p. 50)

The challenge is ambiguous, referring to the poet's refusal to be mastered by a landscape suffused with sexual energy and, subsequently, to his refusal to be "mastered" by language:

> Encompass worlds but never try to encompass me,
> I crowd your noisiest talk by looking toward you.
>
> > (*LG* 1855, p. 51)

Within the democratic economy of *Leaves*, life always exceeds the capacity of language to encompass it: The flesh is prior to the word; the poet prior to the poem.

The main challenge to the poet comes with the onslaught of touch in section 28. This passage records a crisis in which Whitman's hitherto-balanced persona, stimulated by a masturbatory fantasy, is overtaken by the sense of touch:

> Is this then a touch? . . . . quivering me to a new identity,
> Flames and ether making a rush for my veins,
> Treacherous tip of me reaching and crowding to help them,
> My flesh and blood playing out lightning, to strike what is hardly
>     different from myself,
> On all sides prurient provokers stiffening my limbs,
> Straining the udder of my heart for its withheld drip.
>
> > (*LG* 1855, p. 53)

Stimulated and stiffened by the "treacherous" fingertips of himself, the poet loses his bodily balance. Carried away by a solitary act of onanism, he also loses the balance between self and other, body and soul that is part of his democratic posture.

Presented in the language of a violent mass insurrection in which touch, as the "red marauder," usurps the governance of the body, the entire episode has a political nuance:

No consideration, no regard for my draining strength or my anger,
Fetching the rest of the herd around to enjoy them awhile,
Then all uniting to stand on a headland and worry me.

The sentries desert every other part of me,
They have left me helpless to a red marauder,
They all come to the headland to witness and assist against me.
I am given up by traitors;
I talk wildly . . . . I have lost my wits . . . . I and nobody else am the
  greatest traitor,
I went myself first to the headland . . . . my own hands carried me
  there.

(LG 1855, pp. 53–54)

The poet's "worry" in this passage is both personal and political. The vision of insurrection and violence in the poet's democratic body relates not only to the impending crisis of the Civil War but also to the very theory of America itself. If the individual were not capable of self-mastery, if balance were not the natural law of the universe, and if the storms of (homo)sexual passion could usurp the constitution of body and body politic, then the theory of America would be cankered at its source. Just as the insurrection within the body of the poet comes from his own hand, so in the political sphere, the main threat to democracy appeared to come from within the body of the republic itself. The entire sequence links the danger of democracy with the danger of a sexually unruly body. And it is on the level of sex and the body that the poem tests the democratic theory of America.

Whitman symbolically resolves the bodily crisis of his protagonist by linking the onslaught of touch—as a sign of unruliness in body and body politic—with the regenerative energies of the universe:

You villain touch! What are you doing? . . . . my breath is tight in
  its throat;
Unclench your floodgates! you are too much for me.

Blind loving wrestling touch! Sheathed hooded sharptoothed touch!
Did it make you ache so leaving me?
Parting tracked by arriving . . . . perpetual payment of the perpetual
  loan,
Rich showering rain, and recompense richer afterward.

Sprouts take and accumulate . . . . stand by the curb prolific and vital,
Landscapes projected masculine full-sized and golden.

(LG 1855, p. 54)

The moment of sexual release is followed by a restoration of balance as the ejaculatory flow merges with and is naturalized as the regenerative flow of the universe. The parallel lines formally mark the restoration of balance at the same time that they inscribe the process of parting and arriving, efflux and influx that is the generative rhythm of the universe and the main pattern of the poem as the poet advances and retreats, absorbs and bestows.

The body–body politic analogy on which Whitman draws in "Song of Myself" is evident in several of his comments on the state of America. "We have full confidence in the power of the constitution to outlive any gust of passion or feeling," he observed in an article "Hero Presidents" in 1848. "The wisest man is often provoked to anger, and daily weeps his inability to govern his appetites and passions; but the storms of passion are transient, and when they pass away leave his wisdom high and pure like a mountain-top seen in the distance, and serving as a guide for the traveller" (*UPP, I,* 198). The touch sequence in "Song of Myself" enacts a similar pattern of a gust of passion, followed by a restoration of balance and a sense of renewed wisdom.

Read closely, the sequence provides a useful corrective to the popular image of Whitman as the poet of sexual excess. Whitman does not celebrate masturbation in "Song of Myself." On the contrary, his attitude is closer to the antimasturbation tracts published by Fowler and Wells, the distributors of the first edition of *Leaves of Grass.*[12] Whitman presents masturbation as an instance of bodily perturbance—a muted sign, perhaps, of the unruliness of his own homosexual passion—and a trope for disorder in the political sphere. As a figure of democratic unruliness in body and body politic, masturbation becomes the sexual ground on which Whitman tests the democratic theory of America. By demonstrating the restoration of bodily balance after taking democracy to "the verge of the limit" in a masturbatory fit, Whitman tests and enacts poetically the principle of self-regulation in individual and cosmos that is at the base of his democratic faith. But while he successfully manages the onslaught of touch within the symbolic order of the poem, the unruly body—both his own and the bodies of others—would remain a source of anxiety and perturbance in his dream of democracy.

"Is this then a touch? . . . . quivering me to a new identity," the poet asks at the beginning of the touch sequence. In the gloriously regenerative economy of "Song of Myself," (homo)erotic touching is safe and natural, quivering the poet not to a new and marginal identity as a homosexual in heterosexual America but toward an experience of cosmic community. At the outset of the lengthy catalogue in section 33, the poet exclaims:

> Swift wind! Space! My Soul! Now I know it is true what I guessed
>   at;
> What I guessed when I loafed on the grass,
> What I guessed while I lay alone in my bed . . . . and again as I
>   walked the beach under the paling stars of the morning.
>
> (*LG* 1855, p. 57)

The entire sequence confirms the vision of union and harmony revealed to the poet in section 5. But whereas section 5 and subsquent early sections of the poem focus on the self in relation to the human and natural world of the present, in section 33 and later sections of the poem, the self is mobilized, experiencing a sense of identity and relation through time, space, and the cosmos.

"Afoot" with his vision, traveling along, through, over, under, and on the turnpikes, savannas, bayous, huskings, house-raisings, and camp meetings of America, the poet identifies with and participates in the perceptions and emotions of the people he describes:

> I turn the bridegroom out of bed and stay with the bride myself,
> And tighten her all night to my thighs and lips.
>
> My voice is the wife's voice, the screech by the rail of the stairs,
> They fetch my man's body up dripping and drowned.
>
> I understand the large heart of heroes,
> The courage of the present times and all times.
>
> (*LG* 1855, pp. 61–62)

These lines mark the beginning of a series of vignettes in which the poet empathizes with the pain and suffering of others.

"I am the man . . . . I suffered . . . . I was there," he says as he enters a sequence of violent and battle-torn scenes that lead to his second crisis of self-mastery. Once again, the crisis is brought about by a loss of balance, between pride and sympathy, separate person and en masse. But whereas the first crisis is provoked by self-stimulation and a momentary loss of bodily balance, the second crisis is provoked by human suffering and a momentary loss of self in empathetic identification with others. Whitman plotted the drama of this second crisis in an early notebook. Under the title "Poem (idea) 'To struggle is not to suffer,' " he wrote:

> Bold and strong invocation of suffering—to try how much one can stand.
> Overture—a long list of words—the sentiment of suffering, oppression, despair, anguish.
> Collect (rapidly present) terrible scenes of suffering.
> "Then man is a God." Then he walks over all.
>
> (*CW*, X, 25)

The entry suggests a useful distinction in Whitman's verse between moments of symbolic crisis that the poet plotted in advance and moments of unplotted crisis that expose the social anxiety and dissonance he seeks to mask.

In this instance, Whitman appears to lose control of a crisis he plotted in advance. He begins the trial sequence by setting out to describe the large hearts of heroes and the courage of present time and all time, but the events he records, all of which are drawn from the nation's present and past history, have a strong tragic inflection that threatens to overwhelm the democratic protagonist and the national myth he projects. From the wife's screech at the end of section 33 to the poet's crisis of faith in section 38, the entire sequence is laced with the sense of a self and of a nation besieged by its internal contradictions.

In evoking the large hearts of heroes, Whitman focuses first on the persecuted figures of women and blacks:

> The disdain and calmness of martyrs,
> The mother condemned for a witch and burnt with dry wood, and
>     her children gazing on;
> The hounded slave that flags in the race and leans by the fence,
>     blowing and covered with sweat,
> The twinges that sting like needles his legs and neck,
> The murderous buckshot and the bullets,
> All these I feel or am.

(LG 1855, p. 62)

Like Hawthorne reflecting on the sins of the fathers in the Custom House preface to *The Scarlet Letter,* Whitman's reflections on the American past lead him not to moments of national triumph but to moments of social injustice and internal contradiction. Feeling and suffering as the condemned witch and the hounded slave persecuted not from without but on native ground, the poet begins to suffer the sins of the nation. "Agonies are one of my changes of garments," he says, assuming the agonistic garment of America itself (*LG* 1855, p. 62).

As a sign of contradiction in the national ideology, the hounded slave undermines the poet's democratic faith and acceptance, unleashing the demons of hell and despair:

> I am the hounded slave . . . . I wince at the bite of the dogs,
> Hell and despair are upon me . . . . crack and again crack the
>     marksmen,
> I clutch the rails of the fence . . . . my gore dribs thinned with the
>     ooze of my skin,

I fall on the weeds and stones,
The riders spur their unwilling horses and haul close,
They taunt my dizzy ears . . . . they beat me violently over the head
    with their whip-stocks.

                                  (*LG* 1855, p. 62)

The unitary vision of section 15, which "contained" quadroon girls sold at the block and woollypates hoeing in the field, erupts here into a vision of national paradox and American tragedy. As a hounded slave pursued and beaten not by a foreign power but by American marksmen and whippers, the poet suffers firsthand the contradiction of slavery in the democratic republic. Even the horses are "unwilling" participants in the acts of violence daily committed against slaves by booted and spurred American riders. Just as the figure of the caged Negro ruptures the formal structure and idyllic new world vision of Crevecoeur's *Letters,* so in Whitman's *Leaves* it is over and over the issue of slavery that breaks up the text, exposing the contradiction of injustice and violation at the root of the republic.

    The image of a self and a nation under siege is particularly evident in the sequence of battle scenes that begins with the old artillerist: "I am an old artillerist, and tell of some fort's bombardment . . . . and I am there again" (*LG* 1855, p. 63). As the artillerist, the poet enters a generalized national past, in which he witnesses the martyrdom of a dying general: "He gasps through the clot . . . . Mind not me . . . . mind . . . . the entrenchments" (*LG* 1855, p. 64). Once again, however, this moment of heroism is undermined by the artillerist's vision of a fan-shaped explosion of grenades in which human beings become part of the debris of war: "The whizz of limbs heads stone wood and iron high in the air" (*LG* 1855, p. 63).

    The "jetblack sunrise" of the Goliad massacre in section 34 projects a similarly mixed image of triumph and tragedy as the poet recounts in almost journalistic detail "the murder in cold blood of four hundred and twelve young men" in the struggle for Texas independence.[13] Even the tale of John Paul Jones's victory over the British, which Whitman relates in sections 35 and 36, is shrouded in darkness and death:

Formless stacks of bodies and bodies by themselves . . . . dabs of flesh
    upon the masts and spars,
The cut cordage and dangling of rigging . . . . the slight shock of the
    sooth of waves,
Black and impassive guns, and litter of powder-parcels, and the
    strong scent,

Delicate sniffs of the seabreeze . . . . smells of sedgy grass and fields
　　by the shore . . . . death-messages given in charge to survivors,
The hiss of the surgeon's knife and the gnawing teeth of his saw,
The wheeze, the cluck, the swash of falling blood . . . . the short wild
　　scream, the long dull tapering groan,
These so . . . . these irretrievable.

(LG 1855, p. 67)

What is curious about this incident, as well as the artillerist's vision and
the massacre of the 412 young men, is that with the exception of a
reference to "English pluck" in section 35, Whitman does not name either
parties or individuals involved in the battles. These paradoxically ahistorical historical sequences mark the process whereby Whitman translates
history into national myth, representing sites of actual historic struggle in
the American past as generalized events in the seemingly timeless and
universal experience of his democratic protagonist.

Although Whitman's descent into the American past is presumably
intended as a heroic record of personal and national creation, the weight
of human suffering and tragedy in the battles he describes registers anxiety about the blood "falling" over the past and over the future of America.
These scenes of human agony, connected with both the hounded slave
and the horrors of war, trigger the crisis that begins in section 37:

O Christ! My fit is mastering me!
What the rebel said gaily adjusting his throat to the rope-noose,
What the savage at the stump, his eye-sockets empty, his mouth
　　spirting whoops and defiance,
What stills the traveler come to the vault at Mount Vernon,
What sobers the Brooklyn boy as he looks down at the shores of the
　　Wallabout and remembers the prison ships,
What burnt the gums of the redcoat at Saratoga when he surrendered
　　his brigades,
These become mine and me every one, and they are but little,
I become as much more as I like.

(LG 1855, p. 67)

As the poet begins to lose himself in scenes of human suffering and death,
he tries to regain his self-mastery by remembering acts of rebellion associated in particular with the American Revolution.

But these acts of rebellion are once again a paradoxical mixture of
triumph and defeat: Between the rebel "gaily adjusting his throat to the
rope-noose" and the defeat of the British at Saratoga are the tortured
visage of the Indian, the tomb of Washington, and the site of the death of

the American soldiers in British prison ships during the American Revolution. At a time when similar acts of rebellion—whether black or white, North or South—threatened to divide the Union that was the primary guarantee of American life and liberty, Whitman appears both to long for and to fear the rebellious acts he embodies.

The structures of human misery that entrap human life threaten to overwhelm the poet in sections 37 and 38. As he assumes the identity of a prisoner, a mutineer, and a criminal, he becomes static, impotent, caged:

> Askers embody themselves in me, and I am embodied in them,
> I project my hat and sit shamefaced and beg.
>
> (*LG* 1855, p. 68)

No longer "afoot" with his vision, the poet has fallen from the state of democratic grace. His shamefaced beggar is the antithesis of the proud, self-confident persona who straddled continents and cocked his hat as he pleased indoors or out: He has become Jonathan Edwards's man-worm, overwhelmed by the experience of violence, guilt, suffering, and death.

Having lost his democratic balance between self and other, pride and sympathy, the one and the many, the poet undergoes his second major crisis of self-mastery in section 38:

> Somehow I have been stunned. Stand back!
> Give me a little time behind my cuffed head and slumbers and
>     dreams and gaping,
> I discover myself on the verge of the usual mistake.
>
> (*LG* 1855, p. 68)

Whitman's "usual mistake," which has occasioned much critical commentary, is ambiguous. James E. Miller argues that the mistake is the "exclusion of the Divine, the Infinite"; others, including Joel Jay Belson and Sholom J. Kahn, contend that his mistake is his "momentary withdrawal of empathy which he has established with mankind."[14]

It seems fairly clear from the sequence of events that begins with the wife's screech and ends with the shamefaced beggar that the poet is not lacking in personal empathy. On the contrary, his crisis is spurred by the loss of self in empathetic identification with others. His "usual mistake" appears to be his loss of faith in the divine potency of the individual and the regenerative pattern of the whole. Overmastered by scenes of human suffering and death, the poet, like the hounded slave, reaches a nadir of despair. He resolves the crisis by remembering the divinity of Christ:

> That I could forget the mockers and insults!
> That I could forget the trickling tears and the blows of the
>     bludgeons and hammers!
> That I could look with a separate look on my own crucifixion and
>     bloody crowning!
> I remember . . . . I resume the overstaid fraction,
> The grave of rock multiplies what has been confided to it . . . . or to
>     any graves,
> The corpses rise . . . . the gashes heal . . . . the fastenings roll away.
>                                                          (*LG* 1855, pp. 68–69)

The lines clearly allude to Christ's crucifixion, but their meaning is once
again ambiguous. "That I could" might be read as an expression of desire
("How I wish that I could") or regret ("How could I"). However, Whitman's use of the term "I remember" in the subsequent stanza clearly
responds to the regret of having forgotten, rather than the desire to forget,
Christ's crucifixion. His "overstaid fraction" may be temporal, a fraction
of time "overstaid" among scenes of suffering and death. More likely,
however, the "overstaid fraction" refers to Christ. The poet's experience
is Hicksite in that he resumes the "overstaid fraction" of Christ as a living
power existing within himself as part of an eternal present, rather than an
"overstaid" emblem of divine power existing in the past rather than the
present, outside rather than inside the individual.

The sense of the passage is clarified by an early notebook entry that
Whitman made for a lecture "Religion": "America! Why elude—or attempt to elude? I say that unless the Christ you talk so much of is present
in yourselves, here and now, there is no Christ. —The Christ is dead."
Evoking what he calls "a strong description of Crucifixion," he writes:
"The anguish, the treachery, the bloody crowd, the blows of the rods and
mallets, and the piercing of the spears, —long, long ago they ceased; the
sorrowful god the Christ is dead. —And do you seek to revive him by
those Sunday incantations? those bleatings? by a few hollow words? The
effort is vain—it is not in that way you will ever resume the Christ among
you. —"[15] In the "bludgeons and hammers" sequence in "Song of Myself," the poet literally re-members Christ as a living presence, "here and
now," within the self. Having resumed the "overstaid fraction" of Christ
as part of the divine and regenerative potency of himself, the poet is able
to rise out of the structures of human misery—the corpses, gashes, and
fastenings—in which he was entrapped.

The drama of self and nation is completed in images that declare, as
Whitman's notebook prompts, "Man is a God . . . he walks over all."

> I troop forth replenished with supreme power, one of an average
>     unending procession,
> We walk the roads of Ohio and Massachusetts and Virginia and
>     Wisconsin and New York and New Orleans and Texas and
>     Montreal and San Francisco and Charleston and Savannah
>     and Mexico,
> Inland and by the seacoast and boundary lines . . . . and we pass the
>     boundary lines.
>
> Our swift ordinances are on their way over the whole earth,
> The blossoms we wear in our hats are the growth of two thousand
>     years.
>
> (LG 1855, p. 69)

The poet's hat is once again firmly cocked on his head. Christ's lesson of regenerative potency—signified by the "blossoms" in the hat—is transformed into the political base of the American republic and the moral base of a new democratic religion.[16] Restored to democratic grace and poised once again between the one and the many—between the self as "supreme power" and the self as "one of an average unending procession"—the poet projects the ideal of federal union and the expansive dreams of the nation. But as the poet emerges in the figure of a redeemer nation, he becomes a troubling presence. If he bears the hope of democratic regeneration to the whole world, he also marks the tragedy of American diplomacy. As the "I" of the poet merges with the "we" of the nation, his "supreme power" and "unending procession" come to symbolize an imperial policy that passes "the boundary lines"; his democratic dream encompasses not only the present territory of the United States—East, West, North, South—but also Canada and Mexico.

The final section of the poem bears the traces of the revolutionary myths and texts on which Whitman was raised. Metaphors of unhousing, unclothing, and unleashing predominate, as he assumes the guise of a cosmic enlightener endeavoring to release the individual from the oppressive and inhibiting structures of the past. Poetically enacting the idea of Volney, Wright, and Paine that the fear of death and its concomitant, religious orthodoxy, are a primary source of human oppression, Whitman is particularly concerned with pulling down the structures of traditional religion, liberating individuals from the fear of death and the unknown, and releasing them into a vision of the natural law, order, and equability of the universe.

And yet even in the final celebratory passages of "Song of Myself," doubts about the direction of America surface. There are signs that the religion of democracy has become a religion of self-interest and dollar worship as Americans, liberated from classical restraint and Christian humility, blindly run amuck in a laissez-faire economy of capital gains:

> Here and there with dimes on the eyes walking,
> To feed the greed of the belly the brains liberally spooning,
> Tickets buying or taking or selling, but in to the feast never once going;
> Many sweating and ploughing and thrashing, and then the chaff for payment receiving,
> A few idly owning, and they the wheat continually claiming.
>
> (*LG* 1855, p. 73)

This vision of the newly oppressive reign of capital in America, with a "few idly owning" while the many sweat and plow and thrash for a pittance, is at the root of Whitman's anxiety about the future of democracy. The passage registers his fear that the old oppressive order of kingcraft, priestcraft, and aristocracy has been replaced by a new-world order in which capital reigns over labor and rich oppress poor. The capitalist ethos of "buying or taking or selling" separates and alienates people from the abundance of the earth and the products of their labor as brains are liberally spooned to the dollar god, who, like Allen Ginsberg's Moloch, feeds on the body of America.

As one of the "centripetal and centrifugal gang" of new world democrats, Whitman acknowledges his doubts about self and nation:

> Down-hearted doubters, dull and excluded,
> Frivolous sullen moping angry affected disheartened atheistical,
> I know every one of you, and know the unspoken interrogatories,
> By experience I know them.
>
> (*LG* 1855, p. 75)

The "unspoken interrogatories" he knows by experience are related not only to the dominance of capital in the historic landscape of America; his interrogatories are also aroused by what he calls the "untried and afterward," which suggests death, the unknown, the future, and, presumably, the fate of democracy itself.

Whitman overcomes these doubts by a leap of faith in which he affirms his belief in the essential orderliness of the universe:

> I do not know what is untried and afterward,
> But I know it is sure and alive and sufficient.
>
> (*LG* 1855, p. 76)

He supports this affirmation by envisioning the universe not in the image of a divine patriarch or deist machine but in the image of a tender and loving mother:

> Long I was hugged close . . . . long and long.
> Immense have been the preparations for me,
> Faithful and friendly the arms that have helped me.
>
> Cycles ferried my cradle, rowing and rowing like cheerful boatmen;
> For room to me stars kept aside in their own rings,
> They sent influences to look after what was to hold me.
>
> (*LG* 1855, p. 78)

Hugged and coddled and rocked by the cosmic mother, Whitman's persona experiences neither fear nor terror in the face of the great reaches of space and time opened by scientists, geologists, archaeologists, and astronomers; rather, he is empowered by his participation in the processes of universal creation.

Turning and talking "like a man leaving charges before a journey," Whitman struggles in the final section of the poem with the paradox implicit in his role as the poet of democracy: How can he leave his "charges" without making himself and his poem yet another authoritarian structure imposed from without on the body of his readers? This paradox is resolved in section 46 in the figure of the open road:

> I tramp a perpetual journey,
> My signs are a rain-proof coat and good shoes and a staff cut from
>     the woods;
> No friend of mine takes his ease in my chair,
> I have no chair, nor church nor philosophy;
> I lead no man to a dinner-table or library or exchange,
> But each man and each woman of you I lead upon a knoll,
> My left hand hooks you round the waist,
> My right hand points to landscapes of continents, and a plain
>     public road.
>
> (*LG* 1855, p. 80)

What the poet offers the reader is not a closed system but an open road. Although he is a fellow traveler, the poet refuses to make the trip for his readers. In his role as cosmic enlightener, he opens the doors of perception, but it is the reader who must be the see-r. The true test of his poem will be its success in empowering the selves that will overthrow the poet: "He most honors my style who learns under it to destroy the teacher" (*LG* 1855, p. 81). The problem of literary authority is resolved democratically by summoning the self of the reader as the ultimate sovereign of text and world.

Whitman's concern throughout the poem with the problem of self-mastery is related to a more general social anxiety about the increasing centralization of institutional authority, whether in the areas of finance, capital, and trade or in response to the issues of slavery, territorial expansion, and the state of the Union. "You cannot legislate men into morality," Whitman had declared as early as 1842 in an article on popular sovereignty (*Aurora*, p. 100). Later, in an *Eagle* editorial on government, he said: "*Men* must be 'masters unto themselves,' and not look to Presidents and legislative bodies for aid" (*GF*, I, 52). He elaborated his position in *Democratic Vistas:* "That which really balances and conserves the social and political world is not so much legislation, police, treaties, and dread of punishment, as the latent eternal intuitional sense, in humanity, of fairness, manliness, decorum, &c. Indeed, this perennial regulation, control, and oversight, by self-suppliance, is *sine qua non* to democracy; and a highest widest aim of democratic literature may well be to bring forth, cultivate, brace, and strengthen this sense, in individuals and society" (*PW*, II, 421). By imaginatively embodying the individual's capacity for balance—between self and other, body and soul, material and spiritual—and by launching the reader into his or her own self-creation, Whitman tried in his poems to cultivate and strengthen the "perennial regulation, control, and oversight, by self-suppliance" as the "*sine qua non* to democracy" (*PW*, II, 421).

It is this concept of balance as both a principle of self-regulation in humanity and a principle of unity in the cosmos that is the culminating lesson of "Song of Myself." The poet finds the "word unsaid" of the universe in the generative order of creation:

> Do you see O my brothers and sisters?
> It is not chaos or death . . . . it is form and union and plan . . . . it
>     is eternal life . . . . it is happiness.
>
> (*LG* 1855, p. 85)

Like the self-evident truths of "life, liberty, and the pursuit of happiness" enunciated in the Declaration of Independence, Whitman's declaration of faith is rooted in an Enlightenment vision of "form and union and plan" as the natural law of the universe.

Having communicated his lesson of equity and balance, Whitman takes leave of his readers. Moving toward dusk, death, and the future, he acts out his message of faith by joyously dissolving into the elements of earth, air, fire, and water:

> I depart as air . . . . I shake my white locks at the runaway sun,
> I effuse my flesh in eddies and drift it in lacy jags.
>
> I bequeath myself to the dirt to grow from the grass I love,
> If you want me again look for me under your bootsoles.
>
> *(LG* 1855, p. 86)

The death of the poet and the completion of the poem correspond, like the fifty-two weeks of the year and the fifty-two sections in the poem's final version, to the completion of the earth's regenerative cycle. The poet's departure enacts the promise of eternal life not through personal immortality or spiritual transcendence but by merging with the processes of universal creation. Existing under rather than above the "soles" of his readers, the poet becomes the uniform hieroglyphic and sign of democracy he began by contemplating.

The poem returns cyclically to the beginning, with this difference: As the poet had predicted, the reader has now assumed the active and initiatory gaze of the poet contemplating the meaning of the grass:

> You will hardly know who I am or what I mean,
> But I shall be good health to you nevertheless,
> And filter and fibre your blood.
>
> Failing to fetch me at first keep encouraged,
> Missing me one place search another,
> I stop some where waiting for you
>
> *(LG* 1855, p. 86)

Through the use of the present tense and present participles in the final lines, the poet becomes, like the grass, perennially present, waiting in perpetuity somewhere down the road in the future where the reader may encounter him. The lack of a period in the concluding line, an omission that Whitman as the scrupulous printer and editor of his own poems surely intended, further advances the image of an open-ended process. The energy of creation—personal, national, global—is transferred to the indeterminate *you* of the reader, who is like the poet both present and future, singular and plural, male and female, America and Kosmos.

In the succeeding eleven poems of the 1855 *Leaves of Grass,* Whitman continues to invoke a regenerative order of unity and balance as a means of resolving personal and political tensions. He links a diversity of poetic subjects—"A Song for Occupations," "To Think of Time," "The Sleepers," "I Sing the Body Electric," "Faces," "Song of the Answerer," "Europe," and "There Was A Child Went Forth"—with the "rounded never-erased flow" he sees beneath the "creased and cadaverous march" of

humanity (*LG* 1855, p. 125). The weakest poems in the 1855 volume are
those such as "Song of the Answerer," "Who Learns My Lesson Complete," and "Great Are the Myths," in which the potentially divisive forces
in self, nation, and cosmos offer the least resistance to Whitman's unifying
act. His most interesting poems are those such as "To Think of Time," in
which the psychosocial tensions and doubts are closest to the surface,
always threatening to erode the poet's act of artistic resolution. In this
poem, in which Whitman speaks, as it were, from beyond the grave, the
intensity of his anxiety about death and the future and his disquieting
vision of annihilation—of "maggots and rats" and "ashes of dung"—
threaten to overwhelm his rosy assertion that the world has "weight and
form and location," that the "pattern is systematic." "To Think of Time"
suggests that without this sense of systematic pattern in the universe, the
self would become meaningless and the poet suicidal. As the Union began
to dissolve and Whitman's faith in the historic order of democracy corre-
spondingly tottered, a fractured, suicidal, and ontologically insecure self
became an increasingly prominent figure in his poems.

The interrelation between Whitman's personal state and the state of
the nation is particularly evident in "The Sleepers," in which the poet
enters a dream consciousness that is at once private and public, personal
and political. Duplicating in image and structure the ill-assorted, contradictory, and confused impulses of the dream state, the poet whirls into a
timeless and erotically charged dreamscape where social and sexual distinctions break down:

> I am the actor and the actress . . . . the voter . . the politician,
> The emigrant and the exile . . the criminal that stood in the box,
> He who has been famous, and he who shall be famous after today,
> The stammerer . . . . the wellformed person . . the wasted or feeble
>   person.
>
>                                        (*LG* 1855, p. 107)

In this dream world where the bounds between male and female disap-
pear, Whitman releases his own homosexual yearnings in ambiguous but
sexually nuanced images. He imagines himself as the "pet" of a "gay gang
of blackguards with mirthshouting music and wildflapping pennants of
joy" (p. 107), and he assumes the body of a woman aroused by a lover who
becomes one with the darkness:

> My hands are spread forth . . I pass them in all directions,
> I would sound up the shadowy shore to which you are journeying.

> Be careful, darkness . . . . already, what was it touched me?
> I thought my lover had gone . . . . else darkness and he are one,
> I hear the heart-beat . . . . I follow . . I fade away.
>
> (*LG* 1855, p. 107)

This radically democratic and carnivalesque landscape in which the hierarchy and polarity of the daytime world give way to the erotic flow of night is juxtaposed with images of anxiety and shame:

> O hotcheeked and blushing! O foolish hectic!
> O for pity's sake, no one must see me now! . . . . my clothes were
>     stolen while I was abed,
> Now I am thrust forth, where shall I run?
>
> (*LG* 1855, p. 107)

The sense of panic, of being unclothed and unhoused by an ill-assorted flood of erotic desire, climaxes in a passage of sexual ecstasy:

> The cloth laps a first sweet eating and drinking,
> Laps life-swelling yolks . . . . laps ear of rose-corn milky and just
>     ripened:
> The white teeth stay, and the boss-tooth advances in darkness,
> And liquor is spilled on lips and bosoms by touching glasses, and the
>     best liquor afterward.
>
> (*LG* 1855, p. 108)

In this passage, which Whitman later deleted, the images of "life-swelling yolks" and "ear of rose-corn" suggest phallic stimulation, but once again, the sexual boundaries are blurred by the erotic flow of eros. Though the images are orgasmic, the precise nature of the experience—oral, anal, vaginal, heterosexual, or homosexual—is unclear.[17]

As in "Song of Myself," the experience of sexual orgasm leads to a sense of connectedness with the earth's generative cycle:

> I descend my western course . . . . my sinews are flaccid,
> Perfume and youth course through me, and I am their wake.
>
> (*LG* 1855, p. 108)

But this feeling of oneness with the democratic flow of eros is disrupted by an increasingly pained awareness of mortality, as the poet becomes a wrinkled old woman darning her grandson's stockings, then a sleepless widow, and finally a shroud.

"The Sleepers" is deservedly famous as an innovative exploration of the dream state, an experiment in literary free association that anticipates the work of the Surrealists. In the past, the poem has been read as a symbolic enactment of a personal oedipal drama in which the poet seeks to return to an originary union with the mother. However, these psychological readings do not fully account for the poem's fairly marked historic content. The poet's anxiety about the nature of things is linked with anxiety about the democratic dream of America when, midway through the poem, the poet-sleeper enters a kind of political unconscious of the nation.

In section 5, the poet descends into the revolutionary past, remembering the American defeat at the battle of Brooklyn, a defeat that might have meant the defeat of the American cause had it not been for the perseverance of George Washington:

> Now of the old war-days . . the defeat at Brooklyn;
> Washington stands inside the lines . . he stands on the entrenched
>    hills amid a crowd of officers,
> His face is cold and damp . . . . he cannot repress the weeping
>    drops . . . . he lifts the glass perpetually to his eyes . . . . the
>    color is blanched from his cheeks,
> He sees the slaughter of the southern braves confided to him by their
>    parents.
>
>                           (*LG* 1855, p. 110)

At a time when South and North were becoming two separate nations, the image of Southern braves being slaughtered on Northern soil might serve to remind Americans of their common struggle and collective origin in the war for American independence. But in the weeping and blanched visage of Washington, nervously lifting the glass to his eyes, Whitman also registers anxiety about the prospect of another slaughter of Southern braves in a war to preserve the Union.

The tragic burden of American history as seen in Washington's "look" is managed in the poem by combining his defeat at the battle of Brooklyn with a scene of Washington bidding farewell to his troops after peace is declared:

> The same at last and at last when peace is declared,
> He stands in the room of the old tavern . . . . the wellbeloved
>    soldiers all pass through.
>
> The officers speechless and slow draw near in their turns,
> The chief encircles their necks with his arm and kisses them on the
>    cheek,
> He kisses lightly the wet cheeks one after another . . . . he shakes
>    hands and bids goodbye to the army.
>
>                           (*LG* 1855, p. 110)

In this scene and in the scene of the defeat at Brooklyn, both of which are based on passages in *Washington and His Generals* (1847) by Joel Tyler Headley, Whitman presents Washington as a democratic hero, embedded in rather than ruling over the text of the poem. Standing *amid* the crowd of his officers in battle, encircling the necks and kissing the cheeks of his soldiers in peace, Washington is depicted as the benignant father figure popularized in the nineteenth century by Parson Weems. Weeping, kissing, hugging, and compassionating, the figure that Whitman evokes represents a rebellion against more patriarchal models of authority associated with the feudal past.[18]

Like similar scenes in "Song of Myself," however, Whitman's evocation of American victory in "The Sleepers" seems curiously mixed. For one thing, victory is not mentioned, only peace—as if he knew that if there were to be another "family" war, there could be no victor. Like the wet cheeks of the soldiers who file past Washington, the scene is drenched in sorrow, a sorrow that in part reflects the pained awareness of the cost, in lost lives, of victory and peace. Even before the Civil War, Whitman has begun to mourn the war dead. But the passage is also bathed in nostalgia, as if he were also mourning the disbanding of what he called the "Sacred Army" as the initial act in the national retreat from the revolutionary ideals of the founding moment.

The sense of national loss and nostalgia intensifies in section 6 as Whitman descends into his own past to tell the tale of the tender bond between his mother and a native American woman:

> A red squaw came one breakfastime to the old homestead,
> On her back she carried a bundle of rushes for rushbottoming chairs;
> . . . . . . . .
> My mother looked in delight and amazement at the stranger,
> She looked at the beauty of her tallborne face and full and pliant
>   limbs,
> The more she looked upon her she loved her,
> Never before had she seen such wonderful beauty and purity;
> She made her sit on a bench by the jamb of the fireplace . . . . she
>   cooked food for her,
> She had no work to give her but she gave her remembrance and
>   fondness.
>
> (LG 1855, pp. 110–11)

The relationship between mother and squaw suggests a primary, prepatriarchal bond as the foundation of democratic community; once again, the boundaries of gender, race, and class dissolve in the unifying energy of eros. The socially marginalized figures of white mother and red

squaw are linked in a common bond of love, and the bond between them
has a seductively erotic nuance.

The democratic idyll, which is set in the youth of Whitman's mother
and in the time of the "old homestead," implies a departed ideal. The idyll
is interrupted by the squaw's sudden departure:

> The red squaw staid all the forenoon, and toward the middle of the
>     afternoon she went away;
> O my mother was loth to have her go away,
> All the week she thought of her . . . . she watched for her many a
>     month,
> She remembered her many a winter and many a summer,
> But the red squaw never came nor was heard of there again.

> > (*LG* 1855, p. 111)

Like Washington, the squaw is a cultural icon. With her long black hair,
tallborne face, beauty, and purity, she bears the traces of an iconographic
tradition, dating back to the colonial period, in which America was represented as a native American woman. Her departure suggests the loss of a
cultural ideal of America and, ironically, the loss of the native American
race itself with the westward advance of democracy. The loving bond
between the white mother and the red squaw is, indeed, a contrapuntal
figure set against an age that was willy-nilly annihilating native American
culture in the name of civilization and progress.

Whitman penetrates even farther the racial and racist consciousness of
America in the poem's next sequence. Assuming the identity of a
hounded slave, he speaks in the voice of Lucifer:

> Now Lucifer was not dead . . . . or if he was I am his sorrowful
>     terrible heir;
> I have been wronged . . . . I am oppressed . . . . I hate him that
>     oppresses me,
> I will either destroy him, or he shall release me.

> > (*LG* 1855, p. 111)

In an earlier draft of these lines, Whitman described "Black Lucifer" as "a
hell-name and a curse" and the "God of revolt." As the heir of Lucifer,
the slave is a figure of revolutionary defiance and a sign of the evil of
slavery at the foundation of the American republic. Like Frederick Douglass in *My Bondage and My Freedom* (an expanded version of his *Narrative*,
published in the same year as "The Sleepers" in 1855), Whitman's "Black
Lucifer" passage links the logic of slave revolt with the revolutionary
origins of the American republic.[19]

Although Whitman struggled with and sometimes gave in to his culture's notion of black inferiority, he refused to yield to the notion that one man had the right to own another as property: "Every man who claims or takes the power to own another man as his property, stabs me in the heart of my own rights—for they only grow from that vast principle, as a tree grows from the seed" (*DN*, III, 761). The Lucifer passage, which Whitman later deleted from the poem, reveals his fear of slavery as the curse of the land, the fatal flaw of America that threatens to destroy not only the union of the states but also the democratic charter of the republic itself.

In contrast with the egalitarian and essentially matriarchal vision of mother and squaw, the slave master is seen through the eyes of the slave as a cruel patriarch who violates the bonds of blood and family:

> Damn him! how he does defile me,
> How he informs against my brother and sister and takes pay for
>   their blood,
> How he laughs when I look down the bend after the steamboat that
>   carries away my woman.
>
> (LG 1855, p. 111)

The familial violation of the slave master is even more evident in an earlier draft of these lines. What most horrifies Whitman is that given the sexual and racial exploitation of the slave economy, the slave master may indeed be selling his own sons and daughters and the mother who gave them birth:

> Damn him, how he does defile me!
> Hoppler of his own sons; breeder of children and trader of them—
> Selling his daughters and the breast that fed his young.
> Informer against my brother and sister and taking pay for their blood.
> He laughed when I looked from my iron necklace after the steamboat that
>   carried away my woman.[20]

By eliminating the more specific references to black man and slave master, Whitman generalizes the slave's struggle into a more universal struggle of the oppressed against all forms of enslavement—physical, economic, social, and mental. Lucifer's curse becomes a curse on any oppressor, whether Southern slave master or Northern capitalist, who violates the original charter of American liberty.

Horrified at the prospect of national dissolution and wholesale slaughter and yet inhabited by the contradiction of slavery in America, Whitman issues a warning about the prospect of slave revolt in language laced with desire and fear:

Now the vast dusk bulk that is the whale's bulk . . . . it seems mine,
Warily, sportsman! though I lie so sleepy and sluggish, my tap is
    death.
                                          (*LG* 1855, p. 111)

Reflecting what J. C. Furnas has called the "Spartacus complex" of his
age—the belief widely held among Abolitionists that whites were about to
be destroyed in a massive slave revolt—Whitman appears to long for and
invite a slave revolt as a kind of punishment and purgation of personal and
national guilt. His curse of Lucifer anticipates the curse uttered by John
Brown on the day he died: "The crimes of this *guilty, land will* never be
purged *away;* but with Blood."[21]

The specter of death, both personal and national, at the close of the
Lucifer sequence completes the poem's downward spiral. In the curse of
black Lucifer, Whitman reaches the "plague spot" of race and racism in
the very depths of the national psyche. In the concluding passage, the
downward spiral is reversed as the poem moves like the cyclical turn of the
universe, from darkness to light, winter to summer, hate to love. "The
wildest and bloodiest is over and all is peace," the poet declares, finding in
the leveling and restorative power of night and sleep, an earthly analogue
for the "myth of heaven" and the spiritual order of the universe:

The soul is always beautiful,
The universe is duly in order . . . . every thing is in its place,
What is arrived is in its place, and what waits is in its place;
. . . . . . . . .
The sleepers that lived and died wait . . . . the far advanced are to
    go on in their turns, and the far behind are to go on in their
    turns,
The diverse shall be no less diverse, but they shall flow and unite
    . . . . they unite now.
                                          (*LG* 1855, p. 113)

Whitman's declaration of faith is a declaration by fiat. His assertion of
unity is at odds with his confused and ill-assorted dream vision of a
universe in which everything is not in order and diversity does not cohere.
His series of assertions and his rhetorical insistence do not resolve the
personal and political tensions that have erupted in the poem. In fact,
Whitman's metaphoric equation of night and sleep with spiritual regeneration is itself curiously "ill-assorted" in a poem in which sleep brings not
only democratic dreams of an equable and loving order but nightmare
visions of separation and loss, guilt and rage, dredged out of his own and
the nation's psyche.

Whereas in "The Sleepers" Whitman seeks poetic resolution in the baptismal vision of naked sleepers floating "hand in hand over the whole earth from east to west," in "I Sing the Body Electric" he assumes a more active political posture, attempting to uproot the contradictions in the body politic of America by addressing the reader directly. The poem is framed by a question:

> Was it dreamed whether those who corrupted their own live bodies
>     could conceal themselves?
> And whether those who defiled the living were as bad as they who
>     defiled the dead?
>
> *(LG* 1855, p. 116)

The rhetorical ordering of the poem turns on the answer to this question. Here again, however, critics have tended to treat the poem as a fairly tedious enumeration of body parts, failing to note its ominous political prophecy and the fact that the body electric is also black.

The divine and sexually charged body that Whitman celebrates becomes the platform from which he launches an attack on race and class attitudes in America:

> The man's body is sacred and the woman's body is sacred . . . . it is
>     no matter who,
> Is it a slave? Is it one of the dullfaced immigrants just landed on
>     the wharf?
> Each belongs here or anywhere just as much as the welloff . . . . just
>     as much as you,
> Each has his or her place in the procession.
>
> *(LG* 1855, p. 120)

Whitman's questions are intended to engage his readers in a critique that was in part directed at the growth of American nativism, a movement which peaked in 1854 when the organization of antiforeign societies known as the Know-Nothings made substantial gains in state elections, particularly in Massachusetts. In their efforts to restrict immigration, extend the naturalization period, and deny foreigners the vote, the Know-Nothings represented a turn away from the tradition of America as the refuge of the oppressed and the site of a new multinational race.[22] The poem counters this nativist, antiforeign strain in America by celebrating every *body* as a locus of divine and democratic energies and by insisting on the place of every *body* not only in the American procession but in the spiritual progression of the universe.

Beyond the issue of immigration, "I Sing the Body Electric" focuses once again on the whole question of racial inferiority, a question to which Whitman returned over and over again in his notebooks of that time. In one note he reflected on the interrelationship between race, class, and slavery: "But this is an inferior race.    —Well who shall be the judge of inferior and superior races.    —The class of dainty gentlemen think that all servants and laboring people are inferior.    —In all lands, the select few who live and dress richly, make a mean estimate of the body of the people" (*DN*, III, 762). In another notebook entry, he planned a "Poem of Remorse," renouncing his former belief in racial inferiority:

Poem of Remorse

I now look back to the
times when I thought
?          ?
others—slaves—the ignorant
—so much inferior to myself
To have so much less right
(*DN*, III, 791)

Whitman's "Poem of Remorse" was transformed into his poetic celebration of the black person in "I Sing the Body Electric." Echoing his notebook entries on racial inferiority, he asks in the poem: "Do you know so much that you call the slave or the dullfaced ignorant?" (*LG* 1855, p. 121).

The question, this time addressed directly to the *you* of the reader, enlists the audience in the defense of black personhood that is the core of the poem. Thrusting aside the slovenly auctioneer of slaves, the poet mocks the institutions of slavery and the absurdity of buying and selling human lives. Instead, the slave auction becomes an occasion to praise black humanity and black heroism:

> A slave at auction!
> I help the auctioneer . . . . the sloven does not half know his
>     business.
> Gentlemen look on this curious creature,
> Whatever the bids of the bidders they cannot be high enough for
>     him,
> For him the globe lay preparing quintillions of years without one
>     animal or plant,
> For him the revolving cycles truly and steadily rolled.
> (*LG* 1855, p. 121)

In the guise of the auctioneer, Whitman faces his audience directly, delivering, like Frederick Douglass himself, an oration on the glories of black humanity. The body that he celebrates reaches out to include all bodies—red, white, and black—that jostled for freedom and space in the geopolitical landscape of nineteenth-century America:

> In that head the allbaffling brain,
> In it and below it the making of the attributes of heroes.
>
> Examine these limbs, red black or white . . . . they are very cunning
>   in tendon and nerve;
> They shall be stript that you may see them.
>
> <div align="right">(<i>LG</i> 1855, p. 121)</div>

Whitman cracks the culturally imposed mask of black invisibility to expose the same red blood of desire and possibility that links the black person with the democratic commonality:

> Within there runs his blood . . . . the same old blood . . the same
>   red running blood;
> There swells and jets his heart . . . . There all passions and desires
>   . . all reachings and aspirations:
> Do you think they are not there because they are not expressed in
>   parlors and lecture-rooms?
>
> <div align="right">(<i>LG</i> 1855, pp. 121–22)</div>

Here, as throughout the sequence, Whitman continues to interrogate his readers directly, insisting that they probe and resolve the contradictions in their own racial attitudes. To the economics of slavery and Negrophobia of his age, he counters with a vision of blacks as coequal citizens in the process of personal and national creation.

Despite his affirmations, however, the image of the slave auction is a telling reminder of the reality of slavery in the nineteenth-century American marketplace. The poem's concluding lines draw attention to the economics of slavery as the disruptive subtext of both democratic poem and democratic nation. The poem's opening question about defiling the bodies of the living and the dead is answered in its final lines:

> Who degrades or defiles the living human body is cursed,
> Who degrades or defiles the body of the dead is not more cursed.
>
> <div align="right">(<i>LG</i> 1855, p. 123)</div>

Insofar as the defilation of the human body by the institution of slavery remained a reality in nineteenth-century America, the curse with which the poem ends becomes, as in the Lucifer passage in "The Sleepers," a curse on the owners and traders of slaves as well as on America itself. Read in this way, "I Sing the Body Electric" is in its concluding passages a plea for the liberation and regeneration not only of the black person but of white America from the curse of slavery.

Like "Song of Myself," the 1855 edition of *Leaves of Grass* ends on a note of affirmation and faith. But in the concluding poem, "Great Are the Myths," it is an affirmation that is declared rather than earned, as Whitman asserts the greatness of all things from Adam and Eve to the "plunges and throes and triumphs and falls of democracy," from the English language and the English brood to the "eternal equilibrium" and "eternal overthrow of things." His assertions are limp and at odds with the undercurrent of anxiety and doubt, both personal and political, that runs through the volume. The great myths that the poet enunciates have a hollow ring that suggests his growing fear that all may not be great. The universe and along with it the democratic structures of self and republic might be, as for the child in "There Was a Child Went Forth," mere "flashes and specks." Rather than being a final testament of democratic faith, "Great Are the Myths," which Whitman wisely deleted from the final edition of *Leaves of Grass*, is not only anticlimactic and not great; it is also a form of poetic whistling in the dark.

# 6

# The Fractured State

Once, before the war, (Alas! I dare not say how many times the mood has come!) I, too, was fill'd with doubt and gloom. A foreigner, an acute and good man, had impressively said to me, that day—putting in form, indeed, my own observations: "I have travel'd much in the United States, and watch'd their politicians, and listen'd to the speeches of the candidates, and read the journals, and gone into the public houses, and heard the unguarded talk of men. And I have found your vaunted America honeycomb'd from top to toe with infidelism, even to itself and its own programme. I have mark'd the brazen hell-faces of secession and slavery gazing defiantly from all the windows and doorways."

—WHITMAN, *Democratic Vistas*

"I too have at heart Freedom, and the amelioration of the people," Whitman wrote to New York Senator William Henry Seward, the outspoken opponent of compromise on the issue of slavery who had appealed to "a higher law than the Constitution" to keep the new territory free. In his letter, dated December 5, 1855, Whitman asked Seward to send him "public documents, your speeches, and any government, congressional or other publications of general interest, especially statistics, census facts &c."[1] He probably intended to use this material as part of his ongoing plan to take his democratic program on the road; more specifically, he used the material in *The Eighteenth Presidency!* "Where is the real America," he asked in this political pamphlet, which was probably begun amid the controversy over the Fugitive Slave Law and the Kansas–Nebraska Act in the early 1850s and completed in the heat of the presidential campaign of 1856. Subtitled "Voice of Walt Whitman to Each Young Man in the Nation, North, South, East, and West," the pamphlet, which exists in proof but was probably never published, indicates Whitman's continued interest in shaping the course of political events in America by carrying his message to the people (*WWW*, p. 92).

As an impassioned response to the succession of national crises in the 1850s, each of which hinged on the issue of slavery, *The Eighteenth Presidency!* is a masterpiece of political invective. In fierce and scabrous language Whitman presents the national crisis as one of political power in which the current regime of "office-holders, office-seekers, robbers, pimps, exclusives, malignants, conspirators, murderers" and the "three hundred and fifty thousand owners of slaves," committed to adopting slavery as a national policy, seek to impose their will on the thirty million who make up the body of the American people. Split between faith in the "flowing fire of the humanitarianism of the new world" and disillusionment with the machinations of "crawling, serpentine men, the lousy combings and born freedom sellers of the earth," Whitman's pamphlet is a kind of seismograph of the times, measuring the divided mind of a nation committed to a republican ideology of progress and abundance and afflicted by a sense of impending doom (*WWW*, p. 100).

While Whitman's pamphlet is a call to action, it is not, technically, a piece of campaign literature. In the presidential conventions of 1856, the American or Know-Nothing party nominated Millard Fillmore on a nativist platform; the Republicans chose John C. Fremont on an antislavery platform; and the Democrats nominated James Buchanan, who tried to avoid the issue of slavery altogether by denying the right of the federal government to legislate on slavery in the territories. Although Whitman was closest in his views to Fremont's Free-Soil platform, in *The Eighteenth Presidency!* he refuses to identify with any particular political party. "I place no reliance upon any old party, nor upon any new party," he asserts. "Not the so-called democratic, not abolition, opposition to foreigners, nor any other party, should be permitted the exclusive use of the Presidency" (*WWW*, p. 104). Presenting himself as a nonpartisan who speaks for common rather than for special interests, Whitman makes his appeal directly to the people, reminding them of revolutionary principles and the fundamental laws of the land: "The platforms for the Presidency of These States are simply the organic compacts of The States, the Declaration of Independence, the Federal Constitution, the action of the earlier Congresses, the spirit of the fathers and warriors, the official lives of Washington, Jefferson, Madison, and the now well-understood and morally established rights of man, wherever the sun shines, the rain falls, and the grass grows" (*WWW*, p. 105). Invoking the inalienable right to life and liberty set forth in the Declaration of Independence, he urges the people to defy the Fugitive Slave Law and calls for the abolition of slavery, through the spontaneous action of the people from below: "You young men! American mechanics, farmers, boatmen, manufacturers, and all work-people of the South, the same as the North! you are either to abolish slavery, or it will abolish you" (*WWW*, p. 110). The ominous note of Whitman's appeal once again reveals some of the mixed motivation—fear for the Union, fear for the dignity of labor, fear of slave revolt, fear of the black race—that entered antislavery rhetoric in nineteenth-century America.

In the closing passages of his antislavery appeal, Whitman's warning is transmuted into the visionary rhetoric of the revolutionary enlightenment as he presents the struggle against slavery in America not as an isolated event, or as a struggle between blacks and whites only, but as part of the advance of freedom throughout the world: "The times are full of great portents in These States and in the whole world. Freedom against slavery is not issuing here alone, but is issuing everywhere" (*WWW*, p. 112). The agitation against slavery in America, the collapse of traditional frontiers and boundaries, the interlinking of the earth through modern inventions such as the steamship, locomotive, and electric telegraph all are part of the struggle toward a global democratic community: "What historic denouements are these we are approaching?" Whitman asks. "On all sides tyrants tremble, crowns are unsteady, the human race restive, on the watch for some better era, some divine war. No man knows what will happen next, but all know that some such things are to happen as mark the greatest moral convulsions of the earth" (*WWW*, p. 113).

With his ear close to the political ground of the nation, Whitman in *The Eighteenth Presidency!* is uncannily prophetic. Not only does he predict the Civil War that would convulse the nation, but he envisions the legendary president who would rise to save it: "I would be much pleased to see some heroic, shrewd, fully-informed, healthy-bodied, middle-aged, beard-faced American blacksmith or boatman come down from the West across the Alleghanies, and walk into the Presidency, dressed in a clean suit of working attire, and with the tan all over his face, breast, and arms" (*WWW*, p. 93). Whitman's portrait suggests that the figure of a "redeemer president" existed in the popular imagination before Abraham Lincoln rose to play the role. As an unmistakable version of Whitman's democratic rough, the portrait also shows some of his own aspiration to become, if not the president, then at least a political spokesman for the American workers. In the conclusion to his political tract, Whitman addresses the "editors of the independent press" and "rich persons": "Circulate and reprint this Voice of mine for the workingmen's sake," he says. "I am not afraid to say that among them I seek to initiate my name, Walt Whitman, and that I shall in future have much to say to them. I perceive that the best thoughts they have wait unspoken, impatient to be put into shape" (*WWW*, pp. 111–12).

The 1856 edition of *Leaves of Grass*, which was published in late summer at the height of the presidential campaign, is shaped by a similar desire to speak to and for the American people. The poems translate the political jeremiad of *The Eighteenth Presidency!* into a kind of secularized sermon. With the words of Emerson's laudatory letter—"I greet you at the beginning of a great career"—emblazoned in gold on the spine, the volume is split like the age itself between an expansive vision of personal and national fulfillment and a pained awareness of the "dark patches" on the horizon of the American republic. From a historical perspective, the 1856 *Leaves* might well be the most interesting edition of Whitman's poems, for the volume so nearly mirrors the strained countenance of the nation itself in the years immediately preceding the Civil War.

Although Whitman sets out in his letter to Emerson, which appears as a kind of afterword to the poems, to define the role of the American bard, his definition keeps slipping into political invective: "I think there can never be again upon the festive earth more bad-disordered persons deliberately taking seats, as of late in These States, at the heads of the public tables—such corpses' eyes for judges—such a rascal and thief in the Presidency" (*LGC*, p. 737). As in *The Eighteenth Presidency!*, he seeks to incite the people to an act of revolutionary defiance in order to sweep away the current administration, but the people appear to be unequal to the task: "The people, like a lot of large boys, have no determined tastes, are quite unaware of the grandeur of themselves, and of their destiny, and of their immense strides—accept with voracity whatever is presented them in novels, histories, newspapers, poems, schools, lectures, every thing" (*LGC*, p. 737).

The letter to Emerson also registers Whitman's growing fear of political dismemberment: "The union of the parts of the body is not more necessary to their life than the union of These States is to their life," he wrote (*LGC*, p. 735). As the Union began to split, the organic metaphor—the body–body politic trope—became an increasingly insistent figure in Whitman's work. The organic compact of the states was the rather loose organizing principle of the 1856 *Leaves*. "There are Thirty-Two States sketched—the population thirty millions," Whitman noted, suggesting the analogy between the thirty-two poems of the 1856 *Leaves* and the thirty-two states of the Union (*LGC*, p. 738). (Actually, there were only thirty-one states, but he may have been counting on the imminent entry of Kansas).

Whitman's new sense of urgency in addressing the problem of Union is evident in "Poem of Many in One" (later "By Blue Ontario's Shore"), which is, like several of the poems in the 1856 edition, a poetic reworking of the 1855 preface. Whereas in 1855 he envisioned the poet as an incarnator of the nation, in 1856 he moves toward an emphasis on the poet as a fuser of the nation. ,

> By great bards only can series of peoples and
> States be fused into the compact organism of
> one nation.
> To hold men together by paper and seal, or by
> compulsion, is no account,
> That only holds men together which is living
> principles, as the hold of the limbs of the
> body, or the fibres of plants.[2]

Envisioning poetry as a form of political power other than government and militia, "paper and seal," Whitman proposes to join peoples and states by reviving in them the "living principles" of liberty, equality, self-sovereignty, and inalienable right from which the "compact organism" of national union was originally constituted.

In his bid to revive the principles of the founding moment in the political body of the nation, he began in 1856 to date his poems in relation to the formation of the American republic. The 1850 poem "Resurgemus" became "Poem of the Dead Young Men of Europe, the 72nd and 73rd Years of These States"; and his poem on the trial of Anthony Burns became "Poem of Apparitions in Boston, the 78th Year of These States." He even dates his birth and his age at the time *Leaves of Grass* was published in relation to the birth of the nation: in "Lesson Poem" (later "Who Learns My Lesson Complete"), he says he was "born on the last day of May in the Year 43 of America," and he is "a man thirty-six years old in the Year 79 of America" (*LG* 1856, pp. 314–15).

In "Poem of Remembrance for a Girl or a Boy of These States," which was deleted from *Leaves of Grass* after the war, Whitman tries to resolve the problem of national dissolution under the pressure of slavery by appealing directly to his readers, urging them to an act of political "remembrance":

> Remember the organic compact of These
> States!
> Remember the pledge of the Old Thirteen thence-
> forward to the rights, life, liberty, equality, of
> man!

> Remember what was promulged by the founders,
>     ratified by The States, signed in black and
>     white by the Commissioners, read by Wash-
>     ington at the head of the army!
> Remember the purposes of the founders! —Re-
>     member Washington!
>
> (LG 1856, p. 275)

The image of the Union as an "organic compact" resolves on a meta-phoric level the contemporary political debate about whether the Union was an organism or a compact among states and thus dissoluble. In fact, Whitman's representation of the organic compact of the states as both the Declaration of Independence and the living body of the Union—both a political document and the principles of life, liberty, and equality that lived in and through the body of the nation—marks the transition in mid-nineteenth-century America from the Union as political practice, or means, to the political mystification of the Union as absolute, indissoluble, and binding in perpetuity.[3]

Central to Whitman's program of national "remembrance" in 1856 was the creation of a new race of free and equal women. In "Poem of Remembrance" the revolutionary image of America as a potent female republic is translated into the image of a republic of literally potent females:

> Anticipate the best women!
> I say an unnumbered new race of hardy and well-
>     defined women are to spread through all
>     These States,
> I say a girl fit for These States must be free,
>     capable, dauntless, just the same as a boy.
>
> (LG 1856, p. 276)

To empower this new race of dauntless women, the poet enjoins the women of the states to

> Think of womanhood, and you to be a woman!
> The creation is womanhood,
> Have I not said that womanhood involves all?
> Have I not told how the universe has nothing
>     better than the best womanhood!
>
> (LG 1856, p. 278)

Whitman's directive is more than a pious echo of Victorian notions of female moral superiority. His assertion that "the creation is womanhood" reverses the myth of Adam's rib and reclaims the primacy of the female in the order of creation.

This reversal is the subject of "Poem of Women" (later "Unfolded Out of the Folds"), which follows "Poem of Walt Whitman, an American" in 1856. Moving beyond patriarchal inscriptions, the poem celebrates the potency of the female as the creative source of the universe and the source of the poet's own creation:

> Unfolded only out of the folds of the
> woman, man comes unfolded, and is always
> to come unfolded,
> Unfolded only out of the superbest woman of the
> earth is to come the superbest man of the earth,
> . . . . . . . . .
> Unfolded only out of the inimitable poem of
> the woman can come the poems of man—
> only thence have my poems come.
>
> (*LG* 1856, pp. 101–2)

Figuring the male as a term enfolded within the creative body of the female, "Poem of Women" is a female creation myth that radically rewrites the Victorian glorification of the female reproductive capacity.

Whitman's poetic evocation of the fleshy and creative folds of the female body works against the loathing of female flesh, particularly associated with birth, that was part of the Judeo-Christian tradition. "Poem of Women," along with the 1856 "Poem of Procreation" (later "A Woman Waits for Me"), was part of Whitman's attempt to give a literary voice to sex and the body. As he announced in his letter to Emerson, he planned to "celebrate in poems the eternal decency of the amativeness of Nature, the motherhood of all," to work against the "fashionable delusion of the inherent nastiness of sex" and "publicly accept, and publicly name, with specific words, the things on which all existence . . . depend" (*LGC*, pp. 739–40). Writing the body and sex was, in Whitman's view, part of the process of liberating the individual from the political tyranny of the past, and thus part of the democratic creation of America.

In "Poem of Procreation" Whitman publicly names the female body and female sexual desire that had been silenced by literature and culture. Overturning Victorian notions of a closeted and virginal motherhood, he associates maternity with sexual appetite, the open air, and healthy athleticism:

> A woman waits for me—she contains all,
> nothing is lacking,

> Yet all were lacking, if sex were lacking, or if
>     the moisture of the right man were lacking.
>
> Sex contains all,
> Bodies, souls, meanings, proofs, purities, delica-
>     cies, results, promulgations,
> Songs, commands, health, pride, the maternal
>     mystery, the semitic milk,
>
> All hopes, benefactions, bestowals,
> All the passions, loves, beauties, delights of the
>     earth,
> All the governments, judges, gods, followed per-
>     sons of the earth,
> These are contained in sex, as parts of itself
>     and justifications of itself.
>
> Without shame the man I like knows and avows
>     the deliciousness of his sex,
> Without shame the woman I like knows and avows hers.
>                                               (*LG* 1856, p. 240)

As in "Song of Myself," Whitman justifies his new democratic order of
sexual shamelessness by linking the "deliciousness" of an empowered and
empowering sexuality with the amative flow of nature.

The poet refuses the neurasthenic, doll-like, and sexually cloistered
women associated with the cult of true womanhood and the new bourgeois order. As the "robust" husband, he is attracted to women who are
hardy, warm-blooded, and magnetically charged:

> I will dismiss myself from impassive women,
> I will go stay with her who waits for me, and
>     with those women that are warm-blooded and
>     sufficient for me,
> I see that they understand me, and do not deny me,
> I see that they are worthy of me—so I will be
>     the robust husband of those women!
> They are not one jot less than I am,
> They are tanned in the face by shining suns and blowing winds,
> Their flesh has the old divine suppleness and strength,
> They know how to swim, row, ride, wrestle,
>     shoot, run, strike, retreat, advance, resist,
>     defend themselves,
> They are ultimate in their own right—they are
>     calm, clear, well-possessed of themselves.
>                                               (*LG* 1856, p. 241)

Whitman's egalitarian rhetoric, his active, self-possessed, and sexually charged women, and his apparent undoing of traditional male and female spheres have had a particularly powerful appeal for women readers. And yet, read closely, "Poem of Procreation" suggests the extent to which Whitman participated in the sexual ideology of his age even as he sought to challenge and transform it.

Although he succeeds in uncovering a healthy, athletic, and sexually charged female being who had been unnamed by his culture, his aggressively strident and nationalist rhetoric entraps her once again in the codes of patriarchy and the essentially spermatic economy of the bourgeois social order. Drawing on the popular eugenic literature of his time, Whitman presents the male as the carrier of an electric charge that sparks the growth of life in the female body:

> O I will fetch bully breeds of children yet!
> They cannot be fetched, I say, on less terms than
> mine,
> Electric growth from the male, and rich ripe fibre
> from the female, are the terms.
>
> (*LG* 1856, p. 241)

In contrast with the reclamation of female primacy in "Poem of Women," in these lines Eve becomes once again Adam's rib, as the poet himself appropriates the female reproductive capacity.[4]

The image of an electric male impregnating a passive female mass presides over the poem, keeping the female rhetorically prone. "A woman waits" throughout the poem for the phallic male, and when he finally arrives, he takes control, pressing her with "slow rude muscle" into the service of the race:

> I draw you close to me, you women!
> I cannot let you go, I would do you good,
> I am for you, and you are for me, not only for our
> own sake, but for others' sakes,
> Enveloped in you sleep greater heroes and bards,
> They refuse to awake at the touch of any man but
> me.
>
> It is I, you women—I make my way,
> I am stern, acrid, large, undissuadable—but I
> love you,
> I do not hurt you any more than is necessary for
> you,
> I pour the stuff to start sons and daughters fit for
> These States—I press with slow rude muscle,

I brace myself effectually—I listen to no en-
   treaties,
I dare not withdraw till I deposit what has so
   long accumulated in me.
                          (*LG* 1856, pp. 241–42)

As the poet links vigorous (male) sexuality with the creation of a strong
nation, the female is rhetorically plowed under. The male is subject, the
female object; he is active, she is passive. Not only is there no "deliciousness" of female sexual pleasure, but the male appears to take her against
her will. Their union seems less the occasion for a dynamic pairing of
equals and the more the scene of a domestic rape. In a diary entry on the
poem in 1883, Elizabeth Cady Stanton was provoked to comment: "He
speaks as if the female must be forced to the creative act, apparently
ignorant of the great natural fact that a healthy woman has as much
passion as a man, that she needs nothing stronger than the law of attrac-
tion to draw her to the male."[5]

Beneath the tone of strident nationalism that marks "A Woman Waits for
Me" and other 1856 poems is a barely concealed hysteria that self and
world might be coming apart at the seams. Even in "Poem of Many in
One," which is a kind of preface poem to the 1856 *Leaves,* Whitman's
rhetorical insistence on national traditions and the lesson of self-mastery
betrays his growing fear of dissolution: "I swear I will not be outfaced by
irrational things!" he declares toward the end of the poem. And in the
final lines, he faces but does not resolve the forces of wickedness and
unreason that threaten the equilibrium of body and body politic:

I match my spirit against yours, you orbs, growths,
   mountains, brutes,
I will learn why the earth is gross, tantalizing,
   wicked,
I take you to be mine, you beautiful, terrible, rude
   forms.
                          (*LG* 1856, p. 201)

Significantly, in 1867, Whitman revised and improved upon the poem by
refocusing it on the drama of the "throes of Democracy" that culminated
in the victory of the Union in the Civil War, but in the 1856 version the
bard of Democracy appears to have lost his direction.

   The source of his anxiety surfaces in "Poem of the Propositions of
Nakedness" (later "Respondez"), the only new poem to address directly
the political landscape of 1856. Proclaiming his desire to "let that which

was behind advance to the front and speak," Whitman tears away the veils of the republican myth to expose the naked truth of American failure. His propositions present a topsy-turvy world in which the balanced and virtuous republic envisioned by the revolutionary founders has dissolved into a mass of historical contradictions:

> Let contradictions prevail! Let one thing con-
> tradict another! and let one line of my poem
> contradict another!
> Let the people sprawl with yearning aimless
> hands! Let their tongues be broken! Let their
> eyes be discouraged! Let none descend into
> their hearts with the fresh lusciousness of
> love!
> Let the theory of America be management, caste,
> comparison! (Say! what other theory would
> you?)
>
> (*LG* 1856, p. 317)

Like his exclamation at the beginning of the poem "Respondez! Respondez!" the poet's parenthetical questions are rhetorical markers that demand a response from the reader.

Some of the propositions appear absurd: "Let the sun and moon go!" Others look toward the alienation, fragmentation, and global warfare of the twentieth century: "Let the people sprawl with yearning aimless hands!"; "Let there be apathy under the stars!"; "Let there be no God!"; "Let the Asiatic, the African, the European, the American and the Australian, go armed against the murderous stealthiness of each other; Let them sleep armed! Let none believe in goodwill!" (*LG* 1856, pp. 317–20). Other propositions appear to be unreasonable but turn out to be at least half-true:

> Let freedom prove no man's inalienable right!
> Every one who can tyrannize, let him tyran-
> nize to his satisfaction!
>
> .　.　.　.　.　.　.
>
> Let there be money, business, railroads, imports,
> exports, custom, authority, precedents, pallor,
> dyspepsia, smut, ignorance, unbelief!
>
> .　.　.　.　.　.　.
>
> Let there be immense cities—but through any of
> them, not a single poet, saviour, knower, lover!
>
> .　.　.　.　.　.　.

> Let the white person tread the black person under
>    his heel! (Say! which is trodden under
>    heel, after all?)
>                        (*LG* 1856, pp. 318–21)

The juxtaposition of seemingly absurd propositions with propositions that describe the actual conditions in America in 1856 intensifies the impression of fracture and dislocation in the political sphere.

In linking the political failure of America with the collapse of rational order in the universe, "Poem of the Propositions of Nakedness" suggests the connection between the nation's crisis and Whitman's personal crisis of identity in the late 1850s. Whitman associated the American Union with the order of history, even dating his own birth in relation to the birth of the American compact. For the Union to fail would mean the collapse of his own sense of identity and place in the universe and his whole sense of mission as a poet: History would become not patterned progression but meaninglessness, not unity but implacable force. "O seeming! seeming! seeming!" Whitman exclaimed toward the end of the poem as he edged closer to the unhinged and endlessly seeming universe of Melville's *Moby Dick* (1850) and *Pierre* (1851).

Whitman attempted to resolve the crisis of the political moment in poems of visionary democracy that locate the individual and the nation in a unitary and progressive order of history. In "Poem of Wonder at the Resurrection of the Wheat" (later "This Compost"), he counters the specter of personal and national dissolution in the corpse of the land with a vision of the regenerative force of nature. Whitman is himself the unifying figure in "Poem of Salutation" (later "Salut Au Monde!"), in which he links all "ranks, colors, barbarisms, civilizations" in a single democratic embrace:

> My spirit has passed in compassion and deter-
>    mination around the whole earth,
> I have looked for brothers, sisters, lovers, and
>    found them ready for me in all lands.
>                        (*LG* 1856, p. 120)

In "Broad-Axe Poem" (later "Song of the Broad-Axe"), he mythologizes the potentially troubling figure of the ax as a civilizing force, fusing destruction and creation, the past and the future, the advance westward and the advance of the race. Drawing on the progressive ideology of the Enlightenment, Whitman envisions the advance of democracy in America as part of the natural and purposeful evolution of the cosmos. He imagines the realization of the democratic dream in a city where people are sovereign, slavery is abolished, and women participate equally in the government of the nation—a city, in short, that reverses the political order of his time.

The psychic cost of maintaining his exaltingly democratic vision in the face of the imperfect social facts of his time was sometimes too much. At times, even nature itself seemed the source of troubling ambiguity and gnawing doubt. In a notebook of the mid-1850s, Whitman wrote of the paradoxical nature of things:

> O, Nature! impartial, and perfect in imperfection!
> Every precious gift to man is linked with a curse—and each pollution has some sparkle from heaven.
> The mind, raised upward, then holds communion with angels and its reach overtops heaven; and yet then it stays in the meshes of the world too and is stung by a hundred serpents every day. (*UPP*, II, 89)

"Sun-Down Poem" (later "Crossing Brooklyn Ferry"), which appeared for the first time in the 1856 *Leaves*, is Whitman's attempt to come to terms with the "hundred serpents" of every day that "stung" him when he stayed "in the meshes of the world."

Although Whitman had for years been riding the ferry across the East River between Brooklyn and New York, the catalyst for the poem was not these rides but a fit of depression he experienced in the mid-1850s. In a notebook of the time he wrote:

> Everything I have done seems to me blank and suspicious. —I doubt whether my greatest thoughts, as I had supposed them, are not shallow. — and people will most likely laugh at me—My pride is important; my love gets no response—The complacency of nature is hateful—I am filled with restlessness—I am incomplete.[6]

This passage, parts of which were incorporated into the poet's avowal of personal evil in "Sun-Down Poem," reveals the experience of fracture, anomie, and doubt that lay just beneath the rhetorical grandness of the poem's "well-joined scheme."

"Sun-Down Poem" begins not in crisis but in affirmation, as the poet composes the mutable and seemingly blank things of the world into a meaningful pattern, a "simple, compact, well-joined scheme" in which "every thing indicates":

> Flood-Tide of the river, flow on! I watch
> you, face to face,
> Clouds of the west! sun half an hour high! I see
> you also face to face.

> Crowds of men and women attired in the usual
>     costumes, how curious you are to me!
> On the ferry-boats the hundreds and hundreds
>     that cross are more curious to me than you
>     suppose,
> And you that shall cross from shore to shore
>     years hence, are more to me, and more in my
>     meditations, than you might suppose.
>
> (LG 1856, p. 211)

The passage introduces the poem's major figures—the river and the
sunset, the ferry boat and the particularities of daily life, the self and the
crowd, the present generation and the future generation, the *I* of the poet
and the *you* of the reader—all of which are woven into a pattern of
connectedness and continuity through time:

> The simple, compact, well-joined scheme—my-
>     self disintegrated, every one disintegrated,
>     yet part of the scheme,
> The similitudes of the past and those of the
>     future,
> The glories strung like beads on my smallest
>     sights and hearings—on the walk in the
>     street, and the passage over the river,
> The current rushing so swiftly, and swimming
>     with me far away,
> The others that are to follow me, the ties between
>     me and them,
> The certainty of others—the life, love, sight,
>     hearing of others.
>
> (LG 1856, pp. 211–12)

Everything in the poem spins on the axis of Whitman's "well-joined
scheme." The poem continually turns back on itself, taking up and repeating words, images, units, and scenes, creating a pattern of forward motion
and return, unity amid flux. Like the tenses of the poem, which shift from
present, to future, to past and back again, objects are scooped into circular
motion: "Centrifugal spokes of light" radiate like a divine nimbus around
the poet's head; the "whirl" of Manhattan's shipping business takes place
amid the "scallop-edged waves," "ladled cups," and "frolicsome crests"
of the river; and sea gulls edge toward the south in "slow-wheeling circles" that figure the poem's spiraling version of history as a pattern of
repetition and gradual advance.

The discordant, atomizing effects of machine culture in "Philosophy of Ferries" are artistically transfigured in "Sun-Down Poem." The ferry is represented as a richly symbolic figure of human transport, collectivity, and connectedness through time, not in its potentially troubling aspect as a vehicle of commerce. Only a trace remains of the demonic cast of machine power witnessed by Whitman the journalist in "Ten Minutes in the Engine Room of a Brooklyn Ferry Boat":

> On the neighboring shore the fires from the foun-
> dry chimneys burning high and glaringly into
> the night,
> Casting their flicker of black, contrasted with wild
> red and yellow light, over the tops of houses,
> and down into the clefts of streets.
>
> (*LG* 1856, p. 215)

Even this portentous glimpse of the new industrial economy is scooped into the poem's insistently life-affirming scheme.

This poem has usually been read as an attempt to invest the material world with the glow and glory of spiritual significance.[7] But in its overarching concern with the problem of disintegration and union, the poem is also a response to the fact of fracture in self and world at a time when traditional social structures were collapsing under the pressure of the new market economy, when Whitman was experiencing the anomie of being cast off by a seemingly alien and unresponsive world, and when the American Union was itself dissolving. In fact, "Sun-Down Poem" marks a turn in Whitman's work away from an emphasis on personal power toward an increasing focus on the problem of social union.

The poet's affirmation of a "simple, compact, well-joined scheme" is enacted in and through union with others: the "ties between me and them,/The certainty of others—the life, love, sight, hearing of others" (*LG* 1856, p. 212). Whitman demonstrates his assertion of social unity through dramatic interaction with the *you* of future generations of readers. In an apparent effort to overcome the anomie of the literary marketplace, he courts personal intimacy with his readers: "Who knows but I am as good as looking at you now, for all you cannot see me?" (*LG*, 1856, p. 218). At one point he even imagines himself fusing sexually with his readers: "What the push of reading could not start is started by me personally, is it not?" he asks, symbolically impregnating the reader with his meaning (*LG*, 1856, p. 219). The redemption he imagines is achieved not through personal immortality but through the collectivity—through

participation in the rituals of ordinariness that radiate outward from the
daily action of the ferry carrying crowds of people across the East River.

The unity that the poet feels with others is a unity not of pleasurable
experience only but of psychic distress and the shared experience of living
within a wayward, contrary, and seemingly malignant body:

> I too had received identity by my body,
> That I was, I knew was of my body, and what I
>     should be, I knew I should be of my body.
>
> It is not upon you alone the dark patches fall,
> The dark threw patches down upon me also,
> The best I had done seemed to me blank and sus-
>     picious,
> My great thoughts as I supposed them, were they
>     not in reality meagre? Would not people
>     laugh at me?
>
> It is not you alone who know what it is to be
>     evil,
> I am he who knew what it was to be evil,
> I too knitted the old knot of contrariety,
> Blabbed, blushed, resented, lied, stole, grudged,
> Had guile, anger, lust, hot wishes I dared not
>     speak,
> Was wayward, vain, greedy, shallow, sly, a solitary
>     committer, a coward, a malignant person,
> The wolf, the snake, the hog, not wanting in me,
> The cheating look, the frivolous word, the adul-
>     terous wish, not wanting,
> Refusals, hates, postponements, meanness, lazi-
>     ness, none of these wanting.
>                                        (*LG* 1856, pp. 216–17)

The evils that Whitman acknowledges in himself are those he ascribed to
the political state in *The Eighteenth Presidency!*; those in power were robbers, malignants, conspirators, fancy-men, infidels, blowers, bawlers, and
bribers. A cankered self was in republican ideology the source of a cankered nation. By admitting his personal complicity in the evils of his time
and by naming them in the visionary economy of his poem, Whitman
seeks to come to terms with the "old knot of contrariety" in body and body
politic.

But "Sun-Down Poem" is also the site of an increasing conflict between the demands of the body and the demands of the democratic body
of America. The "dark patches" that Whitman acknowledges in himself

are not only those associated with the times; they are also the "dark patches" of his sexual feeling for men—feelings that were in painful conflict with nineteenth-century notions of republican health and virtue:

> I was called by my nighest name by clear loud
>   voices of young men as they saw me ap-
>   proaching or passing,
> Felt their arms on my neck as I stood, or the neg-
>   ligent leaning of their flesh against me as I sat,
> Saw many I loved in the street, or ferry-boat, or
>   public assembly, yet never told them a word,
> Lived the same life with the rest, the same old
>   laughing, gnawing, sleeping,
> Played the part that still looks back on the actor
>   or actress.
>
> (LG 1856, p. 217)

The words are ambiguous, but they suggest Whitman's sense of himself as part of a marginal, dissembling, and (by the standards of his time) abnormal order of loving men. Whitman's autobiographically linked confession represents a new departure in his work, a departure that anticipates the confessional verse of Allen Ginsberg, Robert Lowell, Anne Sexton, and Sylvia Plath a century later. But his confession also points to a potential dissonance between his desire for democracy and what he called the "perturbance" of his homoerotic desire for men. It is only by transforming his homoerotic feelings into bonds of democratic comradeship that Whitman is able to represent these feelings as safe, normal, and nameable in the public sphere. "Gaze, loving and thirsting eyes, in the house or street or public assembly!" Whitman proclaims in the poem's transfiguring final sequence, "Sound out, voices of young men! loudly and musically call me by my nighest name!" (*LG* 1856, p. 220).

In the visionary and "well-joined scheme" of "Sun-Down Poem," the potentially divisive "dark patches" of the body are represented as a source of commonality and a means of strengthening and spreading the bonds of democratic union. "Flow on, river! Flow with the flood-tide, and ebb with the ebb-tide!" the poet declares in the concluding passage, returning to and linking the separate images of the poem in a single exclamatory utterance that commands and affirms the continuity and connectedness of all things. The "beautiful, terrible, rude forms" that challenged the poet at the end of "Poem of Many in One" are transformed into "dumb beautiful ministers" that signify a divine order of union, progress, and spiritual grace. As the poet descends with the reader into the "meshes" of the physical world, he is struck not by its imperfection but by its loveliness:

> We descend upon you and all things, we arrest
>    you all,
> We realize the soul only by you, you faithful solids
>    and fluids,
> Through you color, form, location, sublimity,
>    ideality,
> Through you every proof, comparison, and all the
>    suggestions and determinations of ourselves.
> You have waited, you always wait, you dumb
>    beautiful ministers! you novices!
> We receive you with free sense at last, and are
>    insatiate henceforward,
> Not you any more shall be able to foil us, or with-
>    hold yourselves from us.
>                                  (*LG* 1856, pp. 221–22)

By recasting "all things" in his poem's spiritual and unitary scheme, Whitman seeks, it would seem, to deprive the physical and social world of its potentially corrosive power. The elaborate, ritualistic design of "Sun-Down Poem" is itself a sign of a new formal restraint and caution in his poetry, as if Whitman feared that without a rigorously maintained artistic control, his "simple, compact, well-joined scheme"—and the entire democratic project—would disintegrate under the pressure of time, history, and his own unruly body.

For all Whitman's effort to reach the American people with his message of social union and national "remembrance," the 1856 edition of *Leaves* received little attention. He was visited in 1856 by Bronson Alcott and Henry David Thoreau, and he began to develop a coterie of admirers in art circles, but the great mass of American people remained, and would remain, all but oblivious to his work.

By electing James Buchanan to the presidency in 1856, the people also failed to fulfill Whitman's hope that they would rise up against the anti-democratic direction of American government. He expressed his disillusionment in a notebook of the time: "I have been informed that it is expected that those who address the people will flatter them—I flatter none—I think I could taunt you, rather than flatter you. What have you been about, that you have allowed that scum to be floated into the Presidency?"[8] The election of 1856 also intensified the conflict between North and South, as the major parties split along sectional lines: The Democratic party, which had formerly been a force for union, encompassing a diversity of sectional interests in its ranks, had won the election by carrying a majority of slave states; the Republican party carried no slave states and all but five of the free states.

Events in the political sphere had the immediate effect of spurring Whitman's artistic productivity. Less than a year after he published the 1856 *Leaves*, he wrote to his friend Sarah Tyndale that he had composed sixty-eight new poems toward a third edition of *Leaves of Grass:*

> Fowler & Wells are bad persons for me. They retard my book very much. It is worse than ever. I wish now to bring out a third edition—I have now *a hundred* poems ready (the last edition had thirty-two)—and shall endeavor to make an arrangement with some publisher here to take the plates from F. & W. and make the additions needed, and so bring out the third edition. . . . In the forthcoming Vol. I shall have, as I said, a hundred poems, and no other matter but poems—(no letters to or from Emerson—no notices, or any thing of that sort.) I know well enough, that *that* must be the *true* Leaves of Grass— I think it (the new Vol.) has an aspect of completeness, and makes its case clearer. The old poems are all retained. The difference is in the new character given to the mass, by the additions. (*Corr.*, I, 44)

Although he did not immediately find a publisher for his work, a notebook entry of the same time indicates that he planned more poems, that he planned, in fact, to give the political ideals of America a moral and spiritual base by writing a bible of democracy. "The Great Construction of the New Bible," he wrote in June 1857. "Not to be diverted from the principal object—the main life work—the three hundred and sixty-five. —It ought to be ready in 1859" (*CW*, IX, 6).

At the same time that he planned his bible of democracy, he continued to envision himself in the role of a traveling speaker, carrying his message directly to the American people and those in power. A notebook entry for April 24, 1857, reads:

> The mightiest rule over America could be thus—as for instance, on occasion, at Washington to be, launching from public room, at the opening of the session of Congress—perhaps launching at the President, leading persons, Congressmen, or Judges of the Supreme Court. That to dart hither or thither, as some great emergency might demand—the greatest champion America ever could know, yet holding no office or emolument whatever,— but first in the esteem of men and women. *Not* to direct eyes or thoughts to any of the usual avenues, as of official appointment, or to get such anyway. To put all those aside for good. But always to keep up living interest in public questions—and *always to hold the ear of the people.* (*CW*, IX, 7–8)

In addition to indicating his continued desire to influence the course of political events, the note suggests his interest, at one time or another, in holding some kind of political office.

Among Whitman's manuscripts is a lecture entitled "Slavery—The Slaveholders—The Constitution—The true America and Americans, the laboring persons," which he may have written and delivered around 1858.[9] Like *The Eighteenth Presidency!*, the lecture reveals signs of disenchantment with the state of America. Slavery, Whitman says, is inconsistent with American national law, and yet "Slavery, the greatest undemocratic un-Americanism of all is very live." He comments on the "miserable sight" in Congress of "learned men grovelling; debating whether our government legalized slavery or liberty"; and in an apparent allusion to Chief Justice Roger B. Taney's Dred Scott decision, he asks: "Does the whelp [slavery] fall howling and dead under the blows of an English judge and have his full swing, with meat and drink to boot, from the caressing hand of an American judge?"[10]

The Dred Scott decision, which came in March 1857, intensified Whitman's fear for the future of democracy in America. In ruling against Dred Scott's suit for freedom, Justice Taney not only denied black persons the right of citizenship; he also denied the right of Congress to exclude slavery in the territories. By appealing to the Fifth Amendment guarantee that no person "should be deprived of life, liberty, or property without due process of the law," the chief justice had, in effect, translated the Constitution into an instrument to protect slavery rather than freedom in America.[11]

A similar pessimism about American prospects entered into the articles that Whitman wrote as editor of the Brooklyn *Daily Times* between May 1857 and June 1859. He continued to emphasize the individual as the cure for social ills: "Let everyman look to himself. Then society will take care of itself."[12] And amid the upheavals of the times he still envisioned the possibility of reconstructing things on a new basis: "It is strange how out of evil good continually comes. Such great calamities as that which is now occupying the public mind serve as reminders, as warnings, as lessons. They startle us from our paltry, apathetic selfishness, they elicit feelings better and higher than ordinarily moves us, they link us together, for a time at least, by the bond of a mutual sentiment, they teach us that poor frail human nature can deport itself bravely and well under circumstances the most appalling" (*ISL*, p. 72).

But as conditions worsened and the country glided into a severe economic depression, which began in the fall of 1857 and lasted for over two years, Whitman's mood darkened. He commented on the "unparalleled monetary distress, and the universal want, doubt and apprehension" of the people and the "depressing influences of the time, which are weighing down even the lightest hearts as with a leaden incubus" (*ISL*, pp. 170–71). As America began to experience the convulsive effects of being tied to an international market economy, the land became the very reverse of the American dream of productive and self-sufficient abundance: "The land has been shaken as by an earthquake," Whitman wrote, "and the foundations of industry are dried, the arm of the worker is palsied, the cunning hand is motionless, and the hum and stir of a busy commerce are changed to the dejected silence of a day of national fasting and humiliation." No longer invoking the memory of the revolutionary founders or the lesson of individual power, he expressed panic at the notion of 25,000 laborers in New York and Brooklyn alone "deprived, during the coming winter, by the mere pressure and severity of the times, of all opportunities of employment whatever except the most transient and precarious" (*ISL*, pp. 73–74).

Economic panic, lawlessness, and the continued battle over slavery heightened Whitman's apprehension about the fate of the Union. In an editorial "Prospects of the Slavery Question," he summed up the national dilemma: "The subject is Hydra-headed. No sooner is one phase of it disposed of than another looms up; no sooner do the disputants lay down their weapons and prepare for a season of rest than a fresh alarm is sounded. . . . There is an inexhaustible capital of sectional disputes yet in reserve for them to dilate upon and work with" (*ISL*, pp. 88–89). Recognizing that the annexation of Cuba, the Southwest, and territories in Central America would lead to the extension of slavery, Whitman retreated from the politics of manifest destiny that had led to his earlier support of the Mexican war and President Polk's expansionist policies.

The contradiction of slavery in the American republic became part of a national and highly publicized political debate when in 1858 the Republican Abraham Lincoln opposed the Democrat Stephen Douglas in the Illinois senatorial campaign. "Of the two, Mr. Lincoln seems to have had the advantage thus far in the war of words," Whitman said of the debates that took place between August and October (*ISL*, p. 96). This comment is his first-known reference to the future president who would become bound up with his own sense of identity and destiny as a poet. Although Whitman expressed admiration for Lincoln's eloquence, at this point he had a slight preference for Douglas. He admired Douglas's courage in opposing Buchanan's support of the Lecompton Constitution, which would have admitted Kansas to the Union as a slave state. When Douglas was reelected senator, Whitman congratulated him on his success: "It is a victory of the independent Representative over the party dictator" (*ISL*, p. 98). Amid the "diverging, splitting, forking off" of the Democratic party, Whitman saw two roads: to continue in the current proslavery direction of what he called "the Dead Rabbit Democracy" or "to inaugurate a new era, under new men" (*ISL*, pp. 92, 96). While Douglas might have been one of the men to inaugurate this new era, Whitman also recognized that Douglas's success in opposing Buchanan's proslavery policies would lead ironically to his defeat in the 1860 presidential election by ultra-Southerners "hoping thereby to precipitate a disruption of the Union" (*ISL*, p. 99).

Although Whitman continued his opposition to the extension of slavery, as editor of the Brooklyn *Daily Times* he articulated a position on the black person in America that represented a retreat from the more radical vision of "I Sing the Body Electric." Commenting on the Oregon Constitution which, for both racist and economic reasons, prohibited black persons, slave or free, from entering the state, Whitman asked:

> Who believes that the Whites and Blacks can ever amalgamate in America? Or who wishes it to happen? Nature has set an impassable seal against it. Besides, is not America for the Whites? And is it not better so? As long as the Blacks remain here how can they become anything like an independent and heroic race? There is no chance for it.
>
> Yet we believe there is enough material in the colored race, if they were in some secure and ample part of the earth, where they would have a chance to develop themselves, to gradually form a race, a nation that would take no mean rank among the peoples of the world. (*ISL*, p. 90)

So far does this vision of an all-white America depart from the egalitarian and racially mixed democratic culture of his songs that it is tempting to believe that the article was not really written by Whitman. But the editorial bears the unmistakable impress of his age's racial views. As David Potter says in *The Impending Crisis:* "While slavery was sectional, Negrophobia was national."[13]

Like America itself, Whitman was torn between an ideological commitment to human equality and the social practice of black subordination. His views were shared not only by Jefferson, who articulated a similar view of black people in *Notes on the State of Virginia*, but also by the most prominent political figures of his age. For all his belief in the "irrepressible conflict" between slavery and freedom and the "higher law" of human justice, Senator William Seward supported the Oregon Constitution excluding black persons from the state. When in the Lincoln–Douglas debates, Douglas asked Lincoln, "What shall be done with the free negro?" Lincoln replied that he believed there was "a physical difference" between "the white and the black races" which would "forever forbid their living together upon the footing of perfect equality." He would deny them citizenship and the vote, and like Jefferson, he planned eventually to colonize black people outside the United States.[14]

As economic conditions in the country worsened and the "hydra-headed" subject of slavery continued to threaten the Union's stability, Whitman retreated from the optimism and political radicalism of his early years and all but stopped working on *Leaves of Grass*. According to Fredson Bowers, who carefully examined Whitman's 1860 manuscripts, he may have been composing and revising poems between 1857 and 1859. "But with all allowances," he concludes, "there appears to have been a striking lapse in poetic effort."[15] Although this lapse may be due to his duties as editor, there are signs in Whitman's notes and poems of this period that the nation's political crisis corresponded to an increasing uncertainty about his own identity and destiny as the poet of democracy. In a leather-bound notebook of the period he wrote:

> Walt Whitman stands to-day in the midst of the American people, a promise, a preface, an overture a
> Will he fulfil the half-distinct half-indistinct promise? —Many do not understand him, but there are others, a few who do understand him. Will he justify the great prophecy of Emerson? or will he too, like thousands of others, flaunt out one bright announcement, the result of gathered powers, only to sink back exhausted—or to give himself up to the seductions of (Bowers, p. 56)

Whitman's fragmented and elusive statement and his self-interrogation mirror his growing self-doubt as a poet whose identity was bound up with the destiny of a nation that was itself a "half-distinct half-indistinct promise."

A notebook entry for June 26, 1859, indicates some kind of crisis: "It is now time to *stir* first for *Money* enough, *to live and provide for* M—. *To Stir*—first write stories and get out of this Slough" (*UPP*, II, 91). The *M* may refer to his mother, or it could be a lover or a friend. The "Slough" into which Whitman had dropped had many possible sources. Having resigned his post as editor because of a dispute with local church leaders, Whitman was no doubt depressed by the prospect of being without work during an economic depression.[16] But his depression had other sources as well: His 1856 *Leaves* had received little popular attention, and he was unable to find a publisher for his new poems. Morever, the democratic Union, on which he had staked his identity and fortunes as a poet, was about to dissolve.

His gloominess was exacerbated by the fact that sometime between 1858 and 1859 he appears to have had an unhappy love affair with a man. The primary evidence for this affair exists in a small sheaf of twelve poems, initially entitled "Live Oak with Moss," which he later dispersed among the *Calamus* cluster in the 1860 *Leaves.*[17] In this sequence of poems, which was probably written some time in late 1858 or early 1859, Whitman recounts in pained and agonized terms a personal relationship of love and loss. In poem V of the sequence (later *Calamus* no. 8), he renounces his earlier desire to "strike up the songs of the New World" in order to pursue his personal relationship with his lover:

> For I can be your singer of songs no longer—I have ceased to
>     enjoy them.
> I have found him who loves me, as I him, in perfect love,
> With the rest I dispense—I sever from all that I thought would
>     suffice me, for it does not—it is now empty and tasteless to
>     me,
> I heed knowledge, and the grandeur of The States, and the ex-
>     amples of heroes, no more,
> I am indifferent to my own songs—I am to go with him I love,
>     and he is to go with me,
> It is to be enough for each of us that we are together—We
>     never separate again. —
>
>                                         (Bowers, pp. 80, 82)

This poem is usually cited as evidence that Whitman's crisis of the late 1850s was brought on by a homosexual love relationship. But the poem may also suggest the reverse: that his crisis of faith in the "grandeur of the States" precipitated his turn toward the privacy of love and that the failure of this love relationship exacerbated his sense of personal crisis.

This was not the first time that Whitman had expressed indifference to his songs. In the 1855 poem "There Was a Child Went Forth," he had similarly asked: "Is it all flashes and specks?" And in his 1856 "Sun-Down Poem" he had wondered whether the best he had done was not "blank and suspicious": "My great thoughts, as I supposed them, were they not in reality meagre? Would not people laugh at me?" These questions were not the questions of a lover only but also of a poet who defined himself and his work in relation to a failing democratic order.

The love relationship that Whitman dramatizes in "Live Oak with Moss" is also a sign of his disillusionment with the public culture of democracy. At a time when the South was threatening to secede from the Union and when the country's political administration seemed determined to save the Union by making slavery rather than freedom the law of the land, Whitman enacts a kind of private secession by severing himself from the political sphere: "When I heard at the close of the day how I had been praised in the Capitol, still it was not a happy night for me that followed" (Bowers, p. 86). What makes him happy is his personal relation with his lover:

> I heard the hissing rustle of the liquid and sands, as directed to
>     me, whispering, to congratulate me, —For the friend I love
>     lay sleeping by my side,
> In the stillness his face was inclined towards me, while the
>     moon's clear beams shone,
> And his arm lay lightly over my breast—And that night I was
>     happy.
>
> (Bowers, p. 88)

Renouncing his former identity as the bard of Democracy, Whitman wants his name to be published and his picture to be hung as the "tenderest lover," "Whose happiest days were those, far away through fields, in woods, on hills, he and another, wandering hand in hand, they twain, apart from other men" (Bowers, p. 86). The homosexual order that he celebrates represents a retreat from the city, the crowd, and the public culture of democracy. The city he imagines in "I dreamed in a dream of a city where all the men were like brothers" is not a real city but a version of homosexual pastoral in which men live and love openly:

> O I saw them tenderly love each other—I often saw them, in
>     numbers, walking hand in hand;
> I dreamed that was the city of robust friends—Nothing was
>     greater there than manly love—it led the rest,
> It was seen every hour in the actions of the men of that city, and
>     in all their looks and words. —
>
> (Bowers, p. 114)

The twelve love poems of "Live Oak with Moss" reveal a new separatist impulse in Whitman's verse. In the final poem of the sequence, he envisions himself not as the poet of the many in one but as the poet of an exclusive order of male lovers:

> To the young man, many things to absorb, to engraft, to de-
>     velop, I teach, that he be my eleve,
> But if through him speed not the blood of friendship, hot and
>     red—If he be not silently selected by lovers, and do not
>     silently select lovers—of what use were it for him to seek
>     to become eleve of mine?
>
> (Bowers, p. 118)

Transmuting the national poet into the figure of the "tenderest lover," Whitman is no longer writing to save the Union but to save himself. His urge to confess his hitherto-repressed and displaced homosexual longings supersedes his desire to enlighten the people in the revolutionary ideals of the founding moment. Like the divided nation in which South was breaking from North and private interest was prevailing over public interest, Whitman's separate person had seceded from the democratic en masse.

# 7

# Democracy and (Homo)Sexual Desire

> To this terrible, irrepressible yearning, (surely more or less down underneath in most human souls,)—this never-satisfied appetite for sympathy, and this boundless offering of sympathy—this universal democratic comradeship—this old, eternal, yet ever-new interchange of adhesiveness, so fitly emblematic of America—I have given in that book, undisguisedly, declaredly, the openest expression.
>
> —WHITMAN, 1876 preface

In his first inaugural address of 1861, Abraham Lincoln insisted on the perpetuity of the Union, emphasizing the mystic bonds of memory that united the nation: "Though passion may have strained, it must not break our bonds of affection. The mystic chords of memory, stretching from every battle-field, and patriot grave . . . will yet swell the chorus of the Union, when again touched, as surely they will be, by the better angels of our nature."[1] In the third edition of *Leaves of Grass*, which was published by the Boston firm Thayer & Eldridge in May 1860, Whitman exhibited a similar concern with touching and arousing the "bonds of affection" and the "chords of memory" that would make the Union whole. Through an increased emphasis on comradeship and love, he sought to resolve his personal crisis and the nation's political crisis in poems that were at once private expressions of love and incitements to public affection. "We want satisfiers, joiners, lovers," he wrote in a notebook of the time. "These heated, torn, distracted ages are to be compacted and made whole" (*CW*, IX, 161). Despite the impulse toward union, fusion, and artistic resolution, however, there is in the 1860 *Leaves*—even more than in the 1856 volume—a disquieting undercurrent that registers the throes of personal and public dissolution. Published on the eve of the Civil War, at the same time that the Democratic National Convention was splitting in anticipation of the secession of the South from the Union in the following year, the 1860 *Leaves* is divided between the desire to celebrate and the desire to mourn.

The visual design of the 1860 edition is itself a sign of national uncertainty. Instead of the vegetation that sprouted out of the title on the green cover of the 1855 *Leaves,* the brown-colored 1860 *Leaves* bears the image of a globe of the two Americas set on a cloud. On the back cover is the image of a sun on the sea, and on the spine is the image of a butterfly poised on a hand. Only the image of the butterfly on the hand has a positive implication, suggesting both harmony with the natural world and the hope of regeneration. The cloud–globe and sun–sea images that enclose the poems are more ambiguous: The globe may be passing out of or being engulfed by a cloud, and the sun may be rising out of or falling into the sea (Figure III).

As in the 1855 and 1856 editions, there is no author's name on the title page of the 1860 *Leaves.* In this edition, however, Whitman underscores the identity of his book and the nation by dating the volume in relation to the formation of the Union: *Leaves of Grass.* Boston, Thayer and Eldridge, Year 85 of The States, (1860–61). Once again, he presents himself to his readers through an engraved portrait that appears opposite the title page, but he is no longer the proletarian rough of the 1855 *Leaves.* In the 1860 engraving, which is taken from a painting by Charles Hine, he appears to pose more self-consciously as an artist: His hair and beard are neatly cropped, and a coat and a tie knotted loosely around his open shirt replace the overalls and workshirt of 1855 (Figure IV).

The poet's self-consciously artistic pose is mirrored in the artistic ordering of the 1860 *Leaves.* In presentation and organization there is a mark of finality, implying that Whitman regarded the book as his last will and testament to the American people. He appears to be building an artistic foundation for democracy in America in the event that the actual political edifice dissolve:

> For we support all,
> After the rest is done and gone, we remain,
> There is no final reliance but upon us,
> Democracy rests finally upon us, (I, my brethren,
>     begin it,)
> And our visions sweep through eternity.[2]

The passage represents a significant alteration in Whitman's posture as the poet of democracy: No longer does he present himself as the embodi-

FIGURE III. Imprints of hand, globe, and sun from *Leaves of Grass* (1860). Courtesy of Special Collections, Van Pelt Library, University of Pennsylvania.

FIGURE IV. Engraved frontispiece for *Leaves of Grass* (1860), from a painting by Charles Hine. Courtesty of The Library of Congress.

ment of an already-existing democratic nation. Rather, he presents himself as the initiator, and indeed the sole reliance, of democracy in America—a position that anticipates the role he conceives for the poet in *Democratic Vistas.*

The entire volume represents Whitman's ambition to write the bible of Democracy: "I too, following many, and followed by many, inaugurate a Religion," he says in "Proto-Leaf" ("Starting from Paumanok"). In this poem, which serves as a preface poem in this and all future editions of *Leaves*, Whitman announces his new religious emphasis:

> O I see the following poems are indeed to drop in the
> earth the germs of a greater Religion.
> My comrade!
> For you, to share with me, two greatnesses—And a
> third one, rising inclusive and more resplendent,
> The greatness of Love and Democracy—and the
> greatness of Religion.
>
> (*LG* 1860, p. 13)

The Bible, in Whitman's view, had been the "principal factor in cohering the nations, eras, and paradoxes of the globe, by giving them a common platform of two or three great ideas, a commonality of origin, and projecting cosmic brotherhood, the dream of all hope, all time" (*PW*, II, 548). As the political state hovered on the brink of disaster and the culture of capital tightened its grip on the nation, Whitman came to see the poems of *Leaves of Grass* as a similar axis of unity, commonality, and brotherhood that would bind the nations and "paradoxes of the globe" in the new democratic dispensation of the future.

Whitman began in 1860 to organize his poems into a series of thematic clusters that presage the final ordering of *Leaves of Grass: Chants Democratic and Native American, Leaves of Grass, Enfans d'Adam, Calamus, Messenger Leaves.* In both the clusters and the individual poems that surround the clusters, the poet appears in the guise of a biblical prophet, teaching the principles of the new democratic faith: self-sovereignty, freedom, equality, comradeship, love, community, the divinity of humanity and the universe, and the natural law of unity, regeneration, and perpetual advance.

Throughout the volume, however, there are unresolved tensions that threaten to dissolve the biblical and artistic patterning of the whole. The opening poem, "Proto-Leaf," which was written over several years between 1856 and 1860, is, like the 1860 volume itself, a kind of palimpsest, revealing Whitman's mixed and contradictory impulses at different stages of composition. After presenting himself as a composite figure of the nation in the poem's opening lines—"Boy of the Mannahatta, the city of ships, my city,/Or raised inland, or of the south savannas"—the poet moves toward an isolated stance: "Solitary, singing in the west, I strike up for a new world" (*LG* 1860, pp. 5–6). In an earlier draft of these lines, Whitman said: "I, singing in the West, strike up for The States" (Bowers, p. 36). Besides improving the rhythm of his initial lines, he also alters their sense. His new posture is significant: He no longer envisions himself as part of a democratic commonality at the center of American culture.

Solitary, singing in a west that is both west of "The States" and west of the actual, Whitman now exists on the margins of national discourse. He strikes up for a "new world" that is not an embodiment of but outside the bounds of "The States."

Whitman's new posture is evident in the remainder of the poem, as he slips continually from the center to the margins of American democratic culture. His optimism about American prospects splinters into a frenzy of nationalistic cliches:

> Americanos! Masters!
> Marches humanitarian! Foremost!
> Century marches! Libertad! Masses!
> For you a programme of chants.
>
> (*LG* 1860, p. 7)

Passages of shrill nationalism, in which the poet's "programme of chants" becomes a program of democratic cant, alternate with passages that suggest an actual state of political crisis:

> I will make a song for These States, that no one
>     State may under any circumstances be subjected
>     to another State,
> And I will make a song that there shall be comity by
>     day and by night between all The States, and
>     between any two of them,
> And I will make a song of the organic bargains of
>     These States—And a shrill song of curses on
>     him who would dissever the Union;
> And I will make a song for the ears of the President.
>     full of weapons with menacing points,
> And behind the weapons countless dissatified faces.
>
> (*LG* 1860, p. 10)

In these lines the poet stands face to face with an actual America of political discord, secessionist sentiment, and presidential incompetence—an America that he would rather curse than celebrate.

Whitman's patriotic program of chants is also at odds with a more private homosexual impulse in "Proto-Leaf." At one point, he announces his desire to translate the fire of his passion for men into songs of democratic comradeship:

> I will sing the song of companionship,
> I will show what alone must compact These,
> I believe These are to found their own ideal of manly
>     love, indicating it in me;

> I will therefore let flame from me the burning fires
>     that were threatening to consume me,
> I will lift what has too long kept down those smoul-
>     dering fires,
> I will give them complete abandonment,
> I will write the evangel-poem of comrades and
>     of love,
> (For who but I should understand love, with all its
>     sorrow and joy?
> And who but I should be the poet of comrades?)
>                     (*LG* 1860, pp. 10–11)

The poet's parenthetical questions are of the essence: The burning flame of his homosexual desire becomes, in effect, the authorization of his voice as "the poet of comrades." But despite his desire to come out poetically, to "let flame" the "smouldering fires" of his homosexual feeling by writing the "evangel-poem of comrades and of love," the poem swings wildly between public exhortation—"Take my leaves, America!"—and private address:

> What do you seek, so pensive and silent?
> What do you need, comrade?
> Mon cher! do you think it is love?
>              (*LG* 1860, p. 12)

At the end of the poem, the public and private impulses merge in an address to a *you* that is both fantasy lover and comrade reader:

> O my comrade!
> O you and me at last—and us two only;
> O power, liberty, eternity at last!
> O to be relieved of distinctions! to make as much
>     of vices as virtues!
> O to level occupations and the sexes! O to bring
>     all to common ground! O adhesiveness!
> O the pensive aching to be together—you know not
>     why, and I know not why.
>              (*LG* 1860, p. 22)

With its rush of *O*'s and exclamation points, the passage expresses an urge to collapse all bounds: social, sexual, literary, and metaphysical. In future versions of the poem, Whitman diminished the leveling effect of this passage by eliminating the last four lines. But in the 1860 version, as the poet advances toward the future with his dream lover, it is unclear whether he is advancing toward or away from the public culture of democracy: "O hand in hand—O wholesome pleasure—O one more desirer and lover,/O haste, firm holding—haste, haste on, with me" (*LG* 1860, p. 22).

*Chants Democratic and Native American,* Whitman's first and longest poetic sequence, follows "Walt Whitman" ("Song of Myself") in the 1860 *Leaves.* Although this grouping was deleted in succeeding editions, in 1860, *Chants Democratic* was a major cluster, articulating Whitman's most profound poetic response to the political crisis of democracy.³ It is a poetic counterpart of *The Eighteenth Presidency!* and an early version of *Democratic Vistas.* Like the French-sounding title *Chants Democratic,* which suggests the connection of the songs with the revolutionary traditions of France, the sequence mythologizes democracy as both historic process and native strain.

In poem no. 1 ("By Blue Ontario's Shore") and others in the sequence, including no. 6 ("Poem of Remembrance") and no. 7 ("With Antecedents"), Whitman sets forth the rights and duties of citizenship in the American republic. He represents the development of American democracy as part of an inevitable evolutionary process in no. 2 ("Song of the Broad-Axe"), and he celebrates the glory and dignity of the laborer in no. 3 ("Song for Occupations") and no. 20 ("I Hear America Singing"). In no. 4 ("Our Old Feuillage"), he collects the various regions of the nation in an organic foliage, "all intensely fused to the urgency of compact America," as he said in a letter of the time (*Corr.,* I, 46). He prophesies the future development of democracy in no. 10 ("To a Historian") and no. 11 ("Thoughts"), and in several of the poems—including no. 12 ("Vocalism"), no. 13 ("Laws for Creations"), no. 14 ("Poets to Come"), no. 16 ("Mediums"), no. 17 ("On Journeys Through the States"), and no. 21 ("Vocalism")—he underlines the importance of poets and orators, female and male, in the creation of the democratic future.

*Chants Democratic* represents the social convulsions of the age as part of the parturition process of democracy. "Society waits unformed," Whitman says, "and is between things ended and things begun." In a long line of American writers, extending from the Puritans in the wilderness to Norman Mailer in the postatomic age, Whitman mythologizes the struggle in America as part of a larger metaphysical struggle—"the whirl, the contest, the wrestle of evil with good" (no. 9, "Thoughts," *LG* 1860, p. 180). But while he stands "aplomb in the midst of irrational things" (no. 18, "Me Imperturbe," *LG* 1860, p. 191), his songs reveal an anguished awareness that the democratic edifice might indeed prove to be a sham and a sell. The *Chants Democratic* cluster includes the 1856 "Poem of the

Propositions of Nakedness," to which Whitman added a passage exhorting the people to stop the democratic pretense: "(If it really be as is pretended, how much longer must we go on with our affectations and sneaking?/Let me bring this to a close—I pronounce openly for a new distribution of roles)" (no. 5, *LG* 1860, p. 166).

The low point in Whitman's crisis of faith in self and nation and the darkest moment in the 1860 *Leaves* occurs in poem no. 1 of the *Leaves of Grass* cluster, which immediately follows *Chants Democratic*. This poem, which was first published as "Bardic Symbols" in the *Atlantic Monthly* in April 1860 and later entitled "As I Ebb'd with the Ocean of Life," resonates with a sense of impending doom. "With Whitman," wrote Kenneth Burke, "when all was going well, everything fell into place. It all added up to the literary equivalent of 'mystic unity.' " But, he adds, "if, for whatever cause, the great elation of integrally interrelated symbolic operations abates, the let down can be almost total. One is then but a body in the most desolate sense, a body such that one dies when it does, robbed of all the entrancing humane fullness (the *pleroma*) with which one's symbolically engendered sense of reality had been infused. One then has no place in an ideal community of any sort. One is thrown back incommunicado upon the body; one is in solitary."[4] "As I Ebb'd with the Ocean of Life" marks a crisis similar to the one Burke describes.

The American Union was rooted in a democratic idea of history and natural law that had, in effect, given birth to Whitman as a poet. His siting in the symbolic order of democracy gave him his language, his form, his "bardic symbols," and his sense of identity and purpose as the poet of America. His poetic utterance and his hope for immortality were bound up with the safety and continuity of the Union through time. As the Union disintegrated, Whitman's symbolically engendered sense of reality also collapsed.

"As I Ebb'd" records the loss of democratic ensemble that he had celebrated in "Song of Myself." The poet is thrown back on the separateness of the body; he is a little corpse, a loose windrow on the shore of life:

> As I ebbed with an ebb of the ocean of life,
> As I wended the shores I know,
> As I walked where the sea-ripples wash you, Pau-
>     manok,
> Where they rustle up, hoarse and sibilant,
> Where the fierce old mother endlessly cries for her
>     castaways,

> I, musing, late in the autumn day, gazing off south-
> ward,
> Alone, held by the eternal self of me that threatens
> to get the better of me, and stifle me,
> Was seized by the spirit that trails in the lines
> underfoot,
> In the rim, the sediment, that stands for all the water
> and all the land of the globe.
>
> (*LG* 1860, p. 195)

The sea is no longer as in "Song of Myself" a site of mystic affirmation, where the poet rocks joyously in the "billowy drowse" of the cosmic mother. The sea is discordant, hissing, violent; it is not a cradle but a site of shipwreck, "Where the fierce old mother endlessly cries for her castaways."

The speaker of the poem is himself isolated, a castaway—a fact that is underscored by the break in the rhythmic pulse of the opening lines that occurs in the phrase: "I, musing, late in the autumn day, gazing off southward." The separation of the "I" at the beginning of the line, like the separation of "alone" at the beginning of the next line, stresses the feeling of being cut off. This separate self has something to say that his "eternal self" threatens to stifle; the eternal self appears to be the transcendental self that is seduced by the flow of the ocean of life, a meaning that Whitman made clearer when he later revised the line to read: "Held by this electric self out of the pride of which I utter poems" (*LGC*, p. 253).

What this stifled self wants to utter is the fear of dissolution: a dissolution hinted at by the time of the poem, autumn, the politically resonant image of shipwreck, and the gaze southward in the direction of the nation's political disturbance. It is in fact when the poet gazes southward that he is "seized" by the spirit of the sediment, the "Chaff, straw, splinters of wood, weeds, and the seagluten,/Scum, scales from shining rocks, leaves of salt-lettuce, left by the tide" (*LG* 1860, p. 196). As yet, however, he assigns no significance to these fragments left by the ebbing sea.

As the poet wends toward unknown shores, which are by implication the shores of the future, the sense of impending doom intensifies. The hoarse and sibilant sounds of the sea take shape as "the voices of men and women wrecked." These shipwrecked voices broaden the poem's personal lament into a national elegy, a dirge for the shipwrecked selves of a collapsing culture and a dying world.

Adrift in a world he knows not, Whitman breaks away from his "eternal self" to speak the message of personal insignificance he reads in the "lines underfoot":

> I, too, but signify, at the utmost, a little washed-up
>     drift,
> A few sands and dead leaves to gather,
> Gather, and merge, myself as part of the sands and
>     drift.
>
> (*LG* 1860, p. 196)

The poet is no longer the arrogant, electric self of his earlier poems, in which his utterance was organically linked with the "valved voice" of the cosmos. He has lost his enlightened faith in the balanced structure and orderly procession of the universe; he and the world he perceives have disintegrated into loose windrows, "a little washed-up drift."

Whitman's vision of personal and national shipwreck leads to a crisis of faith in his identity and purpose as the poet of democracy:

> O baffled, balked,
> Bent to the very earth, here preceding what follows,
> Oppressed with myself that I have dared to open my
>     mouth,
> Aware now, that, amid all the blab whose echoes
>     recoil upon me, I have not once had the least
>     idea who or what I am,
> But that before all my insolent poems the real ME
>     still stands untouched, untold, altogether un-
>     reached,
> Withdrawn far, mocking me with mock-congrat-
>     ulatory signs and bows,
> With peals of distant ironical laughter at every word
>     I have written or shall write,
> Striking me with insults till I fall helpless upon the
>     sand.
>
> (*LG* 1860, pp. 196–97)

The short, stressed lines—"O baffled, balked"—halt the flow of the poem as the poet is thrown into a state of confusion. Having lost the vision of ensemble, similitude, and union to which his poems gave voice, Whitman now regards his democratic utterance as mere "blab" that leaves his "real ME" altogether "untold." The poet is, like the nation, torn between democracy and history, between what he appears to be and what he is. His "real ME" is no longer the "compassionating, idle, unitary" self who stands apart from the social pulling and hauling in "Song of Myself." The "real ME" of "As I Ebb'd" is just the opposite: He is the gloomy, doubt-ing, fragmented self whose peals of ironic laughter are heard in "A Boston

Ballad" and "Respondez"; he is the discordant voice of history that continually threatens to disrupt the optative mood of *Leaves of Grass.*

In "As I Ebb'd," the poet doubts not only himself, his poems, and the
nation. His doubts are also epistemological. As in "There Was a Child
Went Forth," he questions the possibility of knowledge itself:

> O I perceive I have not understood anything—not a
>     single object—and that no man ever can.
>
> I perceive Nature here, in sight of the sea, is taking
>     advantage of me, to dart upon me, and sting me,
> Because I was assuming so much,
> And because I have dared to open my mouth to sing
>     at all.
>
> (*LG* 1860, p. 197)

Whitman is once again in the world of Melville's ambiguities, where
contrary to his assertion in "By Blue Ontario's Shore," he has been
"outfaced by irrational things." The world he perceives is an impenetrable
other world of difference and unmeaning. Nature is no longer a lover, but
as he observed in his early notebooks, linked with the sting of "a hundred
serpents" (*UPP*, II, 89). In her he finds not the source of his poetic
utterance but mockery, for having "dared," he says, repeating the phrase,
to open his mouth at all.

In the final movement of the poem Whitman seeks to come to terms
with a world of fragments:

> You oceans both! You tangible land! Nature!
> Be not too rough with me—I submit—I close with
>     you,
> These little shreds shall, indeed, stand for all.
>
> You friable shore, with trails of debris!
> You fish-shaped island! I take what is underfoot;
> What is yours is mine, my father.
>
> I too Paumanok,
> I too have bubbled up, floated the measureless float,
>     and been washed on your shores;
> I too am but a trail of drift and debris,
> I too leave little wrecks upon you, you fish-shaped
>     island.
>
> (*LG* 1860, pp. 197–98)

In his exclamatory address, which encompasses the land and the sea, the
American continent and his native Paumanok, Whitman "closes" with the
world by accepting, instead of organic union—the many in one—the

fragment as a sign for all. His familial address to the land as father and his sequence of lines beginning with "I too" establish a new sense of kinship—a kinship of insignificance and "little shreds."

But Whitman refuses to be a poet of diminution and the fragment. He wants more. He wants the trail of drift and debris to have meaning and significance:

> I throw myself upon your breast, my father,
> I cling to you so that you cannot unloose me,
> I hold you so firm, till you answer me something.
>
> Kiss me, my father,
> Touch me with your lips, as I touch those I love,
> Breathe to me, while I hold you close, the secret of
>     the wondrous murmuring I envy,
> For I fear I shall become crazed, if I cannot emulate
>     it, and utter myself as well as it.
>
> (*LG* 1860, p. 198)

The image of the breast of the father is striking, suggesting the origin of the term *Paumanok* in what Whitman described as "the island with its breast drawn out, and laid against the sea" (*UPP*, II, 274). His impassioned supplication to the father has usually been read as a sign of Whitman's strained relationship with his authoritarian father.[5] Within the context of the poem, however, the passage has a much broader cultural reference. The breast image and the affectionate language—hold, kiss, touch—also suggests the quest for a nurturant paternity that is founded in the revolutionary revolt against patriarchal authority. Experiencing a sense of shipwreck—personal, political, and artistic—Whitman seeks not the authority of a patriarch but the assurance of historic order figured in the caress of a nurturant father(land).

Whitman's representation of his sense of failure as a poet in the culturally resonant image of familial fracture—separation from the heritage of the fathers (the land) and the natural law of the mother (the sea)—underscores the connection between his crisis of faith as a poet and the political crisis of the nation. It is by reuniting with the fatherland that he seeks to renew his political faith and his poetic utterance. The answer he seeks, the secret of the murmuring he craves, is the rhythm, the heartbeat, the pulse of a historic order that will restore his personal sense of identity and the significance of his utterance as the poet of democracy. Without this secret, Whitman remains so radically cut off that, as he says in a line that he deleted after the Civil War, he fears he will become "crazed."

But the poet receives no answer from the fatherland, nor is he able to reconnect with the generative power of the "fierce old mother," the sea:

> Ebb, ocean of life, (the flow will return,)
> Cease not your moaning, you fierce old mother,
> Endlessly cry for your castaways—but fear not,
>     deny not me,
> Rustle not up so hoarse and angry against my feet, as
>     I touch you, or gather from you.
>
> I mean tenderly by you,
> I gather for myself, and for this phantom, looking
>     down where we lead, and following me and
>     mine.
>
>                                   (*LG* 1860, p. 198)

Separated from both father and mother, Whitman becomes, like the ship-wrecked Ishmael, another orphan of the cosmos. Only his plight is worse, for he is not found by any "devious-cruising Rachel . . . in her retracing search after her missing children."[6] And yet as the shipwrecked poet of the nation, he continues to gather fragments, urged on by the phantom eternally seeking "types" and seemingly confident that the ebb in the ocean of his own and the national life will be followed by a period of flow in which these types and fragments will become part of a meaningful pattern.

For the moment, however, the fragment must "stand for all," and thus in the concluding lines of the poem he gathers the drift and debris of his splintered world into a single poetic utterance, a found poem, that he lays at the feet of his readers:

> Me and mine!
> We, loose windrows, little corpses,
> Froth, snowy white, and bubbles,
> (See! from my dead lips the ooze exuding at last!
> See—the prismatic colors, glistening and rolling!)
> Tufts of straw, sands, fragments,
> Buoyed hither from many moods, one contradicting
>     another,
> From the storm, the long calm, the darkness, the
>     swell,
> Musing, pondering, a breath, a briny tear, a dab of
>     liquid or soil,
> Up just as much out of fathomless workings fer-
>     mented and thrown,

A limp blossom or two, torn, just as much over waves
   floating, drifted at random,
. . . . . . . . .
We, capricious, brought hither, we know not whence,
   spread out before You, up there, walking or
   sitting,
Whoever you are—we too lie in drifts at your feet.
                                        (*LG* 1860, p. 199)

The heavily and erratically accented passage, cut over and over again by
marks of punctuation, emphasizes the image of personal and political
fracture. Once again, Whitman reaches in the final lines toward the commonality of the "You, up there," but it is a commonality of waste and
doubt. The final image—"drifts at your feet"—represents an almost self-
conscious reversal of the luxuriantly organic image of the poet at the end
of "Song of Myself," in which as a leaf of grass and a sign of democratic
potency, he promised to sprout eternally under his readers' bootsoles.

   Like T. S. Eliot's drowned Phoenician sailor, whirling in the currents
under sea, Whitman's drowned poet projects the shipwreck of an entire
culture. The posture of Whitman's "me and mine," which is the poet, his
poems, and the voices of shipwrecked men and women cast up by the sea,
is protomodern. Like Ezra Pound in *The Cantos* and Eliot in *The Waste
Land*, he gathers fragments as the only possibility left to the poet in a
world of friable shores and random drift. No longer sustained by the
ensemble of a national democratic order, Whitman sits like Eliot on the
margins of a collapsing civilization, trying to shore up the fragments
against his ruins.

"A Word Out of the Sea" ("Out of the Cradle Endlessly Rocking"), which
was initially published as "A Child's Reminiscence" in the Christmas
issue of the New York *Evening Post* (December 24, 1859), was composed
before "As I Ebb'd," perhaps as early as 1858.[7] In the artistic ordering of
the 1860 *Leaves*, however, "Out of the Cradle" comes after and appears to
respond to the doubts raised by "As I Ebb'd."

   The 1860 version of the poem begins abruptly: "Out of the rocked
cradle." Whitman has frequently been praised for improving these lines to
read in the final version: "Out of the cradle endlessly rocking." Their
present-participial form and the rhythmic progression of dactyl-trochee
are reminiscent of the regular and continuous rocking of the sea/cradle
that is part of the poem's overall message of faith. But this message is
implied rather than stated. The past tense and jolting rhythm of the initial
lines, along with the third line that Whitman later deleted—"Out of the

boy's mother's womb, and from the nipples of her breast"—are closer to the experience of discord, fracture, and separation that informed the 1860 version of the poem. In seeking to improve his poems artistically, Whitman frequently eliminated or toned down passages of crisis, anxiety, and doubt, giving a smoother line to the arc of his own and the nation's development than had in fact been the case. The line "Out of the cradle endlessly rocking," which became the title of the poem in 1871, is at odds with the demonic rumblings of the sea throughout the poem, whereas the 1860 title "A Word Out of the Sea" retains some of the ambiguity and dark mystery of the word that the poet receives from the sea: "Death, Death, Death, Death, Death."

"Once, Paumanok," Whitman says at the outset of his "Reminiscence," giving an American folk quality to his tale of love and loss:

> When the snows had melted, and the Fifth Month
>     grass was growing,
> Up this sea-shore, on some briers,
> Two guests from Alabama—two together,
> And their nest, and four light-green eggs, spotted with
>     brown,
> And every day the he-bird, to and fro, near at hand,
> And every day the she-bird, crouched on her nest,
>     silent, with bright eyes,
> And every day I, a curious boy, never too close, never
>     disturbing them,
> Cautiously peering, absorbing, translating.
>
>                 (*LG* 1860, p. 270)

The he-bird and she-bird exist in a fecund, sun-drenched, and seemingly timeless landscape of love, where they celebrate the union that sustains them against potentially divisive elements:

> *Shine! Shine!*
> *Put down your warmth, great Sun!*
> *While we bask—we two together.*
>
> *Two together!*
> *Winds blow South, or winds blow North,*
> *Day come white, or night come black,*
> *Home, or rivers and mountains from home,*
> *Singing all time, minding no time,*
> *If we two but keep together.*
>
>           (*LG* 1860, pp. 270–71)

This harmonious union is broken when "May-be killed, unknown to her mate," the she-bird disappears one day, never to return.

This story of love and loss has usually been treated as a dramatization of a personal experience.[8] In image and tone, the story seems to relate in particular to the *Calamus* poems and the homosexual love crisis that Whitman records in this sequence. If, however, we read the poem in the specificity of its historical context, we find a democratic elegy written at a time of national crisis that unites all the elements, psychosexual and political. To read the poem in relation to the division of the American Union is not to detract from its significance as a tale of love, loss, and artistic resolution but, rather, to recognize the historical roots of this elegy of dissolution in the state of the nation on the eve of the Civil War.

The poet's tale of two together is a communal idyll, projecting the democratic dream of America that fed the national imagination and spurred Whitman to pour out his own joyous carols. Local Paumanok is a grassy, spring landscape of fertility and generativity, where native American mockingbirds pass their time singing songs of love and union in a version of American pastoral. Whitman evokes their idyllic existence in the vernacular idiom of the locale, using the Quaker term *Fifth Month* for May, and words such as *he-bird* and *she-bird, briers, crouched,* and *peering.*

As birds of passage, the "two guests from Alabama" nesting on the shores of Long Island organically join North and South in a single life-rhythm. The union of he-bird and she-bird sustains them through darkness and light and in the midst of potentially disruptive winds from north and south. When the she-bird disappears, the he-bird looks southward as the source of disunion, invoking the south wind to return his mate to him. All summer long his songs are absorbed by the curious boy:

> Yes, when the stars glistened,
> All night long, on the prong of a moss-scallop'd stake,
> Down, almost amid the slapping waves,
> Sat the lone singer, wonderful, causing tears.
> (*LG* 1860, p. 271)

The fracture of idyllic union transforms the he-bird into a solitary singer of loss and separation. In contrast with the sun-drenched landscape of the two together, the bird is isolated in a nocturnal landscape that appears to be the site of violence and execution. No longer a communal singer of harmony and joy, the bird now comes closer to the neurosis and solipsism of one of Poe's lovelorn characters, tossing himself frantically on the grave of his beloved.

The transformation of the bird from a joyous singer of light and union to an elegiac singer of darkness and separation is similar to the transformation that Whitman himself underwent during the period of heightening

schism in the nation between 1855 and 1860. In fact, Whitman points out
the analogy: Into the past-tense narration from the child's perspective, he
interjects the present-tense voice of the adult poet:

> He called on his mate,
> He poured forth the meanings which I, of all men,
>      know.
> Yes, my brother, I know,
> The rest might not—but I have treasured every note.
>
>                          (*LG* 1860, p. 271–72)

What Whitman knows, he tells us, comes from both shared experience
and the specter of "White arms out in the breakers tirelessly tossing"—
reminding us of similar visions of shipwreck and drowning in "As I
Ebb'd" and other 1860 poems.

The bird's song ends on a forlorn note: "*Loved! Loved! Loved! Loved!
Loved!,*" he repeats, shifting from the present to the past tense, as he
recognizes the fact of "Two together no more" (*LG* 1860, p. 275). As the
bird's song sinks, the poet's song rises in the heart of the boy. "The aria
sinking,/All else continuing," Whitman says as he links the sinking of the
bird's aria with the emergence of the "outsetting bard of love" in a sequence of participial lines that moves beyond the finality of loss and death,
inscribing a unitary pattern of endless process:

> The boy extatic—with his bare feet the waves, with
>      his hair the atmosphere dallying,
> The love in the heart pent, now loose, now at last
>      tumultuously bursting,
> The aria's meaning, the ears, the Soul, swiftly depos-
>      iting,
> The strange tears down the cheeks coursing,
> The colloquy there—the trio—each uttering,
> The undertone—the savage old mother, incessantly
>      crying,
> To the boy's Soul's questions sullenly timing—some
>      drowned secret hissing,
> To the outsetting bard of love.
>
>                          (*LG* 1860, p. 275)

Here for the first time the "fierce old mother" the sea, whose "angry
moans" have surged as a hoarse undercurrent through the poem, joins the
boy and the bird to become a major character in the drama; it is she who
bears the "drowned" secret suspected by the bird, sought by the boy, and
translated by the poet.

Although the poem may say something about the origins of Whitman's art, the interaction between bird and boy is less an enactment of Whitman's emergence as a poet than it is a dramatization of his reemergence as a poet after his crisis of the late 1850s. If the bird projects some of Whitman's despairing sense of personal and national loss, the emergent poet represents the renewed dedication to his art through which Whitman attempted to overcome his crisis of faith. In the final version of the poem, the poet emerged as "the outsetting bard" not the "outsetting bard of love," but the initial line is closer to his concept of his role in 1860 as the lover and fuser of his "heated, torn, distracted" times.

But while the bird's "despairing carols" deepen the boy's awareness and release him into song, the bird's effect is not wholly positive. In the final version of the poem, the bird is addressed as "Demon or bird!," echoing Poe's similar "bird or fiend" addressed to his fateful raven. A demon can be a muse, a genius, or an inspiration, but it can also be an evil spirit, a fiend from the underworld, or a demon like Poe's raven piercing the heart with its beak. The boy's reaction to the bird suggests both senses of the term:

> O throes!
> O you demon, singing by yourself—projecting me,
> O solitary me, listening—never more shall I cease
>     imitating, perpetuating you,
> Never more shall I escape,
> Never more shall the reverberations,
> Never more the cries of unsatisfied love be absent
>     from me,
> Never again leave me to be the peaceful child I was
>     before what there, in the night,
> By the sea, under the yellow and sagging moon,
> The dusky demon aroused—the fire, the sweet hell
>     within,
> The unknown want, the destiny of me.
>                                    (*LG* 1860, p. 276)

Echoing the refrain of "The Raven"—"Nevermore"—the entire sequence has a Poesque ring. The effect of the "dusky demon"—a line Whitman later toned down to "messenger"—is in fact mixed, summed up in the paradox "sweet hell"; sweet because he arouses the flames of desire and hell because this desire can never be satisfied in the world. The distance between the peaceful child and the awakened bard of love marks the distance Whitman traveled between his own visionary songs of 1855 and the elegiac poems of 1860.

Like the poet in "As I Ebb'd," the boy wants to be more than a solitary singer of separation and fracture; he wants a further clue that will allow him to move beyond the tragic perspective of the bird:

> O give me some clew!
> O if I am to have so much, let me have more!
> O a word! O what is my destination?
> O I fear it is henceforth chaos!
> O how joys, dreads, convolutions, human shapes, and
>     all shapes, spring as from graves around me!
> O phantoms! you cover all the land, and all the sea!
> O I cannot see in the dimness whether you smile or
>     frown upon me;
> O vapor, a look, a word! O well-beloved!
> O you dear women's and men's phantoms!
>
> (*LG* 1860, p. 276)

As an intense response to the prospect of dissolution and chaos, the boy's words articulate the poet's mood in 1860: They link Whitman's uncertainty about his identity and destiny as a poet with his doubts about the fate of the nation and the order of the universe. Like the vision of the land as a corpse that he evoked in his antislavery notes and that flits specterlike in and out of his verse, the passage reverses the regenerative myth that is the source of his faith in human and national destiny. The passage registers the fear of some sort of catastrophe, as joys, dreads, convolutions spring at the poet and phantoms cover land and sea. Through the dimness, the poet cannot tell whether he is moving toward light or darkness, regeneration or chaos. In the poem's final version, Whitman deleted all but the first two lines of the boy's desperate address to the sea. The change had the effect of removing from the poem the fact of historic struggle, the sense of panic about human destiny that in 1860 was bound up with the impending dissolution of the nation.

Like the "unsaid word" sought by the poet in "Song of Myself," at the end of "Out of the Cradle" the boy seeks "the word final, superior to all." But the word the boy receives in 1860 is not, as in 1855, "form and union and plan." The word he receives is DEATH:

> Answering, the sea,
> Delaying not, hurrying not,
> Whispered me through the night, and very plainly
>     before daybreak,
> Lisped to me constantly the low and delicious word
>     DEATH,
> And again Death—ever Death, Death, Death,

> Hissing melodious, neither like the bird, nor like my
>     aroused child's heart,
> But edging near, as privately for me, rustling at
>     my feet,
> And creeping thence steadily up to my ears,
> Death, Death, Death, Death, Death.
>
> (*LG* 1860, p. 277)

The fivefold repetition of Death responds to the bird's plaint—"*Loved! Loved! Loved! Loved! Loved!*"—seeming to convey a life-affirming message of continuity and process, a message that is underlined syntactically by the passage's participial flow: answering, delaying, hurrying, hissing, edging, rustling, creeping. But from the child's point of view at least, there is still something "creepy" about Death. Like the monster-sea that overtakes Emily Dickinson on the outskirts of consciousness in "I Started Early Took My Dog," the sea that edges toward the child is not completely reassuring. Lisping and hissing, creeping and rustling like a snake, the sea's word of death is at best ambiguous.

The poem moves in the concluding sequence from past to present, returning to the adult frame of the poet. It is here that Whitman seeks to reconcile the dualities of the poem: life and death, love and loss, child and man, land and sea, sun and moon, day and night, south and north, past and present. The poet's final words are a unifying gesture, articulated in a single phrase that appears as a continuous flow out of the world of the sea and the preceding action of the poem.

> Which I do not forget,
> But fuse the song of two together,
> That was sung to me in the moonlight on Paumanok's
>     gray beach,
> With the thousand responsive songs, at random,
> My own songs, awaked from that hour,
> And with them the key, the word up from the waves,
> The word of the sweetest song, and all songs,
> That strong and delicious word which, creeping to
>     my feet,
> The sea whispered me.
>
> (*LG* 1860, p. 277)

Appearing in his 1860 role as unifier and fuser, Whitman resolves artistically the problem of dissolution by yoking the song of two together, the boy's responsive songs, and the word *death* in a single poetic phrase that encompasses as it inscribes a compensatory rhythm of life and death, love and loss. Beneath and beyond the poem's artistic resolution we still hear the rumbling of a darker sea that floats up the sediment and debris of "As I Ebb'd." But by using an artistic rather than a chronological ordering in the 1860 *Leaves,* Whitman presents "Out of the Cradle" as a progression away from rather than toward the wasted shores of "As I Ebb'd."

As a response to the fact of dissolution in self and world, "Out of the Cradle" marks a turn toward the other-worldly poetics of Whitman's later period. The poet locates the source of his songs not in democratic presence, but in absence and death, in the "unsatified love" and "unknown want" that he seeks to articulate in song but that can never be fully satisfied in the social world.9 If the poem dramatizes Whitman's renewed dedication to his art after his crisis of faith in the late 1850s, it is a dedication that arises out of the disjunction between desire and history, between the poet's democracy of the imagination and the fact of a disintegrating world.

The mythological focus of *Enfans d'Adam* represents a similar progression away from the disquieting facts of social time and American history. Like the *Calamus* poems, the *Enfans d'Adam* poems elaborate on the sexual politics that Whitman announced in his 1856 letter to Emerson: to give new emphasis in democratic literature to the body, sexual love, manly friendship, and the equality of male and female as a means of liberating the individual from the tyrannical structures of the past. "This tepid wash," he says, "this diluted deferential love, as in songs, fictions, and so forth, is enough to make a man vomit; as to manly friendship, everywhere observed in The States, there is not the first breath of it to be observed in print. I say that the body of a man or woman, the main matter, is so far quite unexpressed in poems; but that the body is to be expressed, and sex is" (*LGC,* p. 739). In accordance with his impulse toward balance and artistic ordering in the 1860 *Leaves,* Whitman originally conceived of the *Enfans* cluster as a counterpoise to his sequence of poems on manly love. "A string of Poems (short etc.)," he wrote in his notebook, "embodying the amative love of woman—the same as *Live Oak Leaves* do the passion of friendship for man" (*CW,* X, 18). In another entry he wrote:

> Theory of a Cluster of Poems, the same *to the passion of Woman-Love,*
> as the *Calamus-Leaves* are to adhesiveness, manly love.
> Full of animal-fire, tender, burning—the tremulous ache, delicious
> yet such a torment.
> The swelling elate and vehement, that will not be denied.
> Adam, as a central figure and type.
>
> (*CW,* IX, 150)

The original French title *Enfans d'Adam*, which was later changed to *Children of Adam*, was itself a means of defying the sensibilities of Puritan America by connecting the amative theme of the poems with the sexual freedom and libertarian traditions associated in the popular mind with France.

*Amative* was a term Whitman borrowed from phrenology to describe the love between men and women, just as he used *adhesiveness* to describe the affection of man for man. However, with the exception of poem no. 9 ("Once I Pass'd Through a Populous City"), which was originally addressed to a man, the *Enfans* poems are not really "about" the love relationship between men and women. As part of Whitman's attempt to reclaim the body and sex as a subject of democratic literature, the *Enfans* poems are about the body as a locus of democratic energies and sexuality as personal power and creative force. Although *Enfans* highlights the theme of sexuality as a liberating and democratizing force, the theme is not new to the 1860 *Leaves.*

What is new about the *Enfans* cluster is that Whitman refocuses the sexual theme of earlier poems, including "I Sing the Body Electric" and "A Woman Waits for Me," on Adam as "a central figure and type." Rewriting the myth of Genesis as the culminating myth of the new democratic religion, Whitman presents Adam as a swart, lusty, democratic fertility god, a figure of the poet himself. Eve appears as his organically equal mate, and history is represented as an ascent toward rather than a fall from the garden:

> To the garden, the world anew ascending,
> Potent mates, daughters, sons, preluding,
> The love, the life of their bodies, meaning and being,
> Curious, here behold my resurrection, after slumber,
> The revolving cycles, in their wide sweep, having
>     brought me again,
> Amorous, mature—all beautiful to me—all won-
>     drous,
> My limbs, and the quivering fire that ever plays
>     through them, for reasons, most wondrous;
> Existing, I peer and penetrate still,
> Content with the present—content with the past,
> By my side, or back of me, Eve following,
> Or in front, and I following her just the same.
>                                      (*LG* 1860, p. 287)

As a newly awakened Adam "ascending" toward the new-world garden, the poet embodies the quintessential cultural myth of America as the site of a regenerated republic that will reverse the pattern of history. The hierarchies and polarities of the past collapse as Eve and Adam nimbly switch places—from front to back to side by side—in an egalitarian economy in which the terms female and male are "just the same."

Although Whitman celebrates the sexual union of male and female in *Enfans*, his songs of procreation are not personal love poems but public exhortations to the creation of the democratic future. As in "A Woman Waits for Me," the women he evokes are symbolic embodiments of the reader-nation-future conceived as female, into whom he pours the regenerative seeds of his poems—"the stuff to start sons and daughters fit for These States" (*LG* 1860, p. 303). As a "chanter of Adamic songs," the poet looks to the "new garden, the West" for the realization of the nation's republican dreams (*LG* 1860, p. 313); the children of Adam are the children of the West and the future.

Here again, however, the poet's mythical representation of the West as new-world garden is undercut by the historical formation of America. By 1860, the American West had, in effect, closed; and if the national policy of extending slavery into the territories continued, the West was promising to become not the land of the free but the land of slaves, not a new-world garden but a plantation economy. This contradiction is the subject of poem no. 10 ("Facing West from California's Shores"). As the poet faces west from the shores of the Pacific, recognizing that the circle of civilization's advance has been completed, he asks:

> But where is what I started for, so long ago?
> And why is it yet unfound?
>
> (*LG* 1860, p. 312)

The question represents a supreme moment of cultural recognition, a manifest destiny in reverse, in which the poet realizes that westward expansion has not advanced either civilization or democracy. Rather, it has led to a repetition of the past. The dream of the American republic is bounded by a limited supply of land—by California's shores. In positioning himself on the shores of the Western Sea, "seeking that yet unfound," Whitman is poised for the flight into spiritual seas that became his characteristic renunciatory gesture as America's political failure became increasingly apparent in the post–Civil War period.

If the *Enfans* cluster represents an old theme in new Adamic dress, the poems of the *Calamus* cluster, all of which appeared for the first time in the 1860 *Leaves*, represent a radical departure in both Whitman's work and literary history. Presided over by the phallic Calamus root, a tall and hardy spear of swamp grass, the *Calamus* cluster focuses on the theme of adhesiveness, which Whitman described as "Intense and loving comradeship, the personal attachment of man to man." By interspersing the original twelve love poems of "Live Oak with Moss" among poems of a more public nature, Whitman sought to reconnect his private homosexual feeling with the public culture of democracy. And yet like "Proto-Leaf," the *Calamus* poems are Janus faced, expressing a separatist impulse toward a private homosexual order at the same time that they invoke a national and global community of democratic brotherhood.[10]

This split is evident in poem no. 1 ("In Paths Untrodden"), in which Whitman announces a new direction in his verse:

> In paths untrodden,
> In the growth by margins of pond-waters,
> Escaped from the life that exhibits itself,
> From all the standards hitherto published—from
> the pleasures, profits, conformities,
> Which too long I was offering to feed to my Soul;
> Clear to me now, standards not yet published—
> clear to me that my Soul,
> That the Soul of the man I speak for, feeds, rejoices
> only in comrades.
>
> (*LG* 1860, p. 341)

The new path Whitman travels is away from the material culture of democracy—from "pleasures, profits, conformities"—toward a more spiritual order of democratic comradeship and love. Altering the individualistic emphasis of his earlier poems. Whitman resolves "to sing no songs today but those of manly attachment . . . Bequeathing, hence, types of athletic love." But the poet's escape to the margins of pond-waters, the site of Calamus growth, is also a poetic "coming out." "Escaped from the life that exhibits itself," Whitman resolves to publish the yet-unpublished standard of homosexual love. Alone, away from the "clank of the world," he hints at something secretive, exclusive, and hitherto unuttered by his culture: "No longer abashed—for in this secluded spot I can respond as I would not dare elsewhere" (*LG* 1860, p. 341). Whitman's psychosocial split between homosexual poet and poet of democracy is inscribed in the concluding lines:

> Afternoon, this delicious Ninth Month, in my forty-
> first year,
> I proceed, for all who are, or have been, young
> men,

To tell the secret of my nights and days,
To celebrate the need of comrades.

(*LG* 1860, p. 342)

The ninth-month setting and the reference to his own age suggest a new birth, but the precise nature of this birth is unclear: Speaking now for an exclusive group of young men, the poet hovers between his private desire to confess secrets, to speak the unspeakable, and his public desire to celebrate comrades.

This same divided impulse continues throughout the *Calamus* cluster, as intensely personal poems alternate with poems of public affection. While the focus of the initial "Live Oak" sequence seems private and exclusive, by interspersing these and other personal poems among poems of democratic sentiment, Whitman gave to the entire sequence a more overtly political significance.

> Important as they are in my purpose as emotional expressions for humanity, the special meaning of the *Calamus* cluster of LEAVES OF GRASS . . . mainly resides in its Political significance. In my opinion it is by a fervent, accepted development of Comradeship, the beautiful and sane affection of man for man, latent in all the young fellows, North and South, East and West—it is by this, I say . . . that the United States of the future, (I cannot too often repeat,) are to be most effectually welded together, intercalated, an-neal'd into a Living Union. (*LGC,* p. 753)

Although written at a time when Whitman was attempting to tone down the personal sources of his *Calamus* poems, his words accord with the overall design of the *Calamus* sequence in 1860. At a time when the Union appeared to be falling apart, Whitman ministered to the national crisis by giving "an example of lovers, to take permanent shape and will through The States" (*LG* 1860, p. 343).

Whitman's increased emphasis on adhesiveness was also a response to the deep cultural fear among Northerners and Southerners alike that dismemberment would give rise to a civil or military dictatorship.[11] In poem no. 5 ("For You O Democracy"), Whitman invokes the Union as something more than a legal compact that could be held together by the machinations of lawyers or the use of arms:

> States!
> Were you looking to be held together by the lawyers?
> By an agreement on a paper? Or by arms?
>
> (*LG* 1860, p. 349)

Placing himself in the service of "Democracy . . . ma femme," Whitman announces his intent to "twist and intertwist" the states by circulating "new friendship" throughout the land: "Affection shall solve every one of the problems of freedom," he observes. The problems of freedom to which he refers are the same as those encountered by the framers of the Constitution: how to ensure a maximum of freedom without inviting either a tyranny of the majority or a tyranny of the State. What the founding fathers sought to do through an appeal to republican virtue, the poet seeks to do by arousing the bonds of comradeship and love:

> The dependence of Liberty shall be lovers,
> The continuance of Equality shall be comrades.
>
> These shall tie and band stronger than hoops of iron,
> I, extatic, O partners! O lands! henceforth with the
> love of lovers tie you.
>
> I will make the continent indissoluble,
> I will make the most splendid race the sun ever yet
> shone upon,
> I will make divine magnetic lands.
>
> *(LG* 1860, p. 351)

In this and other poems of the cluster, Whitman moves away from the pond side and back to the center of American culture, legitimizing his calamus emotions as part of the public culture of democracy and as a means of welding the divided nation. His comrades and lovers become, in effect, republican freemen in the affectionate dress of phrenological adhesiveness.

But although democracy and phrenology gave Whitman a positive language in which to name his experience as homosexual, he was still "stifled and choked" by the "smouldering fire" and guilt of what remained unspeakable. It was not until the late nineteenth century that the term *homosexual* was introduced into English, and even then it was used to describe a pathological condition.[12] In poem no. 16, which Whitman deleted from *Leaves* after 1860, he confessed his pain and self-division:

> As if I were not puzzled at myself!
> Or as if I never deride myself! (O conscience-struck!
> O self-convicted!)
> Or as if I do not secretly love strangers! (O tenderly,
> a long time, and never avow it;)
>
> *(LG* 1860, p. 362)

Whitman internalized the homophobia of his culture: His moments of self-persecution and self-doubt fed, and were fed by, his growing doubts about the whole democratic enterprise. Blurring the bounds between personal and political, he experienced his homosexual desire as both a sign of a diseased republican self and a source of his oppression within a putatively healthy democratic order. The real democracy of uncloseted sexual feeling between men was still "that yet unfound" of the poet who faced west from California's shores.

As in "Out of the Cradle," Whitman's personal tale of love and loss in *Calamus* was bound up with the tale of national loss. Poem no. 17 ("Of Him I Love Day and Night") appears to be about the death of a real or fantasy lover:

> Of him I love day and night, I dreamed I heard he
>     was dead,
> And I dreamed I went where they had buried him I
>     love—but he was not in that place,
> And I dreamed I wandered, searching among burial-
>     places, to find him.

But in a passage that anticipates the "Unreal City" of Eliot's *Waste Land*, the poet's dream turns out to be about the death of the nation:

> And I found that every place was a burial-place,
> The houses full of life were equally full of death,
>     (This house is now,)
> The streets, the shipping, the places of amusement,
>     the Chicago, Boston, Philadelphia, the Manna-
>     hatta, were as full of the dead as of the living,
> And fuller, O vastly fuller, of the dead than of the living.
>                                          (*LG* 1860, p. 362)

Like Lincoln's 1858 reference to the nation as a "house divided," Whitman's parenthetical reference to the death of "this house now" has a national resonance: The loss of the poet's comrade and lover comes to figure the loss of America and ultimately of democracy itself.

The complex intermingling of personal and political themes in *Calamus* is partly a function of Whitman's ever-shifting and shifty relationship with the *you* of the reader. At its most public, *you* refers to America and democracy, as in poem no. 5: "For you these, from me, O Democracy, to serve you, ma femme!" (*LG* 1860, p. 351). At its most private, the *you* is an intimate lover, as in the first of the original "Live Oak with Moss" poems: "O love, for friendship, for you" ("Not Heat Flames Up and

Consumes," *LG* 1860, p. 360). Sometimes Whitman uses the *you* to address an apparently exclusive group of homosexual lovers, as in poem no. 3: "Here to put your lips upon mine I permit you" ("Whoever You Are Holding Me Now in Hand," *LG* 1860, p. 345). But he also addresses a more generalized readership, as in "I wish to infuse myself among you till I see it common for you to walk hand in hand" (no. 37, "A Leaf for Hand in Hand," *LG* 1860, p. 375). By conflating these and other uses of the term *you* in the *Calamus* sequence, Whitman interchanges the terms *reader, Democracy,* and *America* with the term *personal lover.* Thus what appears to be a private address to a lover or an exclusive group becomes part of Whitman's democratic strategy, his attempt to create an intimate bond with the reader as a means of depositing seeds of democratic love and comradeship throughout the land.

The *Calamus* cluster is Whitman's most radical sequence personally and politically, in both what it reveals and what it conceals. By placing his personal love poems in a more public, democratic frame and by his ambiguous use of the "compact" *you,* Whitman has it both ways: He is at once the poet of homosexual love and the bard of democracy. As the title of his opening poem suggests, Whitman traveled in "paths untrodden," inscribing homosexuality—the "affection of man for man"—as the condition of his poetic utterance and of democracy itself. His *Calamus* sequence is doubly revolutionary: He infuses the abstractions of democracy with the intensity of erotic passion, giving literature some of its first and most potent images of democratic comradeship; and by linking homoeroticism with a democratic breaking of bounds, he presents one of the most moving and tender accounts of homosexual love in Western literature.

As in "Song of Myself," the 1860 *Leaves* moves toward the poet's emergence in the final sequence, *Messenger Leaves,* in the role of a Christ-like prophet of democracy. But with this difference: Whereas in "Song of Myself" Whitman moved toward a resolution of conflict in a testament of faith, in *Messenger Leaves* the tone of foreboding intensifies as the poet delivers a kind of political jeremiad to his readers. In the first poem "To You, Whoever You Are," he assures his readers of their ability to master the "throes of apparent dissolution" (*LG* 1860, p. 394), and in subsequent poems he prepares for some kind of disaster.

To his "Liberty Poem" of 1856, which was originally part of his encomium on Liberty's triumph in the 1855 preface, he gives a new title, "To a Foil'd Revolter or Revoltress," implying that Liberty has in fact been foiled. He also adds a coda expressing uncertainty about the future of Liberty and his consequent uncertainty about the nature of self and

world: "I do not know what you are for, (I do not know what I am for myself, nor what any thing is for,)." He prepares for defeat and death:

> Did we think victory great?
> So it is—But now it seems to me, when it cannot be
>    helped, that defeat is great,
> And that death and dismay are great.
>
> (*LG* 1860, p. 396)

In the final version of this poem, Whitman addressed these lines to "a Foil'd European Revolutionaire," but in 1860 the lines clearly suggest the foiling of the American Revolution and the imminent defeat and death of the nation.

In several poems of *Messenger Leaves* there is an increased sense of urgency as the poet responds directly to the immediate political crisis. In "To the States, to Identify the 16th, 17th, or 18th Presidentiad," he voices his pessimism about the administrations of Millard Fillmore, Franklin Pierce, and James Buchanan:

> Why reclining, interrogating? Why myself and all
>    drowsing?
> What deepening twilight! Scum floating atop of the
>    waters!
> Who are they, as bats and night-dogs, askant in the
>    Capitol?
> What a filthy Presidentiad! (O south, your torrid
>    suns! O north, your arctic freezings!)
> Are those really Congressmen? Are those the great
>    Judges? Is that the President?
>
> (*LG* 1860, pp. 400–401)

The poet's questions and exclamations, his return to the political invective of his 1850 poems, articulate the disillusionment of an entire decade, a decade in which presidents, congressmen, and judges appeared to cooperate in making the law of the land slavery rather than freedom. The poem suggests that South and North are equally responsible for the widening schism in the Union: If the "torrid" South erred in the direction of fiery passion and rhetoric, the "arctic" North erred in the direction of cold reason and political compromise. All about him, Whitman sees signs of a slumbering nation:

> Then I will sleep a while yet—for I see that These
>    States sleep, for reasons;
> (With gathering murk—with muttering thunder and
>    lambent shoots, we all duly awake,

> South, north, east, west, inland and seaboard, we will
>    surely awake.)
>
> (*LG* 1860, p. 401)

To maintain his democratic optimism in the face of the perturbing social
facts of his time, Whitman had to empty them of historical significance by
refiguring them as natural facts in some larger providential design. As
sleep is followed by waking, so "These States sleep, for reasons" and "will
surely awake"; as storms occur in the physical world, so the gathering
political storm is part of an orderly natural rhythm. Whitman had already
begun to rationalize the Civil War as part of a natural political progression. In fact, his lines appear to prepare the *we* of the nation for the
"muttering thunder" and "lambent shoots" of the Civil War, a prophecy
that Whitman later made good by placing the poem immediately before
the war poems of *Drum-Taps* in the final edition of *Leaves of Grass*.

In another address to the states entitled "Walt Whitman's Caution,"
he voices his fear that the nation will be enslaved by the tyranny of
governmental power:

> To The States, or any one of them, or any city of
>    The States, *Resist much, obey little,*
> Once unquestioning obedience, once fully enslaved,
> Once fully enslaved, no nation, state, city, of this
>    earth, ever afterward resumes its liberty.
>
> (*LG* 1860, p. 401)

Whitman's Jeffersonian directive—*Resist much, obey little*—was delivered
at a time when the traditions of local control were being eroded by an
increasingly centralized and bureaucratized government administration.
In the context of the Union's crisis, however, his advice is double-edged
and seditious. Echoing the political rhetoric of John Calhoun, he encourages the very resistance to national control that was fanning the fires of
Southern secession.

In his urgent response to the political crisis, Whitman addresses the
citizens, cities, and states of the nation, and he also sends a "messenger
leaf" to the president himself. Drawing once again on the Enlightenment
theory of natural law, he enjoins the president to follow the "politics of
nature":

> All you are doing and saying is to America dangled
>    mirages,
> You have not learned of Nature—of the politics of
>    Nature, you have not learned the great ampli-
>    tude, rectitude, impartiality,

> You have not seen that only such as they are for
>    These States,
> And that what is less than they, must sooner or later
>    lift off from These States.
>                     ("To a President," *LG* 1860, p. 402)

These words were probably intended for James Buchanan, who had, among other things, attempted to pacify the fire-eating Southerners by imposing the proslavery Lecompton Constitution on Kansas. But by addressing the poem to an indefinite and nonspecific "President," Whitman encompasses all presidents—past, present, and future—reminding them of the natural laws of liberty, equality, and justice at the political foundation of the American republic.

In the poems that follow *Messenger Leaves*, the movement toward artistic resolution is disrupted by the nation's immediate crisis. "France, the 18th Year of These States" evokes a nightmare vision of the Reign of Terror, linking the birth of liberty in France with the birth of America. The poem is an attempt to justify the "terrible red birth and baptism" of the Terror as the just retribution for years of oppression and suffering. But the example of the French Revolution also suggests that a similar red birth is about to occur in America. Invoking the common cause of liberty and revolution in France and America, the poet declares:

> O Liberty! O mate for me!
> Here too keeps the blaze, the bullet and the axe, in
>    reserve, to fetch them out in case of need,
> Here too, though long deprest, still is not destroyed,
> Here too could rise at last, murdering and extatic,
> Here too would demand full arrears of vengeance.
>                              (*LG* 1860, p. 407)

Like the Lucifer passage in "The Sleepers," the lines anticipate and even appear to invite some cataclysmic uprising as the just "arrears of vengeance" for the logical contradiction of slavery in the American republic. Believing in the long-accrued retribution of Liberty, Whitman sympathized with John Brown when in 1859, "murdering and extatic," he carried out his raid on Harper's Ferry in an effort to liberate Southern slaves. While Whitman was preparing the 1860 *Leaves* for publication in Boston, he attended the trial of Frank Sanborn, who was accused of aiding Brown, and Whitman was one of a group of Abolitionists ready to intervene if the judge decided to hand over Sanborn to federal authorities.[13]

In a series of aphoristic phrases entitled "Says," most of which were

deleted from *Leaves* after the Civil War, Whitman rallies the dying nation by reminding his readers of basic democratic principles. Using biblical epanaphora, he writes a kind of gospel of democracy or poetic bill of rights: "I say man shall not hold property in man" . . . "I say where liberty draws not the blood out of slavery, there slavery draws the blood out of liberty" . . . "I say discuss all and expose all—I am for every topic openly" . . . "I say that every right, in politics or what-not, shall be eligible to that one man or woman, on the same terms as any" (*LG* 1860, pp. 418–20).

The figure of shipwreck presides in the final poems of the 1860 *Leaves*. This sense of impending doom is linked with the nation's uncertain destiny in the penultimate poem "To My Soul" ("As the Time Draws Nigh"):

> As nearing departure,
> As the time draws nigh, glooming from you,
> A cloud—a dread beyond, of I know not what, dark-
>     ens me.
>
> I shall go forth,
> I shall traverse The States—but I cannot tell whither
>     or how long;
> Perhaps soon, some day or night while I am singing,
>     my voice will suddenly cease.
>
> (*LG* 1860, p. 449)

What the poet dreads is not the cessation of himself but his voice which, as the utterance of "The States," may cease with the cessation of the Union. The connection between the cloud that darkens around the poet and the gathering storm of the nation is underscored by the imprint, at the end of the poem, of a cloud engulfing the globe of the two Americas.

Whitman's farewell to his readers in the final poem "So Long!" partakes of the uncertain mood of 1860. No longer conceiving of himself as the "summer-poet" of "consummations," he now looks to the future for realization, "When America does what was promised" (*LG* 1860, p. 451). But as he announces the democratic future—of independent persons, of liberty, equality, and justice, of adhesiveness and the "identity of These States"—his vision collides with the fact of dissolution:

> O thicker and faster!
> O crowding too close upon me!
> I foresee too much—it means more than I thought,
> I appears to me I am dying.
>
> (*LG* 1860, p. 454)

The lines are ambiguous, but they suggest a certain panicked recognition that both the bard of democracy and the nation he embodies are in the throes not of parturition but of death.

To overcome this death, the poet, in a revolutionary gesture, leaps from his *Leaves* into the arms of his readers:

> My songs cease—I abandon them,
> From behind the screen where I hid, I advance per-
>     sonally.
>
> This is no book,
> Who touches this, touches a man,
> (Is it night? Are we here alone?)
> It is I you hold, and who hold you,
> I spring from the pages into your arms—decease
>     calls me forth.
>
> O how your fingers drowse me!
> Your breath falls around me like dew—your pulse
>     lulls the tympans of my ears,
> I feel immerged from head to foot,
> Delicious—enough.
>
>                           (*LG* 1860, p. 455)

Refusing to stay put on the literary page, Whitman once again dissolves the boundary between writer and reader, art and life. In an act of poetic transubstantiation, he dies into the bodies of his readers, to whom he looks for the perpetuation of himself and his democratic vision. Hugging and kissing his reader in the final lines of the poem, he announces his departure: "I am as one disembodied, triumphant, dead" (*LG* 1860, p. 456). These lines are followed in the 1860 edition by an imprint of a butterfly poised for flight, an image that reflects the hope of democratic regeneration toward which the volume moves.

Coming at the end of all future editions of *Leaves of Grass*, "So Long!" completes the volume's cyclical movement from life to death, dawn to dusk, self to other, present to future, and poet to reader, thereby symbolizing the poet's "triumphant" perpetuation in his transubstantiated book, in the body of his reader, and in the regenerative flow of the universe. In the context of 1860 and the "disaster" poems that precede "So Long!" in the 1860 *Leaves*, however, Whitman's triumph seems considerably less assured. His intimate and vernacular "So long" has the sound of a death knell for himself, his book, and the nation. The mood of the poem is closer to Allen Ginsberg's nocturnal encounter with Whitman in "A Supermarket in California":

> Will we stroll dreaming of the lost America of love past
> blue automobiles in driveways, home to our silent cottage?
> Ah, dear father, graybeard, lonely old courage-teacher,
> what America did you have when Charon quit poling his ferry
> and you got out on a smoking bank and stood watching the
> boat disappear on the black waters of Lethe?[14]

Like Ginsberg's vision of the poet in 1955, Whitman, standing on the shores of America on the eve of the Civil War, seems to have sensed that his beloved America was not a place on the map but a place in the imagination that would live and die only with him.

# 8

# The Union War

The Union is proved solid by proof that none can gainsay.
Every State that permits her faction of secessionists to carry
her out, shrivels and wilts at once.... A reign of terror is
inaugurated. All trade, all business stops.... The devils are
unloosed. Theft, outrage, assassination stalk around, not in
the night only but in open days.

—WHITMAN, *Notebooks*

The Civil War took place in part because the American people decided
that it must take place in order to affirm the rational and divine order of
the universe. From the time when slavery was written into the Constitution to Lincoln's second inaugural address, the specter of civil war as the
just retribution for the sin of slavery haunted the American political imagination. While the Constitution was being debated, the Congregational
minister Samuel Hopkins, one of Jonathan Edwards's leading disciples,
expressed his fear that civil war would be the consequence of writing
slavery into the law of the land:

> How does it appear in the light of Heaven, and of all good men, well in-
> formed, that *those* States, who have been fighting for liberty, and consider
> themselves as the highest and most noble example of zeal for it, cannot agree
> in any political constitution, unless it indulge and authorize them to enslave
> their fellow men. I think if this constitution be not adopted, as it is, without
> any alteration, we shall have none, and shall be in a state of anarchy, and
> probably civil war: therefore, I wish to have it adopted: but still, as I said, I
> fear. And perhaps civil war will not be avoided, if it be adopted.[1]

Reflecting on the institution of slavery in *Notes on the State of Virginia*,
Jefferson expressed a similar fear for the well-being of the land: "Indeed I
tremble for my country when I reflect that God is just: that his justice
cannot sleep for ever: that considering numbers, nature and natural

means only, a revolution of the wheel of fortune, an exchange of situation, is among possible events: that it may become probable by supernatural interference! The Almighty has no attribute which can take side with us in such a contest."² Years later, Jefferson's vision of "supernatural interference" against the institution of slavery in America emerged as Lincoln's explanation of the Civil War in his second inaugural address: "If we shall suppose that American Slavery is one of these offences which, in the providence of God, must needs come, but which, having continued through His appointed time, He now wills to remove, and that He gives to both North and South, this terrible war, as the woe due to those by whom the offence came, shall we discern therein any departure from those divine attributes which the believers in a Living God always ascribe to Him?"³

Whitman voiced the liberal Republican sentiments of his time when in 1860 he declared: "I say where liberty draws not the blood out of slavery, there slavery draws the blood out of liberty" (*LG* 1860, p. 418). In the mottled year of 1860, Lincoln appeared to offer a release from the failure of leadership and the series of compromises that had characterized the administrations of Fillmore, Pierce, and Buchanan: "The sixteenth, seventeenth and eighteenth terms of the American Presidency," Whitman said, "have shown that the villainy and shallowness of rulers (back'd by the machinery of great parties) are just as eligible to these States as to any foreign despotism, kingdom, or empire—there is not a bit of difference" (*PW*, II, 429). To Whitman, Lincoln represented a return to revolutionary principles set forth not in the compromised Constitution but in the Declaration of Independence. "Our republican robe is soiled, and trailed in the dust," Lincoln said in his campaign for senator in 1858. "Let us repurify it. Let us turn and wash it white, in the spirit, if not the blood, of the Revolution. . . . Let us readopt the Declaration of Independence."⁴ Lincoln's election to the presidency in 1860 promised to resolve the nation's political crisis through a revival of the ideals of the founding moment: He was the political father, the president as democratic comrade, for whom Whitman longed. "How does this man compare with the acknowledg'd 'Father of his country?' " Whitman asked. Whereas George Washington and Benjamin Franklin were "European," Lincoln was an American original; he represented the republican virtue of the West: "Lincoln, underneath his practicality . . . was quite thoroughly Western, original, essentially non-conventional, and had a sort of out-door or prairie stamp"; he combined "moral and spiritual" qualities with American "*horse-sense*" (*PW*, II, 603).

America has never been known for the quality of the individuals it places at the head of government. And yet as de Tocqueville points out in *Democracy in America,* "when serious dangers threaten the state, the people frequently succeed in selecting the citizens who are the most able to save it."⁵ Whitman was one of the first to embrace Lincoln as the right man for the time. From the time of Lincoln's election there developed a strong bond of affinity between the poet of democratic union and the president who was destined to play an instrumental role in preserving the Union. "Lincoln is particularly my man," Whitman said to Traubel, "particularly belongs to me; yes, and by the same token, I am Lincoln's man; I guess I particularly belong to him; we are afloat in the same stream,—we are rooted in the same ground" (Figure V).⁶

In the 1860 *Leaves,* Whitman had adjured President Buchanan to follow the just and impartial "politics of nature"; so in the period preceding Lincoln's inauguration, Whitman planned to publish a brochure in which he has an imaginary dialogue with Lincoln. A notebook of the period contains the following entry: "Brochure. —Two characters as of a dialogue between A. L._____n and W. Whitman. —as in ? a dream—or better? Lessons for a President elect—Dialogue between W. W. and 'President elect.' "⁷ Although this imaginary dialogue with Lincoln was never written, the note once again reveals the place that Whitman conceived for the poet at the center of American political culture, where he opens and equalizes the power relation between poet and president, governed and government: The note seems almost quaint in its assumption that poets might indeed have lessons to give to presidents and that they should be heeded.

With the election of Lincoln to the presidency, the focus of national politics shifted from the controversy over slavery to the battle to preserve the federal Union. On February 7, 1861, the Confederate States of America was formed, and by the end of the month, seven states of the lower South had left the Union. The atmosphere of Lincoln's inauguration was tense with the threat of assassination and violence. Whitman was himself present when on February 18, Lincoln arrived in New York City on his way to his inauguration in Washington. Secessionist sentiment ran high in New York, particularly among the newly arrived Irish immigrants and the prosperous merchants; and Mayor Fernando Wood was threatening to form a new "Tri-Insula" state of Manhattan, Long Island, and Staten Island.⁸ Whitman remembered Lincoln's "perfect composure and coolness" as he looked out on a sea of potentially hostile faces: "The crowd that hemm'd around consisted I should think of thirty to forty thousand men, not a single one his personal friend—while I have no doubt, (so frenzied were the ferments of the time,) many an assassin's knife and

FIGURE V. Photograph of Abraham Lincoln, 1863, by Alexander Gardner. Courtesy of The Library of Congress.

pistol lurk'd in hip or breast-pocket there, ready, soon as break and riot came" (*PW*, II, 501).

In his first inaugural address, Lincoln asserted the perpetuity of the Union: "I hold that in contemplation of universal law, and of the Constitution, the Union of these States is perpetual." The very idea of secession was, he declared, "the essence of anarchy."[9] In a notebook of the time,

Whitman reflected with similar dread on the consequences of national dissolution: "The Union is proved solid by proof that none can gainsay. Every state that permits her faction of secessionists to carry her out, shrivels and wilts at once. . . . A reign of terror is inaugurated. All trade, all business stops. . . . The devils are unloosed. Theft, outrage, assassination stalk around, not in the night only but in open day."[10] Given the choice among dissolution, compromise, and war, Whitman chose war.

What he most feared in the period immediately preceding the outbreak of hostilities was the possibility that the American people would sacrifice the principle of Union in the interest of continued wealth and prosperity. Whitman dramatized the potential conflict between the long-range benefits of the Union and the more immediate benefits of material wealth in "Song of the Banner at Day-Break," a poem probably begun before the war and intended for inclusion in a volume entitled *Banner at Day-Break*, which was advertised in 1860 by Thayer and Eldridge in the antislavery paper *The Liberator*. Experimenting with a more formal artistic structure, Whitman divided the poem into a colloquy among Poet, Banner, Child, and Father. The poem is a parable of the political debates of the time: The Poet identifies with the Banner flapping in the wind and urges the Child of America to support the cause of Union by slugs and bullets if necessary; the Father, representing the interests of Northern capital, urges the Child to forget the essentially valueless flag for the real value of peace and prosperity:

> Nothing my babe you see in the sky,
> And nothing at all to you it says—but look you my babe,
> Look at these dazzling things in the houses, and see you the
>     money-shops opening,
> And see you the vehicles preparing to crawl along the streets with
>     goods;
> These, ah these, how valued and toil'd for these!
> How envied by all the earth.
>
> > (*LGC*, p. 286)

The Child, however, is a patriot, who sees beyond the dough-faced politics of the Father to the absolute value of the Union:

> O my father I like not the houses,
> They will never to me be any thing, nor do I like money,
> But to mount up there I would like, O father dear, that banner I
>     like,
> That pennant I would be and must be.
>
> > (*LGC*, p. 288)

Even within this emphatically Union poem, one hears the dissenting voice of doubt that would come to dominate later war poems. Support for the cause of liberty and union might, like dissolution, also lead to an unloosing of devils: "Demons and death then I sing," declares the Banner. While the "warlike pennant" undulates "like a snake hissing so curious,/ Out of reach, an idea only," the Poet silences misgivings and potential opposition in a patriotic celebration of the Union as the only guarantee of the future, liberty, and the continued prosperity of the states.

The firing on Fort Sumter by the Palmetto Guard of South Carolina on the morning of April 12, 1861, brought the awakening of the nation for which Whitman longed in his 1860 poem "To the States." In the North, the firing had an electrifying effect, arousing and uniting the divided citizens in support of the Union cause. At the outset of the war, however, the North had what Whitman described as a contemptuous feeling toward "the power and will of the slave States for a strong and continued military resistance to national authority" (*PW*, I, 25–26). He remembered in particular a couple of Brooklyn regiments marching off "with pieces of rope, conspicuously tied to their musket-barrels, with which to bring back each man a prisoner from the audacious South" (*PW*, I, 26).

In the months immediately following the firing on Fort Sumter, no one had any idea how long the war would last or that the war that had begun bloodlessly would be one of the bloodiest in history. Recognition that Union victory would be neither easy nor sure came when the baffled and panic-struck Union troops began to stream back into Washington after their defeat at Bull Run on July 21, 1861. Whitman remembered the defeat as a moment of crisis for the Union: "The dream of humanity, the vaunted Union we thought so strong, so impregnable—lo! it seems already smash'd like a china plate" (*PW*, I, 29). He also remembered the role played by the New York papers in inspiriting the nation "with leaders that rang out over the land with the loudest, most reverberating ring of clearest bugles, full of encouragement, hope, inspiration, unfaltering defiance" (*PW*, I, 31).

Whitman himself became one of these bugles when on September 28, 1861, both *Harper's Weekly* and the New York *Leader* published "Beat! Beat! Drums!" Using stanza division, refrain, incremental repetition, and a fairly regular iambic-anapestic line, Whitman imitates the insistent beat of a war drum summoning the American people away from the normal activities of home, church, school, and business:

> No bargainers' bargains by day—no brokers or speculators—
> would they continue?
> Would the talkers be talking? would the singer attempt to sing?

Would the lawyer rise in the court to state his case before the
    judge?
Then rattle quicker, heavier drums—you bugles wilder blow.

                                                          (*LGC,* p. 283)

In this, as in other mobilization poems such as "Song of the Banner at
Day-Break," "City of Ships," and "Long, Too Long, O Land," Whitman
envisions the war as a dramatic means of drawing the people away from
their self-interested pursuit of business as usual toward the republican
values of patriotism, sacrifice, and courage in defense of the Union. And
yet while the drumbeat of war gathers the separate energies of the people
into the single rhythm of the Union cause, there is in the militaristic
invasion of "Beat! Beat! Drums!" a foreboding sign of the totalizing
structures—institutional and otherwise—that would come to dominate
and control the individual during and after the war.

    Aside from the poems projected for *Banner at Day-Break* and a few
poems published in newspapers, Whitman wrote and published little in
the first two years of the war. In fact, after publishing three editions of
*Leaves of Grass* in five years, he published no major work until after the
war. This may be, as Roger Asselineau suggests, a result of a failure of
inspiration—a sense that he had said all he had to say in the 1860 *Leaves.*[11]
But the sparseness of his poetic output may also reflect his uncertainty
about the role of a national poet in a period of fratricidal war: The role of
warrior-poet sounding the alarum of battle conflicted with his Quaker
instincts and his role as loving comrade and fuser of the nation. Unlike
Whittier, Lowell, and even Melville, Whitman did not see the war as a
moral crusade to liberate the South from the incubus of slavery. Although
Lincoln's Emancipation Proclamation of 1863 outlawed slavery in the
rebel states, in the initial stages of the war, neither Lincoln nor Whitman
looked upon the Civil War as a battle to eradicate the institution of slavery
in the South. Like Lincoln, Whitman saw the war as a struggle to preserve
the Union and to secure the advance of democratic freedom throughout
the world. As a Union poet, he could never fully identify himself as a poet
of North against South. In fact, the argument of states' rights that was the
basis for the Southern secession was a principle with which Whitman had
sympathized in both his verse and his political writings. After years of
defying the ever-stronger foot of federal power, Whitman during the war
years found himself on the side of national, as opposed to state, sovereignty.

    As the Union began to shatter like a china plate, Whitman lost the
sense of identity and power he derived from his democratic muse. In one

of his Civil War notebooks he wrote the following draft of "Quicksand Years":

### Quicksand Years That Whirl Me I Know Not Whither 1861–62

Years that whirl I know not whither
Schemes, politics fail—all is shaken—all gives way
Nothing is sure
Only the theme I sing, the great Soul,
One's-self, that must never be shaken—that out of all is sure,
Out of failures, wars, death—what at last but One's self is sure?
With the Soul I defy you quicksand years, slipping from under my feet.

(Glicksberg, pp. 125–26)

The poem marks a supreme moment of cultural dissonance, as the ground of Whitman's self-protective democratic faith "gives way" under the pressure of time and the uncertain "whirl" of history. As the political scheme and historic order of democracy begin to weaken, everything "gives way," leading Whitman to experience the world as an ever-shifting quicksand of uncertainty and unmeaning. In both the initial and final draft of the poem, he regains control by setting the integrity of the spiritual self against the disintegrative forces of the time. But in a second draft of the poem, he questions the nature of this self:

What do you call One's-Self? (Came one asking me
What is one's-self? (what myself or yours?)
This curious identity
This something that gives me the pow—

(Glicksberg, p. 126)

During the quicksand years of the war, "nothing is sure" for Whitman, not even the ontology of the spiritual self: that something that gives the self *pow-er* appears to be short-circuited by the fact of political "failures, war, death."

These early drafts of "Quicksand Years" are instructive in demonstrating how the political becomes personal in a writer's work. Whitman internalized the war, experiencing the assault on the political Union as an assault on the body and the personality structure of the democratic self. This internalization is evident in the curious entry that he made in his notebook a few days after the attack on Fort Sumter: "I have this hour, this day resolved to inaugurate a sweet, clean-blooded body by ignoring all drinks but water and pure milk—and all fat meats, late suppers—a great body—a purged, cleansed, spiritualized invigorated body."[12] It is almost as if Whitman, by pledging himself to a new bodily regime at the outset of the war, were armoring himself against a sustained assault not only on the integrity of the Union but on the integrity of the (knowable) self. Purging his own body was a means, in some manner, of taking control and preparing for the assault on the larger body of the republic.

Whitman was jolted into a new sense of direction and purpose when in December 1862 he went to the war front in search of his brother George, who had been wounded in the battle of Fredericksburg. On December 19 he arrived at the army camp at Falmouth, Virginia, where he found his brother recovering from a minor face injury. He spent about nine days there, observing the aftermath of battle in the camps and makeshift hospitals opposite Fredericksburg.

"One of the first things that met my eyes in camp, was a heap of feet, arms, legs, &c. under a tree in front a hospital," Whitman wrote to his mother (*Corr.*, I, 59). This was Whitman's nightmare come true: America not as union but amputee, the heritage of the fathers not remembered but dismembered. The sight continued to haunt him as a sign of both the war and the dismembered body of the American republic: Amputation became a recurrent figure in his writings on the war. The heap of amputated arms and legs may have propelled Whitman into a new sense of calling, for he began almost immediately to visit the soldiers in Falmouth's field hospitals. "I go around from one case to another," he wrote. "I do not see that I do much good to these wounded and dying; but I cannot leave them" (*PW,* I, 33). At the camp at Falmouth, he discovered the role he would play during the war: he would minister literally to the wounded body of the republic by visiting the wounded, sick, and dying soldiers in the hospital wards of Washington. And out of this he would compose a record, not of the battles of the Civil War, but of, as he wrote to Emerson, "America, already brought to Hospital in her fair youth—brought and deposited here in this great, whited sepulchre of Washington itself" (*Corr.*, I, 69).

In Washington during the war, Whitman felt that he had touched the pulse of the nation, that he was living at a crossroad of history (Figure VI). He remained there from 1863 to 1873, working part time in the army pay master's office and later in the Bureau of Indian Affairs and the attorney general's office. But his main activity between 1863 and 1865 was visiting the soldiers in Washington's hospitals. "It was a religion with me," he told Traubel. "Every man has a religion . . . something which absorbs him, possesses itself of him, makes him over in its image. . . . That, whatever it is, seized upon me, made me its servant, slave; induced me to set aside the other ambitions: a trail of glory in the heavens, which I followed, followed, with a full heart."[13]

FIGURE VI. Photograph of Whitman, c. 1864, by Alexander Gardner. Courtesy of Walt Whitman Collection (N45-1636-G), Clifton Waller Barrett Library, University of Virginia Library.

In making himself over in the role of a minister to the soldiers during the war, Whitman also in some sense made himself over in the image of a woman. As America during the war moved toward the traditionally masculine polarity of militarism, violence, and aggression, Whitman in his person and his writing moved toward the traditionally feminine polarity of nurturance, compassion, and love. Amid the carnage of war, the athlete and rough of the 1855 *Leaves* assumed the role of a tender and loving nurse and mother. "Whitman was the lover, the healer, the reconciler," wrote his friend John Burroughs in response to Thomas Higginson's charge that Whitman hypocritically urged others to go to war but refused to fight himself. "The only thing in character for him to do in the War was what he did do—nurse the wounded and sick soldiers—Union men and Rebels alike, showing no preference. He was not an athlete, or a rough, but a great tender mother-man, to whom the martial spirit was utterly foreign."[14]

Whitman estimated that he made over six hundred hospital visits during the war, ministering to the needs of eighty thousand to a hundred thousand wounded, sick, and dying soldiers. He would arrive in the hospital wards, freshly bathed, ruddy like St. Nicholas and hairy like a buffalo, carrying a haversack full of gifts for the soldiers. He distributed sweet crackers, preserves, jellies, oysters, tea, butter, condensed milk, plugs of tobacco, wine, brandy, stamps, envelopes, notepaper, the morning paper, undershirts, socks, dressing gowns, and many other articles requested by the soldiers (*PW*, I, 112–13; *Corr.*, I, 153). Sometimes the soldiers merely wanted to be spoken to, read to, touched, and loved, and this was perhaps the most important gift that Whitman brought in his hospital visits: "I believe my profoundest help to these sick & dying men," he wrote to a potential donor, "is probably the soothing invigoration I steadily bear in mind, to infuse in them through affection, cheering love, & the like, between them & me. It has saved more than one life" (*Corr.*, I, 102).

In his hospital visits, Whitman assumed the role of invigorator of the nation he had imagined for himself in "Song of Myself":

> To any one dying . . . . thither I speed and twist the knob of the door,
> Turn the bedclothes toward the foot of the bed,
> Let the physician and the priest go home.
>
> I seize the descending man . . . . I raise him with resistless will.
>
> O despairer, here is my neck,
> By God! you shall not go down! Hang your whole weight upon me.
>
> I dilate you with tremendous breath . . . . I buoy you up;
> Every room of the house do I fill with an armed force . . . . lovers of
>     me, bafflers of graves.
>
> (*LG* 1855, p. 71)

Tramping up and down the aisles of the hospital wards, Whitman came closer to achieving his dream of reaching the democratic masses than he would ever come through his written work. "It has given me my most fervent views of the true *ensemble* and extent of the States," he said of his hospital experiences.

> While I was with wounded and sick in thousands of cases from the New England States, and from New York, New Jersey, and Pennsylvania, and

from Michigan, Wisconsin, Ohio, Indiana, Illinois, and all the Western States, I was with more or less from all the States, North and South, without exception. . . . I was with many rebel officers and men among our wounded, and gave them always what I had, and tried to cheer them the same as any. . . . Among the black soldiers, wounded or sick, and in contraband camps, I also took my way whenever in their neighborhood, and did what I could for them. (*PW*, I, 113–14)

Ministering to a cross section of the democratic nation, North and South, black and white, Whitman literally became the invigorator, comrade, fuser, and reconciler of the American republic he had wanted to become through his writing.

In nursing the war wounded, Whitman found a legitimate social form in which to express his homoeroticism and his desire to mother, love, and nurture men. Describing himself to his good friend Abby Price and her family, he said: "You would all smile to see me among them—many of them like children, ceremony is mostly discarded—they suffer & get exhausted & so weary—lots of them have grown to expect as I leave at night that we should kiss each other, sometimes quite a number, I have to go round—poor boys, there is little petting in a soldier's life in the field, but, Abby, I know what is in their hearts, always waiting, though they may be unconscious of it themselves" (*Corr.*, I, 162). Addressed as "Pa" and "Uncle" by many of the soldiers, Whitman could at least partially realize his desire to father both poems and children (Figure VII).

After what he called the "doubts nauseous" of the prewar years, the war restored Whitman's faith in the democratic masses. Among the sick, wounded, and dying soldiers in the hospitals, he witnessed the extent of human sacrifice for the cause of the Union; here, compassion and sympathy rather than greed and self-interest were the reigning emotions. "I find the best expression of American character I have ever seen or conceived— practically here in these ranks of sick and dying young men," he wrote to Emerson in January 1863, after only a few weeks in the hospital wards. "I find the masses fully justified by closest contact, never vulgar, ever calm, without greediness, no flummery, no frivolity—responding electric and without fail to affection, yet no whining—not the first unmanly whimper have I yet seen or heard." In the throes of death, the soldiers were models of heroism and fortitude. The surgeon of the Patent Office Hospital, Horatio Stone, could not, Whitman observed, remember "one single case of a man's meeting the approach of death, whether sudden or slow, with fear or trembling—but always of these young men meeting their death with steady composure, and often with curious readiness" (*Corr.*, I, 70–71).

The Civil War, which was fought on one side to preserve the Union of

FIGURE VII. Calling card
presented to Whitman by
George Field, one of the many
young men he cared for in the
Washington hospitals. Courtesy
of The Library of Congress.

the republic and on the other to preserve the republican tradition of local
and state sovereignty, was a springboard away from the republican and
essentially agrarian order of the past toward the centralized, industrialized
nation-state of the future.[15] And yet ironically, it was during the war years
and particularly in the hospitals, that America in Whitman's eyes came
closest to realizing the republican dreams of the revolutionary founders:
"Before I went down to the Field, and among the Hospitals," he wrote, "I
had my hours of doubt about These States; but not since. The bulk of the
Army, to me, develop'd, transcended, in personal qualities—and, radically, in moral ones—all that the most enthusiastic Democratic-
Republican ever fancied, idealized in loftiest dreams. And curious as it
may seem the War, to me, *proved* Humanity, and proved America and the
Modern" (*Memoranda,* p. 59).

During the war, however, the battlefields and hospitals were also sites of slaughter and carnage where Whitman's democratic faith was severely tested. The heroism of the soldiers was a flash of light amid what he called the "malignant darkness" of the war. If the war proved humanity and America, it also released a pack of demons that would continue to haunt Whitman and the nation in the postwar years. "Mother, one's heart grows sick of war, after all, when you see what it really is," he wrote a few days after the massive carnage at Gettysburg on July 5, 1863; "every once in a while I feel so horrified & disgusted—it seems to me like a great slaughter-house & the men mutually butchering each other" (*Corr.*, I, 114). Four months later, at the dedication of the National Cemetery on the field at Gettysburg, Lincoln tried to place the unreason of the carnage in a rational frame of history when he presented the war as a testing ground for the ideals of the republic, and the war dead as part of a national sacrifice "that this nation, under God, shall have a new birth of freedom—and that government of the people, by the people, for the people, shall not perish from the earth" (Figure VIII).[16] Whitman pinned Lincoln's address on his wall as a testament of democratic faith. But the carnage of the war continued for almost two more years. In March 1864, as Washington prepared for General Ulysses Grant's spring offensive, Whitman shuddered at the thought of renewed slaughter: "O mother, to think we are to have here soon what I have seen so many times, the awful loads & trains & boat loads of poor bloody & pale & wounded young men again . . . what an awful thing war is—Mother, it seems not men but a lot of devils & butchers butchering each other—" (*Corr.*, I, 204).

Whitman was "tossed from pillar to post" during the war years, for he was equally horrified at the prospect of giving up the contest to preserve the Union. "This country can't be broken up by Jeff Davis, & all his damned crew," he wrote to Tom Sawyer, one of the hospital soldiers with whom he had established a particularly strong bond of affection. To give up the battle would dissolve not only the Union but also Whitman's identity and reason for being: "Tom, I sometimes feel as if I didn't want to live—life would have no charm for me, if this country should fail after all, and be reduced to take a third rate position, to be domineered over by England & France & the haughty nations of Europe &c and we unable to help ourselves" (*Corr.*, I, 92).

Although Whitman sustained his faith in the Union and the democratic masses through the "strange & terrible times" of the Civil War, he exhibited no more faith in the politics of Washington than he had in *The Eighteenth Presidency!* Even as a civil servant during and after the war, he still discovered "bats and night-dogs, askant in the Capitol." Attending the congressional debates on whether to recognize the Confederacy, he

FIGURE VIII. Dead of the 24th Michigan Infantry, Gettysburg, photographed by Tim O'Sullivan. Courtesy of The Library of Congress.

was struck by the low level of ability in the House of Representatives: "It is very curious & melancholy to see such a rate of talent there, such tremendous times as these," he wrote his mother in April 1864 (*Corr.*, I, 211). Washington itself seemed a very symbol of the chaos of the Union. He called it "this huge mess of traitors, loafers, hospitals, axe-grinders, & incompetencies & officials that goes by the name of Washington"; it was "this union Capitol without the first bit of cohesion—this collect of proofs how low and swift a good stock can deteriorate" (*Corr.*, I, 99, 69). Amid the dissolution of Washington and what he called the "wild days of the War," Whitman returned over and over to contemplate J. David D'angers's statue of Jefferson on the White House lawn: "Many, many hours, long and long and long, have I studied that piece of work," he said, as if he were seeking in the author of the Declaration of Independence some reaffirmation of the values being tested in the war (*WWC*, III, 230) (Figure IX).

In Whitman's view, it was President Lincoln above all who revived the spirit of Jefferson and who maintained the steady course of the ship of state amid the rocks and shoals of the time. "I have finally made up my mind that Mr. Lincoln has done as good as a human man could do," he wrote his mother in October 1863. "I still think him a pretty big President—I realize here in Washington that it has been a big thing to have just kept the United States from being thrown down & having its throat cut—& now I have no doubt it will throw down secession & cut its throat—& I have not had any doubt since Gettysburgh" (*Corr.*, I, 174).

For Whitman, as for other American writers, the Civil War was the impetus of a new style that was related to the development of local color and literary realism in the postwar period. The changes wrought by the war—the carnage and physical devastation, the centralization of political and economic power, the growth of an industrial capitalist class, and the emancipation of blacks and in some sense women—had the effect of jolting literary America out of romance and into realism. If in the prewar period Whitman had viewed himself as a poet-prophet, mythically embodying democracy and the revolutionary traditions of the past, during the war years he came to see himself as a kind of poet-historian, preserving a record of the present moment for future generations. In a long line of American writers ranging from Cotton Mather in the colonial period to Norman Mailer in the post-Modern period, Whitman had the sense that events of divine and cosmic significance were being enacted in America and that he as the national poet must record them for posterity. "Is there not, behind all," he asked, "some vast average, sufficiently definite, uniform and unswervable Purpose, in the development of America, (may I not say divine purpose? only all is divine purpose,) which pursues its own will, maybe unconscious of itself" (*Memoranda*, p. 66).

During the war, Whitman kept little notebooks that served as both memoranda books and records of cases treated, sights seen, and stories heard by bedside or in camp. These notebooks, which consisted of sheets of papers folded and secured with a pin, became the germ of his prose memoranda and poems of the war. At first he contemplated writing the

FIGURE IX. Statue of Thomas Jefferson on the White House lawn, 1861, photographed by Mathew Brady. Courtesy of The Library of Congress.

"History of the War, in a great volume, or several Volumes," but realizing the war's "vast complication," he decided to write "a sort of Itinerary" of his hospital experience from "a Democratic point of view" (Glicksberg, p. 166). As in Hemingway's *A Farewell to Arms*, Joseph Heller's *Catch-22*, and Robert Altman's *MASH*, the hospital becomes a metonymic figure of the nation at war: "It seem'd sometimes as if the whole interest of the land, North and South," said Whitman, "was one vast central Hospital, and all the rest of the affair but flanges" (*Memoranda*, p. 5).

Whitman mentioned the project in a letter to Emerson, dated January 16, 1863, only a few weeks after he began visiting the Washington hospitals. Later the same year he wrote to the publisher James Redpath, who had recently printed Louisa May Alcott's *Hospital Sketches* (1863). He proposed to write a volume entitled "Memoranda of a Year," which he described as "something considerably beyond mere hospital sketches . . . memoranda of incidents, persons, places, sights, the past year (mostly jotted down either on the spot or in the spirit of seeing or hearing what is narrated)" (*Corr.*, I, 171). Turned down by Redpath, Whitman contributed several firsthand accounts of his war experiences to New York newspapers, but it was not until a decade after the war that he published, once again at his own expense, *Memoranda During the War* (1875–76).[17] Composed for the most part during the Gilded Age, when the republican ethos released by the war had been replaced by an ethos of public and private greed, Whitman intended his war *Memoranda* to present models of heroism, sacrifice, and the new virtue he called *unionism* to present and future generations: "In the mushy influences of current times," he wrote, "the fervid atmosphere and typical events of those years are in danger of being totally forgotten" (*Memoranda*, p. 5).

Little recognized by critics then or since, the book Whitman proposed to Redpath and wrote in the post–Civil War period was another revolutionary undertaking. In deciding to locate his account of the war in the hospital rather than on the battlefield, he anticipated the model of history offered by the "new social history" of the late twentieth century. With roots in local color and literary realism, his *Memoranda* also looks forward to the genre of new or parajournalism of Norman Mailer, Hunter Thompson, and Tom Wolfe: Like the nonfiction novels of these Vietnam-era writers, Whitman's *Memoranda During the War* is a form of history as fiction, fiction as history. He uses scene-by-scene construction, dialogue, symbolism, and a first-person narrator whose point of view shifts between subjective (and frequently parenthetical) impression and journalistic-style reportage.

Although based on Whitman's war experiences, his reporter-narrator is, like his poetic persona, a fictive creation. His *Memoranda*, he says at the outset, are *"verbatim* renderings" of notes penciled on the spot: "I leave them just as I threw them by during the War, blotch'd here and there with more than one blood-stain, hurriedly written, sometimes at the clinique, not seldom amid the excitement of uncertainty, or defeat, or of action, or getting ready for it, or a march" (*Memoranda*, p. 3). Whitman's Civil War diaries and notebooks may have been written on the spot, and some of them do appear to be blood smeared, but there is little correspondence between them and his published *Memoranda*. That is, his *Memoranda* is not a "verbatim" rendering of his Civil War notebooks, but a remembrance and reinvention of the war written a decade after its close. To give the reader a sense of the immediacy and uncertainty of the war as it was happening, Whitman uses present-tense narration; dated, diarylike entries; a spare, telegraphic style; and an almost cinematic montage of scenes from the war. He invents a narrator who is present, as Whitman was, at the bedside of wounded and dying soldiers, but this same narrator is also present, as Whitman was *not*, at the retreat from Bull Run and the battle of Chancellorsville.

Whitman's point of view is insistently democratic. He focuses not on the feats of the generals but on the bravery of common soldiers: the New York boy who died of a perforated bladder, the Wisconsin boy who had his leg amputated and died, the Yankee sergeant who refused to surrender. "When I think of such things," he says, "all the vast and complicated events of the War on which History dwells and makes its volumes, fall indeed aside, and for the moment at any rate I see nothing but young Calvin Harlowe's figure in the night disdaining to surrender" (*Memoranda*, p. 46).

Like the bodies in the field at Gettysburg, Whitman intended his war *Memoranda* to consecrate the ground of the republic. He presents individual cases as types of the 300,000 unknown soldiers who suffered and died for what Lincoln called a "new birth of freedom." Multiplying the scenes and cases he presents by "many a score—aye, thousands, North and South," he demonstrates the massiveness of the Civil War and the anonymity of modern heroism: "No formal General's report, nor print, nor book in the library, nor column in the paper, embalms the bravest, North or South, East or West. Unnamed, unknown, remain, and still remain, the bravest soldiers" (*Memoranda*, p. 16).

The main substance of Whitman's *Memoranda* is not the issues that provoked the war but, rather, the character displayed by the American people during the war. The issue of slavery appears only as a footnote, and even the cause of the Union is subordinate to the manliness and courage displayed by both sides during the war. "To me," says Whitman, "the points illustrating the latent Personal Character and eligibilities of These States, in the two or three millions of American young and middle-aged men, North and South, embodied in the armies—and especially the one-third or one-fourth of their number, stricken by wounds or disease at some time in the course of the contest—were of more significance even than the Political interests involved" (*Memoranda*, pp. 4–5). It was by representing and mythologizing the war as trial and proof of democratic character that Whitman kept the violence and carnage of the war—and ultimately the entire democratic project—from slipping into the possible chaos and unmeaning of national history.

To foster a national politics of reconciliation, Whitman refused to present the war as one of North against South. The North and South were equally responsible for the war, he argued in a note drawn from *The Eighteenth Presidency!*: The true villain in the national struggle was the Jacksonian one of monopoly and privileged interest, which sought to take control of "municipal, State, and National Politics . . . while the great masses of the people, farmers, mechanics and traders, were helpless in their gripe [*sic*]" (*Memoranda*, p. 64).[18] The war was not between separate peoples but a war for democracy fought within the single body of the Union, a point that Whitman emphasized in his *Memoranda:* "I have myself, in my thought, deliberately come to unite the whole conflict, both sides, the South and North, really into One, and to view it as a struggle going on within One Identity" (*Memoranda*, p. 65).

To stress the single identity within which the war took place, Whitman included the South in his *Memoranda*, with the sequences "A Secesh Brave," "Virginia," "Southern Escapees," and "Calhoun's Real Monument." In "Two Brothers, One South, One North," he presents the Union war in the metonymic figure of two brothers at war: "It was in the same battle both were hit. One was a strong Unionist, the other Secesh; both fought on their respective sides, both badly wounded, and both brought together here after absence of four years. Each died for his cause" (*Memoranda*, p. 53).

Whitman refused to glorify the war. His *Memoranda* is at once a tribute to the heroism of the unknown soldier and a profound antiwar statement. Interwoven in his narrative of the war is the diegesis of "an unending, universal mourning-wail of women, parents, orphans" (*Memoranda*, p. 5). At its worst, the war aroused not republican valor but the devils in human hearts, with "every lurid passion, the wolf's, the lion's lapping thirst for blood—passionate, boiling volcanoes of human revenge for comrades, brothers slain" (*Memoranda*, p. 36). Surveying the camps of the wounded after the battle of Chancellorsville, Whitman asks: "is this indeed *humanity*—these butchers' shambles? There are several of them. There they lie, in the largest, in an open space in the woods, from 500 to 600 poor fellows—the groans and screams—the odor of blood, mixed with the fresh scent of the night, the grass, the trees—that Slaughterhouse!" (*Memoranda*, pp. 14–15).

In Whitman's view it was the military's "feudal" and antidemocratic structure that was largely responsible for the war's massive carnage and waste. Commenting on the need for "A New Army Organization Fit for America," he observed: "In the present struggle . . . probably three-fourths of the losses, men, lives &c., have been sheer superfluity, extravagance, waste" (*PW*, I, 74–75). But Whitman was wrong: The Civil War was not an old-world war of small professional armies, paid mercenaries, and shrewd military maneuver. In the "slaughterhouse" and "waste" of the Civil War, Whitman was witnessing the effects of the first modern war, American style, in which the enemy was defeated by pitting large armies against one another in a war of depletion. What made the Civil War particularly horrid was that this exhaustive trench warfare was fought between citizen and citizen, brother and brother, on native ground. The "butchers' shambles" of the Civil War bled the republic from within, making the war to preserve the Union a war that paradoxically destroyed the possibility of any future Union.

Written a decade after the war's close, Whitman's *Memoranda* is an attempt to comprehend and control the demons released by the Civil War by articulating and representing the war experience as part of a natural historical growth. In a concluding note to his *Memoranda*, Whitman states: "The development of a Nation—of the American Republic, for instance, with all its episodes of peace and war [exhibits] the same regularity of order and exactness, and the same plan of cause and effect, as the crops in the ground, or the rising and setting of the stars" (*Memoranda*, p. 67). To underscore this natural and essentially rational order of history, Whitman interweaves into his war narrative images of grass, moon, and stars as signs of some invisible, redemptive force at work in the world. Amid the "red life-blood oozing out from heads or trunks or limbs" on the battle-field at Chancellorsville is the promise of "green and dew-cool grass" and the "moon shining out full and clear" (*Memoranda*, p. 14). Within this redemptive pattern of history, the blood of the soldiers becomes part of the process of national recreation and the march of humanity toward the democratic future.

The Civil War released into violent and bloody confrontation the tragic contradictoriness of American new-world experience. As in Lincoln's Gettysburg Address, Whitman's urge to naturalize the unnatural bloodiness of the Civil War, by mingling blood and grass in a redemptive teleology, fed willy-nilly the national myth of regeneration through violence that marked, and still does mark, the course of American history.[19] "Strange, (is it not?)" Whitman said in his speech "Death of Abraham Lincoln," "that battles, martyrs, agonies, blood, even assassination, should so condense—perhaps only really, lastingly condense—a Nationality" (*PW*, II, 508). The logic of violence that was part of the "National" policy against native Americans, blacks, and Mexicans had now turned inward against republican freemen and ultimately against the president himself, anticipating Henry Adams's question in *The Education of Henry Adams:* "Was assassination forever to be the last word of Progress?"[20]

At the same time that Whitman was visiting the Washington hospitals and planning his volume of war memoranda, he was also working on the poems of *Drum-Taps* which, along with *Sequel to Drum-Taps*, was published in 1865. In one of his conversations with Traubel, Whitman described his method of incorporating his hospital experiences into his poems: "I took notes as I went along—often as I sat . . . writing while the other fellow told his story . . . I would work in this way when I was out in the crowds, then put the stuff together at home. Drum Taps was all written in that manner—all of it—all put together by fits and starts, on the fields, in the hospitals, as I worked with the soldier boys. Some days I was more emotional than others, then I would suffer all the extra horrors of my experience—I would try to write, blind, blind, with my own tears" (*WWC*, II, 137).

A greater artistic control is evident in the poems of *Drum-Taps* and *Sequel*, in the dramatic presentation of "Song of the Banner at Daybreak" and "Come Up from the Fields Father," the stanzaic form of "Dirge for Two Veterans," the metric regularity of "Beat! Beat! Drums!" and "O Captain! My Captain!" and the imagistic simplicity of "Cavalry Crossing a Ford" and "Bivouac on a Mountain Side." This control may be partly an attempt to distance and aestheticize "all the extra horrors" of the war experience. Formally, *Drum-Taps* was, in Whitman's opinion, superior to *Leaves of Grass:* It was, he wrote to his friend William Douglas O'Connor, "more perfect as a work of art, being adjusted in all its proportions, & its passion having the indispensable merit that though to the ordinary reader let loose with wildest abandon, the true artist can see it is yet under control" (*Corr.*, I, 246). Like Whitman's wound-dresser, his war poems remain cool and impassive on the outside, masking the fire, the "burning flame" within.

If the poetic experiment of *Leaves of Grass* corresponded, as Whitman argued, to the political experiment of the republic, then his return to more conventional forms in *Drum-Taps* and *Sequel* might be read as a sign of the nation's political failure. Like Milton after the collapse of Cromwell's Commonwealth, Blake after the failure of the French Revolution, and Pound after World War I, Whitman moved during the war years away from the visionary poems of his early period toward what Pound would call the "poem including history." This shift may have been signaled by Whitman's initial glimpse of the pile of amputated arms, legs, and feet that entered his field of vision simultaneously with his first experience of the war. Confronted with the reality of the war—its excess and infernal horror—Whitman wanted to leave an artistic record of the time, as a means of comprehending the war imaginatively and of consecrating the war in the national imagination. He becomes a kind of poetic reporter offering a "few stray glimpses" into the dark interiors of the war: His field of poetic vision shifts from psychic event to historic moment, from self to other, from future to present. In the poems of *Drum-Taps* and *Sequel*, Whitman's emphasis is no longer on himself as a model of democratic personality but, as he wrote to O'Connor, "the pending action of this *Time*

*& Land we swim in,* with all their large conflicting fluctuations of despair
& hope, the shiftings, masses, & the whirl & deafening din, (yet over all, as
by invisible hand, a definite purport & idea)—with the unprecedented
anguish of wounded & suffering, the beautiful young men, in wholesale
death & agony, everything sometimes as if in blood color, & dripping
blood" (*Corr.,* I, 246–47).

During the war Whitman was reading Dante's *Inferno* and *The Works
of Virgil* (translated by Joseph Davidson in 1857), perhaps in search of an
epic model for his own war poems.[21] Unlike Melville, however, who draws
on the Bible and *Paradise Lost* in the complex allusive structure of his
*Battle Pieces and Aspects of the War* (1866), Whitman rejects the epic models
of the past for his *Drum-Taps* poems. Whereas Melville assumes a posture
of ironic detachment, organizing his poems around public events, major
battles, and the triumphs and defeats of Northern generals, Whitman's
point of view is democratic and engaged. He enters his poems personally
in the figure of a common soldier, presenting the war in lyric rather than
epic terms: His war scenes could be anywhere, North or South; his heroes
are the masses of ordinary soldiers, particularly the unknown soldiers
whose graves he marks with his own poetic inscription: "*Bold, cautious,
true, and my loving comrade.*"[22] Eschewing the biblical and epic analogy of
Melville's *Battle Pieces,* Whitman finds his symbolic and mythic structure
closer to the democratic ground of America: in the makeshift hospitals of
camp and field; in an army corps on the march; in the death of an
unknown soldier; in the recurrent image of red blood consecrating "the
grass, the ground"; in the figure of the poet as wound-dresser and comrade; and in the female spirit of Democratic Liberty who presides over the
poems as both fierce warrior and cosmic mother.[23]

*Convulsiveness* was the word that Whitman used in his war *Memoranda*
to characterize "the War itself with the temper of society preceding it"
(*Memoranda,* p. 59). In the final ordering of the *Drum-Taps* cluster in the
1881 *Leaves,* Whitman molds the "convulsiveness" of the war into a providential scheme, tracing the movement from militant exultation, to the
actual experience of suffering and death, to demobilization and the final
justification of the war as part of the "throes of Democracy" in its march
toward the future.[24] In the initial version of *Drum-Taps* and *Sequel,* however, the disruption of the war is more apparent: The poems are characterized by radical shifts in style, theme, and mood, as Whitman seeks an
appropriate poetic posture in relation to the whirl and din, fracture and
chaos of the first modern war.

The volume opens with "Drum-Taps" ("First O Songs for a Pre-
lude"), a poem that projects an exultant image of Manhattan arming to the

beat of a drum. In both measure and mood the poem represents the martial spirit of the North at the outset of the war:

> War! an arm'd race is advancing!—the welcome for
> battle—no turning away;
> War! be it weeks, months, or years—an arm'd race is
> advancing to welcome it.
> Mannahatta a-march!—and it's O to sing it well!
> It's O for a manly life in the camp!
>
> *(DT,* p. 7)

Poems of militant, even hysterical exultation in the Union cause alternate with poems of uncertainty and deep mournfulness. The poet's divided mind is evident in "Rise O Days from Your Fathomless Deeps":

> How DEMOCRACY, with desperate vengeful port strides
> on, shown through the dark by those flashes of
> lightning!
> (Yet a mournful wail and low sob I fancied I heard
> through the dark,
> In a lull of the deafening confusion.)
>
> *(DT,* p. 36)

As the poet experiences firsthand the war's carnage and apparent unreason, the rattling of drum corps and sabers modulates into a low sob and mournful wail, and the militant poet in blue becomes the poet of darkness, blood-dripping wounds, and death:

> No poem proud I, chanting, bring to thee—nor mastery's
> rapturous verse;
> But a little book, containing night's darkness, and blood-
> dripping wounds,
> And psalms of the dead.
>
> ("Lo! Victress on the Peaks!")[25]

This elegiac note sounds throughout *Drum-Taps* and *Sequel*, investing the drum taps of the title with multiple significance as taps of recruitment and soldiers on the march and taps for lost lives, lost innocence, and the lost dream of America.

Whitman counterpointed the martial spirit and drum-corps rhythm of "Beat! Beat! Drums!" and "Rise O Days from Your Fathomless Deeps" with the understated tone of poems depicting actual scenes of war. In poems such as "Cavalry Crossing a Ford," "By the Bivouac's Fitful Flame," "Bivouac on a Mountain Side," "An Army on the March," "A March in the Ranks Hard-Prest, and the Road Unknown," "A Sight in

Camp in the Daybreak Gray and Dim," and "The Veteran's Vision," he invents a new style of poetic reportage commensurate with modern and mass warfare. In their focus on realistic detail—"Surgeons operating, attendants holding lights, the smell of ether, the odor of blood"—and in their presentation of the "swarming ranks" and "dense brigades" of men as part of the "resistless" surge of history, the poems anticipate the naturalistic war writing of Stephen Crane, Ernest Hemingway, and Norman Mailer. In their formal precision and economy of means, the poems also anticipate the Imagist poetics announced by Pound in his 1913 "Imagist Manifesto" and summed up by William Carlos Williams's poetic dictum: "No Ideas But in Things."[26]

The immediacy and brutality of the war shocked Whitman into a new realism in which painting and the snapshot rather than oratory and opera became the primary artistic analogues of his poems. If in the poems of 1855 and 1860 he conceived of himself as an orator speaking to present and future generations of readers, during the Civil War he came to see himself as an artist projecting visual images of "this *Time & Land we swim in.*" Whereas his earlier poems nodded in the direction of Elias Hicks and Marietta Alboni, his *Drum-Taps* poems nod in the direction of Winslow Homer's sketches of army life and Mathew Brady's photographs of the Civil War (Figure X).

In Whitman's painterly sketches of war scenes, the vaunting "I" of his earlier poems recedes to the margins of his verse; in some of his sketches, in fact, the "I" is not present at all. "Cavalry Crossing a Ford" projects the tension between individual and en masse through visual notation:

> A line in long array, where they wind betwixt green
>     islands;
> They take a serpentine course—their arms flash in the
>     sun—Hark to the musical clank;
> Behold the silvery river—in it the splashing horses,
>     loitering, stop to drink;
> Behold the brown-faced men—each group, each person,
>     a picture—the negligent rest on the saddles;
> Some emerge on the opposite bank—others are just
>     entering the ford;
> The guidon flags flutter gaily in the wind.
>
>                                                           (*DT*, p. 8)

Like the single sentence of the poem itself, the men form a single line. In the opening lines the anonymous "they" is subsumed in the larger figure of the "line in long array" winding its "serpentine course"; the men are

FIGURE X. *Infantry Column on the March,* drawing by Winslow Homer. Courtesy of the Cooper-Hewitt Museum, Smithsonian Institution/Art Resource, New York.

distinguished only by the image of arms—not flesh but metal—flashing in the sun. In the river the en masse breaks into individual "pictures" as the poet's camera eye moves, as it does throughout the volume, from panorama to close-up. The images of "splashing horses" loitering and the "brown-faced men" resting "negligently" on their saddles contradict traditional notions of military order, discipline, and hierarchy, thereby projecting the figure of a democratic army. And yet like the pattern of alliteration, repetition, and internal rhyme that links the poem's separate images, the cavalry's apparently random motion is part of a single pattern. The separate brown-faced men are part of the line of cavalry that stretches along both sides of the river and part, too, of the democratic masses advancing under the guidon flags of the Union. This harmonious image of a democratic army, in which individual and national will are merged under the banner of the Union cause, splinters in subsequent poems into a discordant vision of individuals atomized and commodified under the dehumanizing press of a war spun out of control.

Some of Whitman's painterly war poems, including "A March in the Ranks Hard-Prest, and the Road Unknown," "A Sight in Camp in the Day-Break Grey and Dim," and "The Veteran's Vision," are based on firsthand accounts of the war that he recorded in his Civil War notebooks. During his stay at Camp Falmouth, where the Army of the Potomac was recovering from its defeat in the battle of Fredericksburg, Whitman made the following entry in his notebook:

*Sight at daybreak*—in camp in front of the hospital tent on a stretcher, (three dead men lying,) each with a blanket spread over him—I lift one and look at the young man's face, calm and yellow—'tis strange!
(Young man: I think this face of yours the face of my dead Christ!)
(Glicksberg, p. 79)

His vision becomes in "A Sight in Camp in the Day-Break Grey and Dim" the base of a sustained action, a sequence of "uncoverings" in which the poet-reporter faces a scene of suffering and death, responds with compassion and love, and tries to place the horror of the war into a redemptive pattern of history. As in other *Drum-Taps* poems, he performs a kind of rescue mission, recuperating the body of the unknown soldier from the anonymity of mass warfare and mass death.

Like Lincoln consecrating the field at Gettysburg, Whitman consecrates the ground of the war by focusing not on "soldiers' perils or soldiers' joys" but on the wounded "Where they lie on the ground, after the battle brought in;/Where their priceless blood reddens the grass, the ground" ("The Dresser," *DT*, p. 32). In only one *Drum-Taps* poem, "The Veteran's Vision" ("The Artilleryman's Vision"), does Whitman describe a scene of battle. An early draft of this poem, which may be based on an account of a battle told to Whitman by one of the soldiers, appears in his 1862 notebook. Whereas the initial draft of the poem, under the title "A Battle. (Scenes, Sound, &c.)," presents the battle directly, in the poem's final version, the psychological complexity is heightened by presenting the battle as a nightmare vision that haunts the veteran after the close of the war (Glicksberg, pp. 121–23).

During his service in the Washington hospitals, Whitman frequently remarked on the fact that the sights he witnessed in the daytime returned to haunt him at night, and years after the war's close he continued to be troubled by memories of what he called the "fearful scenes of the war." "I don't think the war seemed so horrible to me at the time, when I was busy in the midst of its barbarism, as it does now, in retrospect," he told Traubel in 1888 (*WWC*, I, 198). This experience of the retrospective effects of war is the subject of "The Veteran's Vision." Framed as a veteran's dream, the poem is both a firsthand account of battle and a realistic probe into the long-term effects of the war on the individual and the national psyche.

The vision of the war that presses on the veteran as he lies with his wife and child sleeping peacefully by his side is rendered as an explosive volley of "shots," punctuated by parenthetical passages that give emotional and psychological resonance to the realistic presentation:

The skirmishers begin—they crawl cautiously ahead—
I hear the irregular snap! snap!
I hear the sounds of the different missiles—the short
*t-h-t! t-h-t!* of the rifle balls;
I see the shells exploding, leaving small white clouds—
I hear the great shells shrieking as they pass;
The grape, like the hum and whirr of wind through the
trees, (quick, tumultuous, now the contest rages!)

(*DT*, p. 55)

In "fantasy unreal" the veteran reexperiences the pride and heroism of the infantrymen amid the noise and confusion of anonymous combat; the sound of the cannon, he confesses parenthetically, arouses even in dreams "a devilish exultation, and all the old mad joy, in the depths of my soul" (*DT*, p. 56). But the veteran's "devilish exultation" also suggests the underlying malignity of the war.

In his notebook draft of the poem, Whitman commented directly on the hell of war:

Then after the battle, what a scene! O my sick soul!
how the dead lie,
The wounded—the surgeons and ambulances—
O the hideous hell, the damned hell of war
Were the preachers preaching of hell?
O there is no hell more damned than this hell of war.

(Glicksberg, p. 123)

In the final version of the poem, Whitman suppressed these lines, perhaps out of his desire to focus on the artilleryman's almost nostalgic longing for the "devilish" excitement of combat. The effect of the cover-up is an increase in subtlety. Beneath its flashes of heroism and "mad joy," the veteran's vision reveals the chaos and conflagration of a war in which life and death become meaningless:

—Elsewhere I hear the cry of a regiment charging—
(the young colonel leads himself this time, with
brandish'd sword;)
I see the gaps cut by the enemy's volleys, (quickly
fill'd up—no delay;)
I breath the suffocating smoke—then the flat clouds
hover low, concealing all.

(*DT*, p. 55)

The veteran's vision bears witness to the technology of modern warfare, in which mass armies confront each other in anonymous, mechanical combat. Soldiers appear as infinitely replaceable goods filling the gaps in the assembly line of war.[27] The mechanical production of the Civil War, with its heavy artillery and machine gun weaponry, commodifies and literally amputates the individual in a manner that anticipates the dehumanizing effects of machine technology in the postwar industrial economy.

Under the pressure of battle, men become automatons, stripped of compassion and humanity:

> And ever the hastening of infantry shifting positions—
>   batteries, cavalry, moving hither and thither;
> (The falling, dying, I heed not—the wounded, dripping
>   and red, I heed not—some to the rear are hob-
>   bling;)
>
>                                          (*DT*, p. 56)

The veteran who "heeds not" is himself a psychic fatality of the war; lacking the engaged perspective of Whitman's compassionating wound-dresser, his vision represents the war's debilitating human effects.

The poem is built on the complex psychic tension generated by the emotional stripping and demonic unleashing that occur as the veteran relives scenes from the war, and the emotions of love, tenderness, and familial union associated with the domestic, postwar context in which his vision occurs. What "The Veteran's Vision" suggests is that the enduring wounds of the Civil War were mainly psychic: that the demons unleashed by the war defied the notions of personal virtue, individual freedom, and a rationally ordered universe on which the democratic republic was built. The demons that continue to haunt the veteran after his return to home and family suggest the disruptive and potentially devastating psychic effects of the war not only on individual citizens, including Whitman, but also on the political nation after its much-vaunted return to family union.

It was partly in response to these psychic wounds of the nation that Whitman emerged in the role of wound-dresser during the war years. In "The Dresser" ("The Wound-Dresser"), he presents himself as a male nurse literally binding the wounds of the American soldiers:

> From the stump of the arm, the amputated hand,
> I undo the clotted lint, remove the slough, wash off the
>   matter and blood;
> Back on his pillow the soldier bends, with curv'd neck,
>   and side-falling head;

> His eyes are closed, his face is pale, he dares not look on
>     the bloody stump.
> And has not yet looked on it.
>
> > > > (*DT*, p. 33)

Harold W. Blodgett and Sculley Bradley, the editors of the New York University edition of *Leaves of Grass*, describe the content of this poem as "a faithful description of Walt Whitman's ministrations to the war-wounded in Washington hospitals" (*LGC*, p. 309). Rarely, however, did Whitman actually dress the wounds of soldiers during his service in the Washington hospitals: rather, the image of wound-dresser is a metaphor for the role he played both in the hospitals and in his war poems as the soother, reconciler, and psychic healer who dared to look—as others could not—on the dismembered bodies and bloody corpses produced by the war.

As a figurative dresser of the amputated body and wounded psyche of the American republic, Whitman seeks in poems such as "The Centenarian's Story," "Pioneers! O Pioneers!," "Rise O Days from Your Fathomless Deeps," "Years of the Unperform'd," "Turn O Libertad," and "Lo! Victress on the Peaks" to explain the war to the American people. He represents the war as a trial of the Union and of the democratic masses and thus of the whole theory of America. Standing prophetically above the geography of his time, he seeks to place his blood-dripping pictures in a soothing and rational pattern of democratic advance:

> Years of the unperform'd! your horizon rises—I see it
>     parting away for more august dramas;
> I see not America only—I see not only Liberty's nation,
>     but other nations preparing;
> I see tremendous entrances and exits—I see new com-
>     binations—I see the solidarity of races;
> I see that force advancing with irresistible power on the
>     world's stage.
> > > ("Years of the Unperform'd," *DT*, p. 53)

Whitman mythologizes the Civil War as part of a cosmic drama presaging the birth of a new era of popular democracy and international brotherhood: "Are all nations communing? is there going to be but one heart to the globe?" he asks, "Is humanity forming, en-masse?" (*DT*, p. 54). Rather than question the "irresistible power" of democracy and thus admit the possible futility of the war's blood sacrifice, Whitman, like President Lincoln, committed himself to the creation and dissemination of an almost willfully legendary version of the Civil War as a triumphant battle to preserve democracy worldwide.

This prophetic note of democratic triumph is supported in *Drum-Taps* and *Sequel* by what Whitman called "an undertone of sweetest comradeship & human love, threading its steady thread inside the chaos, & heard at every lull & interstice thereof" (*Corr.*, I, 247). The intense bonds of compassion, comradeship, and love that Whitman witnessed and formed among the soldiers were a source of democratic sustenance amid the blood-drenched scenes of war. These loving bonds formed by men at war also gave Whitman a positive language and social form in which to experience and articulate his own homosexual desire.[28] In the poems of *Drum-Taps* the lover of the *Calamus* poems becomes the soldier-comrade and wound-dresser ("Many a soldier's kiss dwells on these bearded lips"), thus heightening the lyric intensity and emotional immediacy of several of the war poems (*DT*, p. 34).

Whitman's "undertone of sweetest comradeship & human love" is particularly strong in the elegy "Vigil Strange I Kept on the Field One Night," where in the starlight illuminating the darkened landscape of war, the poet-soldier buries his "dear comrade" and "son of responding kisses" in a private ritual of mourning and love. Modulating formal control with a tone of uttermost woe, Whitman's "strange" vigil suggests that it was the loving affection among men—released and allowed in a wartime context—that enabled him to rise from the "chill ground" of the battlefield and conduct his own burial of the dead in the poems of *Drum-Taps* and *Sequel*.

"*Bold, cautious, true, and my loving comrade,*" Whitman's common soldier is more than a soldier of war in *Drum-Taps*. He is a figure of democratic—and homosexual—humanity marching the "untried roads" of the future. "I know my words are weapons, full of danger, full of death," Whitman asserts in "As I Lay with My Head in Your Lap, Camerado," urging his readers to join him in the democratic struggle:

> For I confront peace, security, and all the settled laws, to
>     unsettle them;
> I am more resolute because all have denied me, than I could
>     ever have been had all accepted me;
> I heed not, and have never heeded, either experience, cau-
>     tions, majorities, nor ridicule;
> And the threat of what is call'd hell is little or nothing to
>     me;
> And the lure of what is call'd heaven is little or nothing
>     to me;

> . . . Dear camerado! I confess I have urged you onward
> with me, and still urge you, without the least idea
> what is our destination,
> Or whether we shall be victorious, or utterly quell'd and
> defeated.
>
> *(Sequel,* p. 19)

It was ironically in the fields and hospitals of the Civil War that Whitman came closest to realizing his democratic and homosexual dream of a "new City of Friends." Included among the poems of demobilization, "As I Lay" registers uneasiness as the poet moves away from the true democracy of wartime comradeship toward the potentially oppressive structures of a peacetime—and heterosexual—economy. Addressing a *you* who is, as in *Calamus,* both reader and lover, the poet expresses renewed dedication to a boundless democratic "destination" that will include the homosexual as a fully free and human being.

In stressing the ideological origins of the Civil War as a war to save the Union nationally and democracy globally, Whitman avoided the tone of moral indignation, evident in the poems of John Greenleaf Whittier and Oliver Wendell Holmes, who presented the war as a self-righteous struggle of Northern freemen against demonic Southern slave masters. But he also avoided the issue of slavery which was the root cause of the war. In the initial edition of *Drum-Taps* and *Sequel,* Whitman makes no allusion to the specific political issue of slavery. Slavery and the black person exist as a potent ellipsis, an unvoiced subject in the text of his war poems. Only in the 1867 poem "Ethiopia Saluting the Colors" does Whitman link the issue of slavery with the war; but it was not until 1881, in the final edition of *Leaves of Grass,* that he incorporated "Ethiopia" into the *Drum-Taps* grouping.

The unvoiced question of black emancipation is one of many questions about the consequences of the war that "convulsed" the democratic text of *Drum-Taps* and *Sequel* in 1865. The sense of reeling that Whitman expresses in "Year That Trembled and Reel'd Beneath Me," which was probably composed in 1863 after the Union defeats at Charleston and Chancellorsville, he experienced as a result of defeat in battle and the disintegrative forces unleashed by the war. Like the "solid, ironical, rolling orb" in the poem of the same title, the war brought to "practical, vulgar tests" all of Whitman's "ideal dreams" of democracy. If the war affirmed his faith in the democratic masses and at times appeared to body forth his dream of an ideal republic and a "new City of Friends," the war also aroused further "doubts nauseous" about the state of the Union and the nature of things.

The low sob and mournful wail prevails in the *Drum-Taps* poems, revealing a sense of irreparable national loss and damage. The poems return again and again to the politically resonant image of family fracture. "Dirge for Two Veterans" mourns the death of both father and son in a manner that suggests the end of the national line. "Come Up from the Fields Father," in which Whitman imagines the receipt of the kind of letter he himself wrote for the soldiers in the hospitals, expresses a similar sense of permanent familial fracture. Situated in the West, "amid all teeming and wealthy Ohio," the poem registers in the figure of the mother wanting only to die after the loss of her son in the war, the sense of depletion amid plenty brought by the war. In "Reconciliation" Whitman, like Lincoln, attempts to bind up the wounds of the nation by encouraging a spirit of reconciliation, but the reconciliation he envisions occurs not in life but in death, as "the hands of the sisters Death and Night, incessantly softly wash again, and ever again, this soil'd world" (*Sequel*, p. 23). As in Melville's "Magnanimity Baffled," in which the Northern conqueror reaches out to a Southern hand only to find it dead, so in "Reconciliation" the enemy that the poet bends down to touch with his lips is a corpse.

Presided over by the moon rather than the sun, the poems of *Drum-Taps* and *Sequel* present a darkened prospect of the nation. Despite Whitman's emphasis on himself in the role of wound-dresser and the war as a redemptive blood sacrifice, his war poems, particularly in the initial editions of *Drum-Taps* and *Sequel*, are full of perturbations and questions that reflect a diminished sense of self and world. In "Not My Enemies Ever Invade Me," which Whitman deleted in future editions of *Leaves of Grass*, he presents himself "grovelling on the ground," overmastered by his lovers. This poem is followed in *Drum-Taps* by another, "O Me! O Life" in which he questions the "endless trains of the faithless," the "cities fill'd with the foolish," the "poor results of all," the "plodding and sordid crowds," and the "empty and useless years of the rest." The answer he receives, like the response of the fierce old mother in "As I Ebb'd," is minimalist, revealing a world of diminished possibility:

> *Answer.*
>
> That you are here—that life exists, and identity;
> That the powerful play goes on, and you will contribute a
>     verse.
>
> (Sequel, p. 18)

The soldier of war and peace becomes in "Weave In, Weave In, My Hardy Life" a mere strand in a pattern of history that he neither knows nor understands:

> (We know not what the use, O life? nor know the aim,
>      the end—nor really aught we know;
> But know the work, the need goes on, and shall go
>      on—the death-envelop'd march of peace as well
>      as war, goes on;)
> For great campaigns of peace the same, the wiry
>      threads to weave;
> We know not why or what, yet weave, forever weave.
>
> (*DT*, p. 69)

Although Whitman presents the war as a mass action and part of an inexorable march toward the democratic future, in several of his poems the power and autonomy of the individual human agent appear to be lost as the masses become anonymous actors in a drama of history that they no longer produce or control. In "An Army on the March" ("An Army Corps on the March"), the "swarming ranks" rise and fall "to the undulations of the ground" as the "army resistless advances" (*Sequel*, p. 20). Anticipating the snakelike curling and writhing of Stephen Crane's troops in *The Red Badge of Courage*, the men are not in the foreground in this poem but are viewed from a distance, diminished by the landscape and the swarming masses that engulf them.

The individual is similarly swept along by the "resistless" force of history in "A March in the Ranks Hard-Prest, and the Road Unknown." Based on an account of the battle of White Oaks Church related to Whitman by a soldier in one of the hospital wards, the poem is framed by the image of an army marching in darkness along an unknown road. When the army comes to a clearing in the forest, the "we" of the army drops out of formation to become the "I" of the soldier-poet:

> We come to an open space in the woods, and halt by the
>      dim-lighted building;
> 'Tis a large old church, at the crossing roads—'tis now
>      an impromptu hospital;
> —Entering but for a minute, I see a sight beyond all
>      the pictures and poems ever made:
> Shadows of deepest, deepest black, just lit by moving
>      candles and lamps,
> And by one great pitchy torch, stationary, with wild red
>      flame, and clouds of smoke.
>
> (*DT*, p. 44)

The church-made-hospital, where pews become beds for wounded soldiers, the gleams of light amid "shadows of deepest, deepest black," and the hellish cast of flame and smoke all reflect an ambivalent response to the war as a site of redemption and a descent into hell. Before being drawn back into his march in the ranks, the soldier stops to minister to the wounds of a fellow soldier:

> At my feet more distinctly, a soldier, a mere lad, in
>     danger of bleeding to death, (he is shot in the ab-
>     domen;)
> I staunch the blood temporarily, (the youngster's face is
>     white as a lily;)
>
> *(DT,* p. 44)

The lily white of the youngster's face contrasts with the hellish glow of the scene, suggesting a Christ-like suffering and redemption. The soldier appears to be a countervailing figure, relieving the scene of "deepest, deepest black" by an act of compassion and love. But the gesture is only temporary, perhaps meaningless.

Beyond "all the pictures and poems ever made," the bloody "sight" of war that Whitman describes almost overwhelms him, making it impossible to write. His poet-soldier tries to "absorb" the scene, but the "postures beyond description" come out as a chaos of sight, sound, and smell:

> Faces, varieties, postures beyond description, most in
>     obscurity, some of them dead;
> Surgeons operating, attendants holding lights, the smell
>     of ether, the odor of blood;
> The crowd, O the crowd of the bloody forms of soldiers
>     —the yard outside also fill'd;
> Some on the bare ground, some on planks or stretchers,
>     some in the death-spasm sweating;
> An occasional scream or cry, the doctor's shouted orders
>     or calls;
> The glisten of the little steel instruments catching the
>     glint of the torches;
> These I resume as I chant—I see again the forms, I
>     smell the odor.
>
> *(DT,* pp. 44–45)

The presentation of the soldiers as a crowd of "bloody forms" and the insistent repetition of the indefinite pronoun *some* emphasize both the numbers and the anonymity of the war dead.

Unrelieved by any larger teleology that would give meaning and significance to the "bloody forms" of war, the soldier is swept back into the ranks marching in darkness along an unknown road:

> Then hear outside the orders given, *Fall in, my men,*
>   *Fall in;*
> But first I bend to the dying lad—his eyes open—a
>   half-smile gives he me;
> Then the eyes close, calmly close, and I speed forth to
>   the darkness,
> Resuming, marching, as ever in darkness marching, on
>   in the ranks,
> The unknown road still marching.
>
> (*DT*, p. 45)

The half-smile of the dying lad represents a sustaining gesture of comradeship, love, and human affirmation, shooting its light into the surrounding darkness as the soldier falls back into line and speeds onward into the night. The almost-hypnotic repetition of the word *marching* impels the final lines forward with the resistless motion and beat of an army corps on the march. As the controlling image of Whitman's Civil War poems, the figure of the march has a dual suggestiveness as both an army march and the march of humanity. But in the war poems, the march is no longer along the sun-drenched vistas of "Song of the Open Road." In the poems of *Drum-Taps* and *Sequel,* as in America in the post–Civil War period, the direction of the march is uncertain and the road is unknown.

# 9

## *Burying President Lincoln*

(Nor for you, for one, alone;
Blossoms and branches green to coffins all I bring:
For fresh as the morning—thus would I chant a song for
you, O sane and sacred death.

All over bouquets of roses,
O death! I cover you over with roses and early lilies;
But mostly and now the lilac that blooms the first,
Copious, I break, I break the sprigs from the bushes:
With loaded arms I come, pouring for you,
For you and the coffins all of you, O death.)
                    —WHITMAN, "When Lilacs Last in the Door-yard Bloom'd"

When General Robert E. Lee surrendered at Appomattox Courthouse on
April 9, 1865, the event seemed to confirm the universe of "form and
union and plan" that Whitman had celebrated in "Song of Myself." "And
could it really be, then?" he said of the Union victory, "Out of all the
affairs of this world of woe and failure and disorder, was there really come
the confirm'd, unerring sign of plan, like a shaft of pure light—of rightful
rule—of God?" (*PW*, II, 503). This apparent sign of an unerring plan was
once again thrown into question when only a few days later on April 14,
Abraham Lincoln was shot by John Wilkes Booth while attending a play at
Ford's Theater in Washington. The assassination negated the prospect of
"rightful rule," suggesting that "woe and failure and disorder" were indeed the only "confirm'd" plan of the world.

   In his second inaugural address, delivered a month before his death,
Lincoln had urged a spirit of national reconciliation: "With malice toward none; with charity for all; with firmness in the right, as God gives
us to see the right, let us strive on to finish the work we are in; to bind
up the nation's wounds; to care for him who shall have borne the battle,
and for his widow, and his orphan, —to do all which may achieve and

cherish a just, and a lasting peace, among ourselves, and with all nations."¹ Even before Lincoln's death, however, his conciliatory attitude
was challenged by radical Republicans who tried to pass legislation excluding from the business of reconstruction Southerners who had supported the Confederacy. The assassination heightened the animus
against the South and dealt a final blow to the cause of national reunifica-
tion. Moreover, as the first act of violence committed against an American president, the assassination—which was called a "black horror" by
Lincoln's attending physician—undermined the institutions of democracy and raised the specter of a reign of terror in which violence would
become the primary means of effecting political change.

In New York on April 15, Whitman described the day of Lincoln's
death as a black day, full of horror and gloom among the people:

> In the forenoon, the news had not more than been rec'd. All Broadway is
> black with mourning—the facades of the houses are festooned with black—
> great flags with wide & heavy fringes of dead black, give a pensive effect—
> towards noon the sky darkened & it began to rain. Drip, drip, & heavy moist
> black weather—the stores are all closed—the rain sent women from the
> street & black clothed men only remained—black clouds driving overhead—
> the horror, fever, uncertainty, alarm in the public. . . .
>   Lincoln's death—black, black, black—as you look toward the sky—long
> broad black like great serpents slowly undulating in every direction. (Glicks-
> berg, pp. 174–75)

This initial response to Lincoln's assassination as a sign of "black, black,
black" on the horizon of the nation informs Whitman's first poetic response
in "Hush'd Be the Camps To-day." Speaking from the communal point of
view of the soldiers, Whitman "celebrates" Lincoln's death as a release
from "life's stormy conflicts"—from "time's dark events,/Charging like
ceaseless clouds across the sky" (*DT*, p. 69). The poem was inserted into
*Drum-Taps*, which Whitman was having printed in New York at the time of
Lincoln's assassination. Recognizing the symbolic significance of Lincoln's
death in the total pattern of the war, however, Whitman decided at the last
minute to hold up the publication of *Drum-Taps* in order to incorporate a
sequel, which included his famous Lincoln elegies "When Lilacs Last in
the Door-yard Bloom'd" and "O Captain! My Captain!"²

In "When Lilacs Last in the Door-yard Bloom'd," which is featured
on the title page of *Sequel to Drum-Taps* and opens the volume, the poet
attempts to resolve for himself and the nation the panic-struck vision of
Lincoln's assassination as a black horror, a demonically charged darkness
blocking the nation's democratic vista. But while Lincoln's death is its

occasion, the poem operates on several different, continually overlapping levels. Whitman's attempt to reconcile himself to Lincoln's death becomes as well a reconciliation to death on a personal, national, and global plane. More specifically, Lincoln's death becomes an occasion for the poet as wound-dresser to grieve publicly in an effort to make poetic and historic sense out of the seeming waste of the Civil War.

Although Whitman employs some of the conventions of pastoral elegy, including the mourning of nature and pathetic fallacy, the funeral procession, the placing of flowers on the coffin, the juxtaposition of death with nature's renewal, and apotheosis, "Lilacs" is, like his other Civil War poems, cast in a local and democratic mold.[3] Refusing to name Lincoln as a subject in this or any other of his Lincoln poems, Whitman assumes a community of knowledge, a familiarity with the events surrounding Lincoln's death, that becomes part of his poem's unvoiced text. "Lilacs" is, on this level, a kind of civil ritual in which the poet remembers what the nation already knows.

By refusing to name Lincoln, to single him out, Whitman also emphasizes his representative status. Lincoln is embedded in the poem as a figure of the American people: The one is continually balanced with the many, the separate person with the en masse, as Whitman places Lincoln and the particularities of his death in a poetic pattern that is at once national and universal. Although the poem moves toward the apotheosis of Lincoln in the silver face of the western star, that apotheosis does not take place apart from the American people; the western star is in its largest configuration a star of the American Union. What Whitman presents, finally, is a politicopoetic myth to counter Booth's cry on the night of the assassination—*Sic Semper Tyrannis*—and the increasingly popular image of Lincoln as a dictatorial leader bent on abrogating rather than preserving basic American liberties.

In the opening lines of the poem, Whitman remembers a loss that appears to have occurred not a few months previously but in some timeless order of nature:

> When lilacs last in the door-yard bloom'd,
> And the great star early droop'd in the western sky in the
>     night,
> I mourn'd . . . and yet shall mourn with ever-returning
>     spring.
>
> O ever-returning spring! trinity sure to me you bring;
> Lilac blooming perennial, and drooping star in the west,
> And thought of him I love.
>
> *(Sequel,* p. 3)

Although Lincoln was shot on Good Friday and died the following day, Whitman avoids the obvious Lincoln–Christ symbolism, preferring instead the local symbolism of lilac and star, which were associated in his imagination with the time of Lincoln's death.[4] He gives the poem a religious suggestiveness by linking Lincoln's death with the resurrection of spring and by weaving throughout the poem trinities formed of senses and sounds, lines and images.

At the time that Lincoln was shot, the marketplace was flooded with poems calling for blood vengeance. By deciding to present Lincoln's death as the culminating sacrifice of the tragedy of the Civil War—an occasion for national mourning and national unity—Whitman chose another way, a way that Lincoln himself might have chosen. He mourns his death in the same redemptive terms that informed Lincoln's political speeches during the Civil War. Nevertheless, his failure to refer to the actual circumstances of Lincoln's death—the fact that his death was not an act of nature but an unnatural act of violence against the state—remains a significant silence in the poem. This silence, with its unvoiced horror of rupture, blood, and democratic anarchy, exerts a subtle pressure in the text, undermining Whitman's own act of poetic and political resolution.

This pressure is evident in the frenzied plaint of section 2:

> O powerful, western, fallen star!
> O shades of night! O moody, tearful night!
> O great star disappear'd! O the black murk that hides the
> star!
> O cruel hands that hold me powerless! O helpless soul of
> me!
> O harsh surrounding cloud that will not free my soul!
> *(Sequel, p. 3)*

The despair the poet expresses in this passage is more than a response to Lincoln's death. The mournful O's of grief and nothingness, the frantic exclamations, the "black murk" and "shades of night," the "cruel hands" and "harsh surrounding clouds" gather into a single passage the "doubts nauseous" and "black horror" of an entire decade. Just as during the war years the poet reeled as the Union reeled, so in the postwar period the self is left helpless and impotent by a "powerful, western, fallen star" associated with Lincoln's death and with the fallen star of America itself.

The first four sections of the poem serve as a kind of overture, introducing the major motifs of lilac, star, and bird that are played and replayed sonatalike throughout the poem. These sections operate symbolistically, requiring the reader to make the imaginative links between figures that are introduced abruptly with no logical line of development. Through-out the poem, the reader is summoned to participate in the ritual of unification and reconciliation that is its overall design.

The symbolistic evocation of the order of nature in the overture is linked with the order of history in sections 5 and 6, in which a narrative sequence plots the circuitous route of a coffin journeying through the land:

> Over the breast of the spring, the land, amid cities,
> Amid lanes, and through old woods, (where lately the
>     violets, peep'd from the ground, spotting the gray
>     debris;)
> Amid the grass in the fields each side of the lanes—passing
>     the endless grass;
> Passing the yellow-spear'd wheat, every grain from its
>     shroud in the dark-brown fields uprising;
> Passing the apple-tree blows of white and pink in the
>     orchards;
> Carrying a corpse to where it shall rest in the grave,
> Night and day journeys a coffin.
>
> *(Sequel,* p. 4)

The journey alludes to the actual funeral procession that carried Lincoln's body from Washington to Springfield, Illinois, where he was buried on May 4, 1865. But by envisioning a journey through old woods where "lately" violets spotted gray debris (suggesting the blue and gray of the Union and Confederate soldiers) and through fields where wheat rises out of the shroud of the land, Whitman imaginatively encompasses the corpses of all the war dead. The coffin that "journeys" in the present tense memorializes the circumstances of Lincoln's death in a perpetual present that links that death with the coffins of the war dead and all the dead that journey end-lessly through time. The possibility of consolation is embedded in the very grammar of the passage: Just as the figure of the journeying coffin appears at the end of a long periodic sentence and an elaborate sequence of preposi-tional and participial phrases, so the particularity of the single death is contained in a larger regenerative cycle of life and death.

The journeying coffin assumes in section 6 a more specifically na-tional reference as the poet envisions the states, personified as crape-veiled women, united in mourning Lincoln's death:

> Coffin that passes through lanes and streets,
> Through day and night, with the great cloud darkening the
>     land,

> With the pomp of the inloop'd flags, with the cities draped
>    in black,
> With the show of the States themselves, as of crape-veil'd
>    women, standing,
> With processions long and winding, and the flambeaus of
>    the night,
> With the countless torches lit—with the silent sea of faces,
>    and the unbared heads,
> With the waiting depot, the arriving coffin, and the sombre
>    faces,
> With dirges through the night, with the thousand voices
>    rising strong and solemn.
>
> (*Sequel*, p. 5)

The journeying coffin is once again both singular and representative,
marking the path of Lincoln's coffin and the paths of all the war dead that
journeyed in coffins throughout the states to depots full of mourning
faces. In an act that expresses the collective grief of the people, the poet
does what he is expected to do. He places a sprig of lilac on the coffin as a
sign presumably of perpetual renewal and the unity of life: "Here! coffin
that slowly passes, I give you my sprig of lilac."

In the poem's next sequence, Whitman articulates parenthetically the
communal nature of his action:

> (Nor for you, for one, alone;
> Blossoms and branches green to coffins all I bring:
> For fresh as the morning—thus would I chant a song for
>    you, O sane and sacred death.
>
> All over bouquets of roses,
> O death! I cover you over with roses and early lilies;
> But mostly and now the lilac that blooms the first,
> Copious, I break, I break the sprigs from the bushes:
> With loaded arms I come, pouring for you,
> For you and the coffins all of you, O death.)
>
> (*Sequel*, p. 5)

Here for the first time the perennial bloom of the lilacs is associated with
the poet's act of creation. His desire to deck the coffins of all with chants
of "sane and sacred death" is a half-note that will become a full note in
the bird's "song of death." But in this passage the desire to sing of "sane
and sacred death" is at odds with the frenzied action of the subsequent
stanza: As the poet unloads arms full of flowers on the coffins of death, he
seems impelled by an urge not only to *cover over* but to *cover up* death. If
Whitman's flowers are blossoms of renewal and of art, then this action

suggests what has hitherto been unvoiced in the poem: that the elaborate artistry of "Lilacs" is an attempt to "cover over" not only death but the horror of a fratricidal war that culminated in blood violence against the president and the institutions of democracy itself.

Although the poet "would" chant a song for "sane and sacred death," he is still held by the particularity of his grief. In section 8 he plunges even further into the past, remembering the despair that he felt when he watched the western star sink into the night:

> O western orb, sailing the heaven!
> Now I know what you must have meant, as a month since
>     we walk'd,
> . . . . . . .
> As I saw you had something to tell, as you bent to me night
>     after night,
> As you droop'd from the sky low down, as if to my side,
>     (while the other stars all look'd on;)
> . . . . . . .
> As I watched where you pass'd and was lost in the nether-
>     ward black of the night,
> As my soul, in its trouble, dissatified, sank, as where you,
>     sad orb,
> Concluded, dropt in the night, and was gone.
>
>                      (*Sequel*, p. 6)

Here the western star, Venus, that hovered portentously over the nation at the time of Lincoln's second inauguration in 1864 is transformed into a multivalent symbol, associated with Lincoln's death, the poet's plunge into despair, and the western orb of America sinking into the black of night.

The movement toward consolation begins at midpoint in the poem as the poet, seeking an aesthetic form in which to contain his sorrow, reflects on the process of composing a democratic elegy:

> O what shall I hang on the chamber walls?
> And what shall the pictures be that I hang on the walls,
> To adorn the burial-house of him I love?
>
>                      (*Sequel*, p. 7)

The lines suggest once again the pictorial analogue of Whitman's war poetry. While he alludes to the ancient Egyptian custom of decorating the tombs of the pharaohs with hieroglyphics, the pictures he hangs in the burial house of his verse are democratic. He draws idyllic pictures of an essentially agrarian American republic: "Pictures of growing spring, and farms, and homes . . . and the city at hand . . . and the workshops, and the workmen homeward returning." His pictures fuse North, South, East, and West in a national vision of promise and abundance:

> Mighty Manhattan, with spires, and the sparkling and hur-
> rying tides, and the ships;
> The varied and ample land—the South and the North in
> the light—Ohio's Shores, and flashing Missouri,
> And ever the far-spreading prairies, cover'd with grass and
> corn.
>
> (*Sequel*, p. 7)

But these scenes of an idyllic republic achieving its fullest realization in the fertile prairies—the grass and corn of the West—conflict with the twilight in which the scenes are cast. The republic of farms and homes, cities and workshops is presented at sundown "with floods of the yellow gold of the gorgeous, indolent, sinking sun, burning, expanding the air"; and the poet's vista of America from "Mighty-Manhattan" to the "far-spreading prairies" moves toward night. Like the falling western star, this America of a sinking rather than a rising sun lends particular significance to the burial house as poem in which the poet hangs his democratic pictures. For his act of picture hanging wavers between commemoration and entombment: It is unclear whether he is memorializing the death of the president with pictures of the republic that he preserved or burying an American republic that was, along with Lincoln and the lives of 600,000 soldiers, the major casualty of the war.

Still held by his grief, the poet advances in section 14 toward the "sacred" knowledge that will enable him to move from sorrow to consolation, dramatized in the poem as a move away from the star and lilac toward the song of the hermit thrush. This "sacred" knowledge comes as the poet sits once again in a twilight landscape—"in the close of day"— observing the ongoing life of a generally prosperous world. But in the midst of this prosperity, a dark shadow falls on the land and the people:

> Falling among them all, and upon them all, enveloping me
> with the rest,
> Appear'd the cloud, appear'd the long black trail;
> And I knew Death, its thought, and the sacred knowledge
> of death.
>
> (*Sequel*, p. 8)

The causal link between the vision of material prosperity and the thought and knowledge of death is unclear, but the passage suggests that the thought of death in the midst of the ongoing activity of the world leads the poet to the "sacred" knowledge of death as part of a regenerative cycle.

This knowledge allows the poet to move toward accepting both the particularity of Lincoln's death and the death of everybody.

In a startling democratic transformation of the trinity, Whitman represents this movement toward a philosophical acceptance of death in the image of the poet walking with the knowledge of death (process) on one side of him and the thought of death (loss) on the other. Girded by these figures, he turns away from the star and lilac toward the consoling song of the hermit thrush:

> Then with the knowledge of death as walking one side of
>     me,
> And the thought of death close-walking the other side of me,
> And I in the middle, as with companions, and as holding the
>     hands of companions,
> I fled forth to the hiding receiving night, that talks not,
> Down to the shores of the water, the path by the swamp in
>     the dimness,
> To the solemn shadowy cedars, and ghostly pines so still.
>
> And the singer so shy to the rest receiv'd me;
> The gray-brown bird I know, receiv'd us comrades three;
> And he sang what seem'd the song of death, a verse for
>     him I love.
>
>                                 *(Sequel,* p. 9)

The repetition of *s* sounds gives the passage a sibilance that suggests the almost-religious hush of the setting and the soothing that the poet receives as his spirit tallies the song of the bird.

The song of the hermit thrush, which offers the possibility of consolation in the poem, is not, finally, a song of the regenerative flow of life and death. Rather, death is celebrated as a release from the fallen world and the "fathomless universe" evoked in the poem:

> Come, lovely and soothing Death,
> Undulate round the world, serenely arriving, arriving,
> In the day, in the night, to all, to each,
> Sooner or later, delicate Death.
>
> Prais'd be the fathomless universe,
> For life and joy, and for objects and knowledge curious;
> And for love, sweet love—But praise! O praise and praise,
> For the sure-enwinding arms of cool-enfolding Death.
>
>                                 *(Sequel,* p. 9)

Fusing sound and sense, the bird's song mimics in the systematic repetition of open vowel sounds the enfolding and encompassing action of death. In the bird's carol, the "Dark Mother" death replaces the mother republic as Whitman's sustaining myth. The "husky whispering wave" of death no longer threatens the poet as it did in his prewar lyrics; death is seductive rather than fearful, lovingly invited rather than evaded. Sung away from the settlements in "secluded recesses," the bird's song marks the poet's flight away from the desolate post–Civil War American landscape toward death as a source of release from the dark events of time and history.

What the bird's song expresses is the consoling vision of death that Whitman learned during the war, when death became a "strong Deliveress" for the wounded and suffering soldiers. In the context of the poem, it is only after the poet has received this consoling knowledge that he, and along with him the reader, is able to gaze with "unclosed" eyes on the untold horror of the war. Even more than Lincoln's death, this horror is perhaps the real subject of Whitman's elegy. As his spirit continues to tally the bird's consoling song, the poet plunges even deeper into his own and the nation's memory, dredging up scenes of the war years: "My sight that was bound in my eyes unclosed/As to long panoramas of visions." Unlike the "loud" song of the bird that spreads over the night "that talks not," the poet's war "visions" are "noiseless dreams," suggesting a nightmare reality that cannot finally be articulated in the poem's grammar of consolation.

The poet sees first a kind of mimed enactment of the wounding and maiming of the American republic:

> I saw the vision of armies;
> And I saw, as in noiseless dreams, hundreds of battle-flags;
> Borne through the smoke of the battles, and pierc'd with
>     missiles, I saw them,
> And carried hither and yon through the smoke, and torn
>     and bloody;
> And at last but a few shreds of the flags left on the staffs,
>     (and all in silence,)
> And the staffs all splinter'd and broken.
>
> (*Sequel*, p. 11)

As emblems of the nation, the battle flags dramatize the devastating effects of the war on the American republic, which was, like the flags, "pierc'd with missiles" and ultimately left "torn and bloody" on staffs "all splinter'd and broken."

Repeating the term "I saw" as a rhythmic base that insists on the act of looking at the war's reality, the poet finally "uncloses" his vision to the masses of battle corpses left by the war:

> I saw battle-corpses, myriads of them,
> And the white skeletons of young men—I saw them;
> I saw the debris and debris of all dead soldiers;
> But I saw they were not as was thought;
> They themselves were fully at rest—they suffer'd not;
> The living remain'd and suffer'd—the mother suffer'd,
> And the wife and the child, and the musing comrade suf-
>   fer'd,
> And the armies that remain'd suffer'd.
>
> *(Sequel,* p. 11)

The myriads of white skeletons and the debris of dead soldiers suggest both the massive waste of the war and its ultimate meaninglessness. Like the "debris" thrown up by the sea in "As I Ebb'd," the "debris and debris of all dead soldiers" exists apart from any rational order of history; their death is, at best, merely an escape from the dis-order of the world. The ultimate wound of the war, Whitman suggests, was delivered not to the war dead but to the living body of the nation. The dead are released from pain; it is the living who continue to suffer.

Insofar as Whitman's consoling vision of death represents a coming to terms with Lincoln's death and the total sacrifice of the war, the poem could have ended here. But he must have recognized that in his vision of the war he had reached a nadir of darkness and negation that needed to be reversed, or at least balanced, by a turn backward toward life. In the poem's final movement, he envisions himself "passing" beyond the scenes of war and death and beyond the death carol of the hermit thrush, as he seeks to propel himself and the poem outward toward life and the future:

> Passing the visions, passing the night;
> Passing, unloosing the hold of my comrades' hands;
> Passing the song of the hermit bird, and the tallying song
>   of my soul.
>
> *(Sequel,* p. 11)

In the poem's final version, Whitman presents this movement away from the land of the dead in a sequence of declarative sentences: "I leave thee lilac with heart-shaped leaves. . . . I cease from my song for thee." But in the poem's initial version, his movement is more tentative, presented in the form of a question rather than a declaration:

> Must I leave thee, lilac with heart-shaped leaves?
> Must I leave thee there in the door-yard, blooming return-
>   ing with spring?

> Must I pass from my song for thee;
> From my gaze on thee in the west, fronting the west, com-
>      muning with thee,
> O comrade lustrous, with silver face in the night?
>
> *(Sequel,* p. 12)

The poet seems reluctant to leave his thoughts of death, associated here with Lincoln and the West. But as in other war poems, he also seems reluctant to leave the formal ordering of the poem itself.

Whitman overcomes his reluctance by realizing that in leaving his thoughts of Lincoln and the war dead in a song, he will also *leave* them in his own and the national memory: "Yet each I keep and all," he says in the concluding lines of the poem:

> Comrades mine, and I in the midst, and their memory ever
>      I keep—for the dead I loved so well;
> For the sweetest, wisest soul of all my days and lands . . .
>      and this for his dear sake;
> Lilac and star and bird, twined with the chant of my soul,
> With the holders holding my hand, nearing the call of the
>      bird,
> There in the fragrant pines, and the cedars dusk and dim.
>
> *(Sequel,* p. 12)

The final consolation offered to self and nation comes not from the song of the bird but from the poet's song as an act of life and an act of creation that will memorialize Lincoln's death and the total sacrifice of the war.

And yet the sight of the torn and bloody republic and the "myriads" of corpses left by the war resists Whitman's act of artistic consolation. His elaborately orchestrated resolution seems at best an effort to "cover over" the personal and national loss that are at the very source of his elegy. Moreover, although he offers "Lilacs" as a sign of life and renewal, in the poem's final lines he does not in his usual manner turn toward the reader and the future. Rather, he retreats back into the poem toward the landscape of death, a movement that is particularly evident in the poem's initial version in which Whitman presents himself "nearing the call of the bird." His move back into the poem toward the "cedars dusk and dim" represents not only a return to the self-protective ordering of art; his move backward toward the bird's carol of death also anticipates the turn in Whitman's postwar lyrics away from the "present and the real" toward an increasing emphasis on death, night, and the spiritual world.[5]

The formal regularity of "O Captain! My Captain!", which comes after "Lilacs" in *Sequel to Drum-Taps* and other groupings of the Lincoln poems, is a further sign of the artistic control that Whitman had to exercise in order to "cover over" the sense of "horror, fever, uncertainty, alarm in the public" aroused by Lincoln's assassination. Speaking in the voice of the civil servant, the poet refuses really to engage the feelings unleashed by Lincoln's violent death. Rather than the personal and local symbolism of lilac, star, and bird, he uses the more conventional image of the ship of state:

> O Captain! my captain! our fearful trip is done;
> The ship has weather'd every rack, the prize we sought is
>     won;
> The port is near, the bells I hear, the people all exulting,
> While follow eyes the steady keel, the vessel grim and daring:
>     But O heart! heart! heart!
>         Leave you not the little spot,
>             Where on the deck my captain lies,
>             Fallen cold and dead.
>
> *(Sequel,* p. 13)

Through the rigid deployment of rhyme, meter, refrain, and regularly patterned stanzas, Whitman keeps Lincoln's death distant, contained, and safe, as he memorializes the president in his more public and legendary dimension as the martyr of the cause of national union.

As it turned out, the American reading public liked Whitman best when he was being most traditional. "If Walt Whitman had written a volume of My Captains," wrote a contemporary reviewer, "instead of filling a scrapbasket with waste and calling it a book the world would be better off today and Walt Whitman would have some excuse for living." Recited by schoolchildren across the land, "O Captain! My Captain!" is both Whitman's most conventional poem and the only one to reach the masses of people he envisioned as the audience for his poems. The irony was not lost on Whitman: "I'm honest when I say, damn My Captain and all the My Captains in my book! . . . I'm almost sorry I ever wrote that poem. . . . I say that if I'd written a whole volume of My Captains I'd deserve to be spanked and sent to bed with the world's compliments— which would be generous treatment, considering what a lame duck book such a book would have been!" (*WWC*, II, 304).

In the poems of what he would later call *Memories of President Lincoln,* as in his annual lecture on "The Death of Lincoln," Whitman tried to diffuse the "black, black, black" of Lincoln's assassination through aesthetic trans- figuration. Lincoln's murder was, he said in his lecture, the culminating act of "lightning-illumination" through which "a long and varied series of contradictory events arrives at last at its highest poetic, single, central, pictorial denouement." Representing Lincoln's assassination as the final act of the tragedy of the Civil War, Whitman invests his death and the war itself with the value and meaning of an artistic performance. Like his "Lilacs" elegy, his Lincoln lecture is impelled by the desire to make the meaningless seem necessary and natural as the fact of Lincoln's assassination is refashioned into a myth of national regeneration. Whitman articulates even as he shapes a saving national vision of Lincoln's death as a redemptive blood sacrifice that saved the Union and delivered the republic from the internal contradiction of slavery. His death was "that seal of the emancipation of three million slaves—that parturition and delivery of our at last really free Republic, born again, henceforth to commence its career of genuine homogeneous Union, compact, consistent with itself" (*PW*, II, 508).

Whitman discovered in Abraham Lincoln a flesh-and-blood figure whose life and death appeared to confirm the historic order of democracy. As such, he was "Dear to the Muse—thrice dear to Nationality—to the whole human race—precious to this Union—precious to Democracy— unspeakably and forever precious—their first great Martyr Chief" (*PW*, II, 509). (Re)presenting the circumstances and significance of Lincoln's death in a lecture he delivered annually between 1879 and 1890 was Whitman's manner of *saying over* the "contradictory events" of his time that history was an action within rather than a fact without the democratic design of his own artistic creation.

# 10

## *"Who Bridle Leviathan?"*

> Pride, competition, segregation, vicious wilfulness, and license
> beyond example, brood already upon us. Unwieldy and im-
> mense, who shall hold in behemoth? who bridle leviathan?
> Flaunt it as we choose, athwart and over the roads of progress
> loom huge uncertainty, and dreadful, threatening gloom.
>
> —WHITMAN, *Democratic Vistas*

"Washington is filled with *darkies*," Whitman wrote his mother in the
spring of 1867. "The men & children & wenches swarm in all directions—(I am not sure but the North is like the man that won the
elephant in a raffle)" (*Corr.*, I, 323). Although he hailed the "absolute
extirpation and erasure of slavery from the States" as "the greatest
revolutionary step in the history of the United States, (perhaps the
greatest of the world, our century)," in the postwar period he no longer
dreamed the egalitarian dream of "I Sing the Body Electric" (*PW*, II,
509). Like Lincoln and the majority of the American people, he was not
fully prepared to integrate the black person into his vision of a free and
equal America. With the exception of the "hardly human" black woman
in "Ethiopia Saluting the Colors," black people are absent from his
poetry of the postwar years, and in his letters and journals of the time,
blacks remain on the periphery of his vision as sources of dread and
emblems of retribution.

Whitman supported the principle of universal suffrage—"the widest
possible opening of the doors"—but he subscribed to the popular prejudice against "Negro rule" (*PW*, II, 531). When several thousand black
persons turned out in June 1868 to celebrate the election of Samuel B.
Bowen as mayor of Washington, the procession seemed to him the very
image of bedlam and riot: "The men were all armed with clubs and
pistols," he wrote to his mother. "Besides the procession in the street,
there was a string went along the side-walk in single file with bludgeons &

sticks, yelling & gesticulating like madmen—it was quite comical, yet very disgusting & alarming in some respects—They were very insolent, & altogether it was a strange sight—they looked like so many wild brutes let loose" (*Corr.*, II, 35) In Whitman's view, the "black domination, but little above the beasts" in South Carolina, Mississippi, and Louisiana was not a permanent condition but "a temporary, deserv'd punishment for their Slavery and Secession sins" (*Memoranda*, p. 66).

As Melville said of slavery in his supplement to *Battle Pieces*, emancipation had "ridded the country of the reproach, but not wholly of the calamity," and he predicted another Civil War that would engage the "hatred of race toward race."[1] If the bloodbath of the Civil War had resolved the paradox of slavery in the land of the free, it had not solved the problem of black freedom and black civil rights. In Whitman's struggle to reconcile his culture-bound prejudice against blacks with the ideal of democratic equality, he, like Melville, anticipated the racial struggle that would mark the American future. "Did the vast mass of the blacks, in Slavery in the United States, present a terrible and deeply complicated problem through the just ending century?" he asked in a note to his *Memoranda During the War*. "But how if the mass of the blacks in freedom in the U.S. all through the ensuing century, should present a yet more terrible and deeply complicated problem?" (*Memoranda*, p. 66).

"Ethiopia Saluting the Colors," which was composed in 1867 under the title "Ethiopia Commenting," is Whitman's attempt to come to terms publicly with the meaning of black liberation in America. The poet presents himself as a soldier observing a black woman who stands by the roadside while General Sherman's army passes by on its famous march from Atlanta to the sea. Using a highly conventional form, with regular stanzaic structure, a seven-stress syllabic line, and an elaborate scheme of internal and terminal rhyme, alliteration, and clearly marked caesuras, Whitman keeps the black woman safely at a distance. Even the poem's title, with its incongruent image of Ethiopia saluting the American flag, reveals the failure of America and the national literary imagination to assimilate the black race. The black woman appears as a kind of racial mother, an allegorical figure of black America, but as Ethiopia, she is the figure of *another* country, an exotic alien foreign to the native colors of America.

The poem begins with a series of questions that indicate uncertainty about Ethiopia's identity and the meaning of her rising to greet the colors of America:

> Who are you dusky woman, so ancient hardly human,
> With your woolly-white and turban'd head, and bare bony feet?

Why rising by the roadside here, do you the colors greet?
                                                      (*LGC*, p. 318)

Reflecting popular nineteenth-century notions of blacks as less far along on the evolutionary scale, Whitman suggests that their "rising" will be brought about by events transpiring in America. Stepping forth from her hovel door, Ethiopia singles out the soldier-poet, to whom she explains her presence in America:

> *Me master years a hundred since from my parents sunder'd,*
> *A little child, they caught me as the savage beast is caught,*
> *Then hither me across the sea the cruel slaver brought.*
>                                                      (*LGC*, p. 318)

In this rather crude attempt to imitate black dialect, Whitman once again registers his protest against the inhumanity and barbarism of the slave trade. As in Phillis Wheatley's "On Being Brought from Africa to America" (1773), he tries to place the violations of slavery into a redemptive and democratic pattern of history. But for all Whitman's effort to make imaginative sense of the slave's presence in America, the poem perpetuates the politics of racial violation by representing Ethiopia as a "hardly human" figure, wagging her turbaned head, rolling her "darkling eye" and curtsying to "doughty" Sherman's regiment on its "liberating" march to the sea.

Rather than answering the question of black identity and black freedom raised by the poem, Whitman ends with another series of questions:

> What is it fateful woman, so blear, hardly human?
> Why wag your head with turban bound, yellow, red and green?
> Are the things so strange and marvelous you see or have seen?
>                                                      (*LGC*, p. 319)

Although the poet's questions express wonder at the "strange and marvelous" social revolution caused by the war, they also reveal his anxiety about the effect of black liberation on the future of America. The black person's "hardly human" figure is "blear"—something indistinct, dim, and out of focus in Whitman's vision of America. She is a "fateful woman" whose liberation and rising are part of the world's fated march toward the democratic future, but in her fatefulness, she is also a figure of fate, a possibly ominous presence who will affect the country's destiny in uncertain ways.

The questions about the consequences of the war that frame "Ethiopia Saluting the Colors" are questions about the future of democracy that Whitman continued to ask in the postwar period. Along with the specter

of racial war, the Civil War had unloosed a horde of psychic and socioeconomic demons that continued to haunt the national dream of democracy.
Even before the war, that dream seemed tainted by the dominant business
interests of the time. To Whitman the war represented a spark of democratic idealism in the midst of the vulgar commercialism of the time:
"Amid the whole sordidness—the entire devotion of America, at any
price, to pecuniary success, merchandise—disregarding all but business
and profit—this war for a bare idea and abstraction—a mere, at bottom,
heroic dream and reminiscence—burst forth in its great devouring flame
and conflagration" (*PW*, II, 706).

Ultimately, however, the conflagration of the war served not to purge
but to strengthen the country's business interests. In a diary entry for
1863, Whitman noted the general prosperity of the North: "It looks anything else but war—everybody well drest, plenty of money, markets boundless & of the best, factories all busy—" (Glicksberg, p. 139). He was also
conscious of a certain transposition of values during the war years, when
infidelity seemed to prosper and the true heroes were shot down. In his
war diary he described the case of William Grove, who, after serving
loyally for two years, was shot for desertion "amid all this show of general
stars, & the bars of the captains & lieutenants—amid all the wind &
puffing and infidelity—amid the swarms of contractors and their endless
contracts, & the paper money." The ironic inversion of values penetrated
Whitman's public show of confidence, arousing the doubts of "Respondez": "O the horrid contrast & the sarcasm of this life—to know who
they really are that sit on judges benches, & who they perched on the
criminal's box—to know" (Glicksberg, p. 127).

By the end of the war, republican idealism had itself taken a strong
economic and material turn as the concepts of liberty, equality, individualism, and laissez-faire were put in the service of one of the most aggressively capitalist economies in history. Even Whitman was led to doubt the
shibboleth of democratic individualism divorced from any sense of communal responsibility: "About this business of Democracy & human rights
&c, often comes the query—as one sees the shallowness and miserable
selfism of these crowds of men, with all their minds so blank of high
humanity and aspiration—then comes the terrible query, and will not be
denied, *Is not Democracy of human rights humbug after all*—Are these flip-
pant people with hearts of rags and souls of chalk, are these worth preaching for & dying for upon the cross? May be not—may be it is indeed a
dream—" (italics added).[2]

The conditions of material abundance by which America had from the first defined itself became in the postwar period a vehicle of potential destruction. Whitman engaged this paradox of destructive abundance in "A Carol of Harvest, for 1867" ("The Return of the Heroes"), which was published in the New York *Galaxy* in September 1867. Equating the Union as Mother with a boundlessly fertile land, he celebrates the return to a peacetime order of fecund abundance. But the poet's vision of plenty is ambiguous: America groans under the weight of her "great possessions," entrapped and choked by the material wealth that is the source of her power:

> Fecund America—to-day,
> Thou art all over set in births and joys!
> Thou groan'st with riches, thy wealth clothes thee as a swathing-
>     garment,
> Thou laughest loud with the ache of great possessions,
> A myriad-twining life like interlacing vines binds all thy vast
>     demesne,
> .  .  .  .  .  .  .  .  .
> Thou, bathed, choked, swimming in plenty.
>
>                                                   (*LGC*, p. 359)

At the end of the war, despite the death of 600,000 men and the handshake at Appomattox, the Union remained split. Indeed, rather than preserving the Union, the war had sowed a deep sectional antagonism that would become an enduring tension not only in the political life of the nation, but more subterraneously in its artistic and cultural production. In the period immediately following the war, the antagonism between North and South was rehearsed in the conflict between the president and Congress. Like Lincoln, Andrew Johnson wanted to restore the Southern states to the Union as quickly as possible, but guided by the philosophy of Jefferson and Jackson, he wanted to leave to the states the question of reconstruction and black civil and political rights. But the radical republicans opposed both his leniency and his attempt to return to the old state-centered Union.[3]

As a clerk in the attorney general's office, Whitman participated directly in the process of reconstruction during the postwar period. Positioned amid the streams of Southerners who flowed into the office seeking special pardons, he was occasionally commissioned to respond for the president. "The President," he noted in September 1865, "has fix'd his mind on a very generous and forgiving course toward the return'd secessionists" (*PW*, II, 611).

As the conflict between the president and Congress intensified, Whitman attended the debates in Congress, in which by the summer of 1866 the radicals had gained control. "The Radicals have passed their principal measures over the President's vetos," he wrote to his mother in March 1867, referring to the passage that year of the Reconstruction Act, the Army Appropriations Act, and the Tenure of Office Act (*Corr.*, I, 316). The struggle between the executive and legislative branches culminated in 1868 with the move to impeach President Johnson for "high crimes and misdemeanors in office." This move was, in Whitman's view, a "shaky," "doubtful business," but had it succeeded it would have institutionalized impeachment as a political rather than a judicial proceeding, thus destroying the constitutional system of checks and balances.

Whitman followed the debates over Reconstruction and the growing crisis of power between president and Congress, and on the surface at least, he appeared confident that balance would eventually be restored. Only a few days before the impeachment proceedings began, he wrote to his English admirer Moncure Conway: "Our American politics, as you notice, are in an unusually effervescent condition—with perhaps (to the mere eye-observations from a distance) diverse alarming & deadly portending shows & signals. Yet we old stagers take things very coolly, & count on coming out all right in due time." Like many in the country, Whitman hoped for a retreat from the more radical premises of Republican reconstruction and a restoration of balance through the election of Ulysses S. Grant: "The Republicans have exploited the negro too intensely, & there comes a reaction. But that is going to be provided for. According to present appearances the good, worthy, non-demonstrative, average-representing Grant will be chosen President next fall" (*Corr.*, II, 15).

Of the alarming "shows & signals" in American politics in the Reconstruction period, perhaps the most "deadly portending" to the republic was the emergence, under the pressures of a wartime economy, of a leviathan state. In his *Battle-Pieces* poem "Conflict of Convictions," Melville had foreseen the irony that in using national power to quell a local rebellion, the ultimate victim of the Civil War might be the "Founders' dream." Out of the Union victory would emerge a new power state, symbolized by the iron dome that replaced the original wooden dome of the Capitol during the Civil War:

> Power unanointed may come—
> Dominion (unsought by the free)
>   And the Iron Dome,
> Stronger for stress and strain,
> Fling her huge shadow athwart the main;
> But the Founders' dream shall flee.[4]

During the war years, "power unanointed" had come. Whitman might well celebrate the disbanding of the army in his "Carol of Harvest, for 1867," for under the presidencies of both Lincoln and Johnson there was widespread fear of a military coup. "I have little hope of any man or any community of men, that looks to some civil or military power to defend its vital rights," Whitman wrote in his lecture notes on Democracy (*WWW*, p. 58). But in the post–Civil War period, the nation was becoming increasingly dependent on civil and military power to enforce what were deemed "vital rights." In fact, under the terms of the Reconstruction Act of 1867, ten unreconstructed southern states were placed under military rule.

The gravitation of power from the periphery to the center during the Civil War represented a radical shift from the state-centered union of the past: American was no longer a union of states but a national union. Without the opposition of the South during the war years, government power was exercised increasingly in the interests of the urban and industrial North; the economic legislation of the period—a protective tariff, a national banking system, railroad subsidies, and the release of national resources—reflected the new nationalism and the entrepreneurial interests of an industrial capitalist class. Under the "stress and strain" of the war the "Founders' dream" had fled in the shadow of what Noam Chomsky later called "military-based state capitalism."[5]

The consolidation of political and economic power in the national government threatened to bring a standardization of American life that to Whitman, among others, was another of the deadly "signals" of the postwar years. In fact, he himself felt the moral impact of the new power state when he became one of the targets of Secretary of the Interior James Harlan's attempt to purge the Department of the Interior in the name of the "rules of decorum & propriety prescribed by a Christian Civilization."[6] When on June 30, 1865, Whitman was relieved of his duties as clerk in the Bureau of Indian Affairs, he may well have been the first "literary" victim of the new genteel morality that, as Henry Adams suggested in *Democracy* (1880), all too frequently masked crass motivation and unbridled acquisitiveness.

The iconography that emerged in the wake of William Douglas O'Connor's ardent defense of Whitman in *The Good Gray Poet* (1866) has blurred the extent to which Whitman as both social critic and poet placed himself at the center of the political debates over Reconstruction in the postwar period. In his writings, Whitman struggled with what he called the "behemoth" of racial, sectional, socioeconomic, and psychic tensions in post–Civil War America. His first major attempt to address the problem of democracy in the period immediately following the Civil War came in response to Thomas Carlyle's 1867 critique of democracy in "Shooting Niagara: And After?"[7] Responding to Benjamin Disraeli's proposed reform bill of 1867, which would extend the vote to the working class, Carlyle characterized the bill as a devil-inspired leap into the Niagara of "completed Democracy" brought about by the "swarmery" of the masses.

Although England was the prime object of Carlyle's political jeremiad, he found "by far the notablest case of *Swarmery*, in these times, is that of the late American War, with Settlement of the Nigger Question for result."[8] Almighty God, said Carlyle, had intended the blacks to be servants in perpetuity. And yet having "got into *Swarmery*" over the question of black slavery, "a continent of the earth has been submerged, for certain years, by deluges as from the Pit of Hell; half a million . . . of excellent White Men, full of gifts and faculty, have torn and slashed one another into horrid death, in a temporary humour, which will leave centuries of remembrance fierce enough: and three million absurd Blacks, men and brothers (of a sort), are completely 'emancipated'; launched into the career of improvement, —likely to be 'improved off the face of the earth' in a generation or two!"[9]

Carlyle's representation of democratic enfranchisement in the Calvinistic image of the devil unchained must have aroused some of Whitman's own reservations about the hell pit of the Civil War, the consequences of black freedom, and the extension of the vote. Within a few weeks after Carlyle's essay was carried in the New York *Tribune* on August 16, 1867, he had prepared an essay "Democracy" as "in some sort a counterblast or rejoinder to Carlyle's late piece, *Shooting Niagara*" (*Corr.*, I, 342). "Democracy" was published in the *Galaxy* in December 1867. In May 1868 the *Galaxy* also published Whitman's essay "Personalism," which, as Whitman says, "sketches" an ideal democratic male and female personality and "overhauls the Culture theory" measured against "any grand, practical Democratic test" (*Corr.*, II, 19). In the same year he completed an essay on democratic literature entitled "Orbic Literature," which was submitted to the *Galaxy* but never published (*Corr.*, II, 32).[10]

Whitman later incorporated his essays "Democracy," "Personalism," and "Orbic Literature" into an eighty-four-page pamphlet published in 1871 under the title *Democratic Vistas*. Although his *Vistas*, or, as he called them, "speculations," originated in an effort to "counterblast" Carlyle's attack, Whitman quickly realized that he shared Carlyle's diagnosis of the diseases of democracy. In a footnote regarding "Shooting Niagara," he admitted:

I was at first roused to much anger and abuse by this essay from Mr. Carlyle, so insulting to the theory of America—but happening to think afterwards how I had more than once been in the like mood, during which his essay was evidently cast, and seen persons and things in the same light, (indeed some might say there are signs of the same feeling in these Vistas)—I have since read it again, not only as a study, expressing as it does certain judgments from the highest feudal point of view, but have read it with respect as coming from an earnest soul, and as contributing certain sharp-cutting metallic grains, which, if not gold or silver, may be good hard, honest iron. (*PW*, II, 375–76)

Under the stimulus of Carlyle's attack, Whitman undertook his most sustained and eloquent meditation on the problems and contradictions of democracy in America. Written in the post–Civil War world of collapsing values, *Democratic Vistas* was impelled by Whitman's will to heal, to make whole, what had been damaged by the war. But the essay is not only an attempt to lay the demons released by the Civil War. It is also an attempt to lay the intermeshed demons of self and republic that had been spooking Whitman's democratic dream of America for over three decades. Moving between the political invective of *The Eighteenth Presidency!* and the prophetic vision of *Leaves of Grass*, *Democratic Vistas* is at once an *ars republica* and an *ars poetica* in which Whitman seeks to justify the "democratic republican principle" as the theory of America and the political ground of his own life and work.

In his attempt to reconcile the ideals of democracy with the carnage of the war and the corruptions of post–Civil War America, Whitman found philosophical support in the ideas of Hegel and the German idealistic thinkers. "Only Hegel is fit for America," he wrote in a series of lecture notes on Kant, Hegel, Fichte, and Schelling prepared in the late 1860s or early 1870s (*CW*, IX, 170).[11] Hegel's idea of a triplicate process through which opposites are merged into a higher synthesis, and his vision of history as a manifestation of spirit rationalized Whitman's own vision of American democracy progressing through the evils and contradictions of the present toward divine ends. In Hegel's system, Whitman wrote, "the varieties, contradictions and paradoxes of the world and of life, and even good and evil, so baffling to the superficial observer, and so often leading to despair, sullenness or infidelity, become a series of infinite radiations and waves of the one sea-like universe of divine action and progress, never stopping, never hasting" (*CW*, IX, 171).

The structure of *Democratic Vistas* is itself Hegelian, working through oppositions and contradictions toward some higher synthesis. At the outset, Whitman calls attention to the dialectical procedure of his essay: "Though it may be open to the charge of one part contradicting another—for there are opposite sides to the great question of democracy, as to every great question—I feel the parts harmoniously blended in my own realization and convictions, and present them to be read only in such oneness, each page and each claim and assertion modified and temper'd by the others" (*PW*, II, 362–63). Throughout the essay the political realist does battle with the visionary poet, as Whitman modifies the claims of each in some overarching unity. The essay moves back and forth between present and future, fact and vision, individual and aggregate, urban and rural, culture and nature, matter and spirit, science and religion, working through these dialectical oppositions toward a dynamic democratic synthesis.[12]

Like a physician diagnosing the ills of his time, Whitman begins his *Vistas* by acknowledging the dangers of democracy. "I will not gloss over the appaling [*sic*] dangers of universal suffrage in the United States. In fact, it is to admit and face these dangers I am writing." His *Vistas* are addressed to those who are, like himself, split between the ideal and reality of democracy in America—"to him or her within whose thought rages the battle, advancing, retreating, between democracy's convictions, aspirations, and the people's crudeness, vice, caprices" (*PW*, II, 363). Accepting Carlyle's portrait of a world in crisis, Whitman admits: "The spectacle is appaling [*sic*]. We live in an atmosphere of hypocrisy throughout." Even before the worst scandals of the Grant administration were exposed, he presents an image of America saturated with corruption and greed from the national to the local level:

> The depravity of the business classes of our country is not less than has been supposed, but infinitely greater. The official services of America, national, state, and municipal, in all their branches and departments, except the judiciary, are saturated in corruption, bribery, falsehood, maladministration; and the judiciary is tainted. The great cities reek with respectable as much as non-respectable robbery and scoundrelism. . . . In business, (this all-devouring modern word, business,) the one sole object is, by any means, pecuniary gain. The magician's serpent in the fable ate up all the other serpents; and money-making is our magician's serpent, remaining to-day sole master of the field. . . . I say that our New World democracy, however great a success in uplifting the masses out of their sloughs, in materialistic development, products, and in a certain highly-deceptive superficial popular intellectuality, is, so far, an almost complete failure in its social aspects, and in really grand religious, moral, literary, and esthetic results. . . . It is as if we were somehow being endow'd with a vast and more and more thoroughly-appointed body, and then left with little or no soul. (*PW*, II, 369–70)

In this vision of an America driven by hypocrisy, infidelity, superficiality, and greed, Whitman had already discovered the distinctive characteristics of what Mark Twain a few years later would call the "Gilded Age" in the book that named the period.[13]

Although Whitman looked beyond the evils of the present toward the future for what he called the "justification and success" of America, in *Democratic Vistas* even the visionary gleam of the future was clouded in darkness. "What prospect have we?" he asked. "We sail a dangerous sea of seething currents, cross and under-currents, vortices—all so dark, untried—and whither shall we turn?" Among the problems of the present age, he notes the lack of leadership and the "plentiful meanness and vulgarity of the ostensible masses" (*PW*, II, 422). And in language reminiscent of the Jacksonian rhetoric of his newspaper days, he assails the "system of inflated paper-money currency, (cause of all conceivable swindles, false standards of value, and principal breeder and bottom of those enormous fortunes for the few, and of poverty for the million)—with that other plausible and sugar-coated delusion, the theory and practice of a protective tariff, still clung to by many" (*PW*, II, 753).

If in the 1840s and 1850s the agitation against slavery drew attention away from the emergent conflict between labor and capital, in the post–Civil War period, Whitman returned to the anticapitalist rhetoric of his early years.[14] It is in the growing conflict between labor and capital, rich and poor, that he sees the most dangerous current of the future. In a footnote regarding "THE LABOR QUESTION," which he later deleted from *Democratic Vistas*, Whitman wrote:

> The immense problem of the relation, adjustment, conflict between Labor and its status and pay, on the one side, and the Capital of employers on the other side—looming up over These States like an ominous, limitless, murky cloud, perhaps before long to overshadow us all; —the many thousands of decent working-people, through the cities and elsewhere, trying to keep up a good appearance, but living by daily toil, from hand to mouth, with nothing ahead, and no owned homes—the increasing aggregation of capital in the hands of a few—the chaotic confusion of labor in the Southern States, consequent on the abrogation of slavery—the Asiatic immigration on our Pacific side—the advent of new machinery, dispensing more and more with hand-work—the growing, alarming spectacle of countless squads of vagabond children, roaming everywhere the streets and wharves of the great cities, getting trained for thievery and prostitution—the hideousness and squalor of certain quarters of the cities. (*PW*, II, 753)

What Whitman was seeing in the streets and cities of post–Civil War America was the dissolution of the traditional values of the republic as the country came fully under the power of the new lords of capital. He not only described but predicted the struggles over working conditions, urban poverty, immigration, automation, alienation, and the ghettoization of America that would mark the progress toward the future. After the Civil War, America had, in effect, already entered the twentieth century.

Anticipating the work of Brooks and Henry Adams, Ignatius Donnelly, and Jack London later in the century, Whitman was one of the first major writers to chart the potentially downward spiral of American history in the postwar years. Naming the new age of unbridled forces and disintegrating values—of "pride, competition, segregation, vicious willfulness, and license beyond example"—he was one of the first to grapple with the problems not only of democracy but also of machine technology, incorporation, modernization, and unlimited production that would mark the twentieth century. "Unwieldy and immense, who shall hold in behemoth?" he asked, "who bridle leviathan?" (*PW*, II, 422).

His answer to this question is at the very center of his *Vistas:* It is to literature, and to the poet in particular, that Whitman looked for the power that would bridle the gilded monsters to which America, and perhaps the theory of America, had given birth. Carlyle, too, had called for a literary *Aristos* of "inspired speakers and seers" to "deliver the world from its swarmeries." But as the poet of these democratic "swarmeries," Whitman turned Carlyle on his head: He called for a class of native American writers, "a new and greater literatus order" that would lead the people out of the feudal past and into the democratic future: "Our fundamental want to-day in the United States, with closest, amplest reference to present conditions, and to the future, is of a class . . . of native authors, literatuses . . . permeating the whole mass of American mentality, taste, belief, breathing into it a new breath of life, giving it decision, affecting politics far more than the popular superficial suffrage, with results inside and underneath the elections of Presidents or Congresses" (*PW*, II, 365). The true politics of democracy would manifest itself neither on the level of government nor through material advance but in the emergence of a new state of mind that would bring about a total transformation of American life. In "reconstructing, democratizing society," Whitman said, the true "revolution" would be of "the interior life"; and in bringing about this democratic revolution, the poet would play the leading role by providing the language, commonality, and myths by which America named itself (*PW*, II, 410). By "fusing contributions, races, far localities, &c., together," the poet had the power to "give more compaction and more moral identity, (the quality to-day most needed,) to these States, than all its Constitutions, legislative and judicial ties, and all its hitherto political, warlike, or materialistic experiences" (*PW*, II, 368).

It was in *Democratic Vistas* that the word and the concept of culture came first into Whitman's writing as he set about defining what he called a "programme of culture" for America. In claiming for the poet the authority to define America and thus name reality, Whitman placed himself at the center of the ideological debates about the meaning and direction of America in the post–Civil War period. But in presenting the poet as a cultural authority and bridler of leviathan, he was saying nothing new. In *Culture and Anarchy* (1869), Matthew Arnold had also found in culture "a principle of authority, to counteract the tendency to anarchy."[15] What made Whitman's "culture theory" different was that he challenged the hegemonic dominance of an elite class, calling for "a radical change of category, in the distribution of precedence" (*PW*, II, 396). Arnold located culture in the best that had been thought and said, whereas Whitman located the sources of culture in the common life of the masses, who had been neither known nor named in the artistic productions of the past: "I should demand a programme of culture, drawn out, not for a single class alone, or for the parlors or lecture-rooms, but with an eye to practical life, the west, the working-men, the facts of farms and jack-planes and engineers, and of the broad range of the women also of the middle and working strata, and with reference to the perfect equality of women, and of a grand and powerful motherhood" (*PW*, II, 396).

Whitman proposed a democratic "programme of culture" different from the genteel model of bourgeois culture that was becoming the American standard: "To prune, gather, trim, conform, and ever cram and stuff, and be genteel and proper, is the pressure of our days" (*PW*, II, 394). He launched a full-scale attack on genteel culture as a system of power no less class based than the feudal products of Europe. Reflecting on the absence among American writers of a "single image-making work of the people," he asked: "Do you call those genteel little creatures American poets? Do you term that perpetual, pistareen, paste-pot work, American art, American drama, taste, verse?" (*PW*, II, 388–89).

At the core of *Democratic Vistas* is a proposal for a radical reconstruction of literature as it had been traditionally understood. Anticipating postmodern investigations into the ideological bases of literature, literacy, and literary value and the corresponding call for a reconstruction of American literature along ethnic, feminist, and ultimately democratic lines, Whitman proposed that "almost everything that has been written, sung, or stated, of old . . . be rewritten, re-sung, re-stated, in terms consistent with the institution of these States" (*PW*, II, 425). He not only named the politics of past literature, he also called for the repoliticization of the literature of the future.

But for all the apparent radicalism of Whitman's democratic project, his attack on American politics and culture is an attack launched from within. He turns the terms of democratic ideology to a critique of American culture as it was manifesting itself in the capitalist and bourgeois production of his time, but he never questions the American system itself. As a work of political criticism, *Democratic Vistas* illustrates the particular power of American democracy as a political ideology that simultaneously creates and contains the terms within which its movements of protest occur.[16]

Whitman did not repudiate the material prosperity or capitalist enterprise of the post–Civil War period. "My theory includes riches, and the getting of riches," he wrote, "and the amplest products, power, activity, inventions, movements, &c." (*PW*, II, 385). However, like other critics of the Gilded Age, including Mark Twain, Rebecca Harding Davis, John de Forest, and Henry Adams, Whitman recognized the need for some higher system of moral or spiritual value to counter the centrifugal force of material and technological production. Perhaps under the influence of Hegel, he came to see American history as a sequence of stages, moving from the political founding through a stage of material development to a higher stage of spiritual realization.

But in his attempt to give democracy a spiritual base, separate from the institutions of church and state, Whitman came more and more to deny the material conditions of American life. Describing the actual state of America, which included an emphasis on "sense, science, flesh, incomes, farms, merchandise . . . buildings of brick and iron," he dismissed these conditions as "illusions! apparitions! figments all!" (*PW*, II, 417–18). Whereas in his early work he balanced matter with spirit, in his later work, as the material production of America seemed to spin out of control and the action of the individual was reduced to a clockwork motion, Whitman placed an increased emphasis on transcendence and the higher law of religiospiritual value. His pursuit of spiritual union, his attempt to oppose the "growing excess and arrogance of realism" with the "central divine idea of All," became finally a push away from the historic conditions of Gilded Age America.

In addressing the problem of democracy in *Democratic Vistas*, Whitman never doubted the fact of democratic advance. Appealing neither to the Rousseauistic doctrine of natural good nor even to the Jeffersonian notion of inalienable right, he represents democracy as an irreversible movement of history. "The great word Solidarity has arisen," he says. "Of all dangers to a nation, as things exist in our day, there can be no greater one than having certain portions of the people set off from the rest by a line drawn—they not privileged as others, but degraded, humiliated, made of no account" (*PW*, II, 382). Whereas Carlyle sought to stem the tide of democracy through a reassertion of aristocratic power and the imposition of a rigorous militaristic order, in Whitman's view the class system contained the seeds of its own destruction; the coming of democracy was inevitable. He rationalizes democracy not as a possible system but as the only possible system. The question is not, he states, "whether to hold on, attempting to lean back and monarchize, or to look forward and democratize—but *how*, and in what degree and part, most prudently to democratize" (*PW*, II, 383).

A good part of Whitman's *Vistas* is taken up with the question of how and in what degree to democratize, a question that, like the ideological debates of the time, revolved around the use and distribution of power. In taking up this question here as in "Song of Myself," Whitman returned to the essentially constitutional theory of a balance and separation of powers. He presents the reconstructed nation as a political balance wheel that merges what he called the "conflicting and irreconcilable interiors" of America into a dynamic synthesis: The power of the self is balanced with the power of the aggregate, the state with the nation, the South with the North, country with city, self-interest with social love, matter with spirit, science with religion, money with soul.

Although Whitman recognized the "wounds and diseases" of the new market economy, his reconstructed politics did not recognize the possible need for more vigorous government action to protect the people from themselves. To Carlyle's call for an authoritarian state, Whitman countered with the democratic notion of a government for the people in the largest sense of the term: "We believe the ulterior object of political and all other government . . . to develop, to open up to cultivation, to encourage the possibilities of all beneficent and manly outcroppage, and of that aspiration for independence, and the pride and self-respect latent in all character" (*PW*, II, 379). At the same time, however, he recognized the potential conflict between unleashed individualism and the good of the republic: "Must not the virtue of modern Individualism, continually enlarging, usurping all, seriously affect, perhaps keep down entirely, in America, the like of the ancient virtue of Patriotism, the fervid and absorbing love of general country?" (*PW*, II, 373). In this "serious problem and paradox in the United States," Whitman struggled with the conflict at the foundation of the American republic: how to reconcile the desire for personal liberty with the demands of social union.

In addressing the problem of liberty and union, Whitman worked toward two dynamic syntheses. The first was pragmatic and Madisonian:

It is in the individual interest, he argues, to maintain national union, for individual freedom cannot exist apart from a strong political community. Whitman's second synthesis was utopian and visionary. Countering the revolutionary movement away from the tyrannical structures of the past toward the sovereign power of the individual, he postulates a universal force that binds and fuses humanity. There is "not that half only, individualism, which isolates. There is another half, which is adhesiveness or love, that fuses, ties and aggregates, making the races comrades, and fraternizing all." Although Whitman does not, as did the political founders, advocate a doctrine of self-sacrifice, in his vision of a democratizing force that "ever seeks to bind, all nations, all men, of however various and distant lands, into a brotherhood, a family" (*PW*, II, 381), he does propose a kind of republican virtue in which individual power is continually balanced by a bond of sympathy, love, and relatedness to a larger community. Swerving from the individualistic ethos of Emerson and Thoreau, Whitman advocates a form of freedom that exists not in an isolated self or a romanticized state of nature but in relation to others. While he did not, like Hegel, subordinate the interests of the individual to those of an ethical state, in his increased concern with the need for solidarity, he was moving closer to the socialist concept of the individual finding her or his greatest freedom within a political community.

Like de Tocqueville, Whitman recognized the potentially leveling effects of democracy, effects that were evident in the increased mechanization and standardization of American life in the postwar period. And thus to "democracy, the leveler, the unyielding principle of the average," he opposed another law of nature, the principle of "individuality, the pride and centripetal isolation of a human being in himself—identity—personalism" (*PW*, II, 391). In his doctrine of personalism, a concept that Bronson Alcott appropriated to describe his own philosophical system, Whitman found a counterpart to "the mortal dangers of republicanism"—a "compensating balance-wheel of the successful working machinery of aggregate America" (*PW*, II, 392).

Whitman's doctrine of personalism had ideological roots in the land-based agrarian economy of the past: "Singleness and normal simplicity and separation, amid this more and more complex, more and more artificialized state of society—how pensively we yearn for them! how we would welcome their return!" The key word here is *return:* The model of democratic personality that Whitman imagines represents a return to the "healthy rudeness" and "savage virtue" advocated by Jefferson (*PW*, II, 394). The growth of personalism and indeed Whitman's hope for democracy are rooted in the agrarian ideal of universal landownership and a general distribution of wealth: "The true gravitation-hold of liberalism in the United States will be a more universal ownership of property, general homesteads, general comfort—a vast, intertwining reticulation of wealth" (*PW*, II, 383).

Whitman's vision of a land-based democratic economy is contradicted by the fact of growing impoverishment in American cities. Presenting himself as a kind of new-world Gulliver among Lilliputians, he confesses: "Using the moral microscope upon humanity, a sort of dry and flat Sahara appears, these cities, crowded with petty grotesques, malformations, phantoms, playing meaningless antics" (*PW*, II, 371–72). Rather than resolve the conflict between the ideal of republican abundance and the actual waste of post–Civil War America, Whitman projected his dream of democracy onto the future and the West. Like *Leaves of Grass*, *Democratic Vistas* turns on a paradox generally prevalent in nineteenth-century America. While Whitman looks to the future for the realization of democracy, the reconstructed nation he imagines represents a return to the past and the ideal republic of Jefferson:

> I can conceive a community, to-day and here, in which, on a sufficient scale, the perfect personalities, without noise meet; say in some pleasant western settlement or town, where a couple of hundred best men and women, of ordinary worldly status, have by luck been drawn together, with nothing extra of genius or wealth, but virtuous, chaste, industrious, cheerful, resolute, friendly and devout. I can conceive such a community organized in running order, powers judiciously delegated—farming, building, trade, courts, mails, schools, elections, all attended to; and then the rest of life, the main thing, freely branching and blossoming in each individual, and bearing golden fruit. (*PW*, II, 402)

Whitman never reconciles the conflict between dream and reality, the virtuous citizens of the West and the petty grotesques of the city: The Garden and the Sahara remain emblems of a poet and a nation split between the republican ideals of the past and the dislocations of a modern market economy.

Whitman arrives at no final synthesis of the values he juggles in *Democratic Vistas*. His *Vistas* conclude with a series of contradictions that await some higher synthesis in the future. Amid the modernizing and standardizing whirl of America, where "with steam-engine speed" generations of humanity are turned out "like uniform iron castings," Whitman recognized that the road to the democratic future might be, as Carlyle had predicted, the road to hell.

I say of all this tremendous and dominant play of solely materialistic bearings upon current life in the United States, with the results as already seen, accumulating, and reaching far into the future, that they must either be confronted and met by at least an equally subtle and tremendous force-infusion for purposes of spiritualization, for the pure conscience, for genuine esthetics, and for absolute and primal manliness and womanliness—or else our modern civilization, with all its improvements, is in vain, and we are on the road to a destiny, a status, equivalent, in its real world, to that of the fabled damned. (*PW*, II, 424)

The conclusion of *Democratic Vistas* is open-ended. Without the "force-infusion" of a "native literary and artistic formulation," America "will flounder about, and her other, however imposing, eminent great-ness, prove merely a passing gleam." With such an artistic formulation, "she will understand herself, live nobly, nobly contribute, emanate, and, swinging, poised safely on herself, illumin'd and illuming, become a full-form'd world, and divine Mother not only of material but spiritual worlds, in ceaseless succession through time" (*PW*, II, 426).

It is not surprising that Whitman's *Democratic Vistas* concludes with the figure of America as a "divine Mother" swinging self-poised into the democratic future. Torn between growing opportunity and increased re-straint during the antebellum period, the female was another of the fig-ures who was in some sense unchained by the Civil War. During the war, women entered the work force in increased numbers as factory workers, teachers, nurses, and civil servants, and in the post–Civil War debates over equal rights and black suffrage, women united under the leadership of Elizabeth Cady Stanton and Susan B. Anthony to demand not only the vote but also an equal place and voice in the reconstructed nation.

Whitman acknowledged and indeed encouraged this new presence of women in American life. Central to *Democratic Vistas* is "the idea of the women of America, (extricated from this daze, this fossil and unhealthy air which hangs about the word *lady*,), develop'd, raised to become the robust equals, workers, and, it may be, even practical and political deciders with the men" (*PW*, II, 389). Women, he says, are the equals of men in "all departments" or at least are "capable of being so, soon as they realize it, and can bring themselves to give up toys and fictions, and launch forth, as men do, amid real, independent, stormy life" (*PW*, II, 389). Reflecting the agitation for women's rights and women's suffrage after the war, Whitman foresees a time when women will fully participate in the political gover-nance of the nation: "The day is coming when the deep questions of woman's entrance amid the arenas of practical life, politics, the suffrage,

&c., will not only be argued all around us, but may be put to decision, and real experiment" (*PW*, II, 401).

In "I'm ceded—I've stopped being Theirs—," Emily Dickinson renounces, in the language of political secession, her dolls, her childhood, and her "string of spools" in order to assume the "rank" of poet. Like her, Whitman saw in literature a primary means of giving utterance to women, who had been silenced both politically and culturally. By recasting past models of female character, by reinventing womanhood, literature might also counter the genteel model of femininity that had become the standard in nineteenth-century America. Literature, in Whitman's view, had the power to achieve "the entire redemption of woman out of these incredible holds and webs of silliness, millinery, and every kind of dyspeptic depletion" (*PW*, II, 372). As part of his own effort to create empowering types of a new athletic womanhood, he included among his *Vistas* models of women in the traditional roles of wife and mother and also in their new roles as worker and business woman outside the home.

In *Democratic Vistas*, as in *Leaves of Grass*, Whitman returned over and over again to the image of "divine maternity" as a woman's "towering, emblematic attribute." One of the main dangers he saw in post–Civil War America was the lack of moral fiber through all society and the "appaling [*sic*] depletion of women in their powers of sane athletic maternity, their crowning attribute, and ever making the woman, in loftiest spheres, superior to the man" (*PW*, II, 372). Although Whitman insisted on the superiority of the mother, he did not limit the female to a maternal *role*, or trap her in what Simone de Beauvoir would later call biological "immanence." Like the sexually charged and spiritually unifying figure of the Virgin that Henry Adams opposed to the divisive and ultimately destructive power of the dynamo, Whitman sought to revive the mother not as a biological function only but as a creative and intellectual force.

A "new birth" of women is at the base of the reconstructed nation that Whitman envisioned in *Democratic Vistas:* "I have sometimes thought, indeed, that the sole avenue and means of a reconstructed sociology depended, primarily, on a new birth, elevation, expansion, invigoration of woman, affording, for races to come, (as the conditions that antedate birth are indispensable,) a perfect motherhood. Great, great, indeed, far greater than they know, is the sphere of women" (*PW*, II, 372). Whitman had little patience with the Christian benevolent model of female behavior propagated by Catherine Beecher in *A Treatise on Domestic Economy* (1842). His mothers do not exist as wives in relation to individual husbands, nor are they pious, pure, domestic, or self-sacrificing in any limited sense of the terms. Like feminist works ranging from Margaret Fuller's

*Woman in the Nineteenth Century* (1845) to Charlotte Perkins Gilman's *Herland* (1915) to Adrienne Rich's *Of Woman Born: Motherhood As Experience and Institution* (1976), Whitman sought to remove motherhood from the private sphere and release the values of nurturance, love, generativity, and community into the culture at large. Exceeding the bounds of home, marriage, and the isolate family, Whitman's "perfect motherhood" is motherhood raised to the height of solicitude for the future of the race.

# 11

## *The Poetics of Reconstruction*

LEAVES OF GRASS, already published, is, in its intentions, the song of a great composite *Democratic Individual*, male or female. And following on and amplifying the same purpose, I suppose I have in my mind to run through the chants of this Volume (if ever completed,) the thread-voice, more or less audible, of an aggregated, inseparable, unprecedented, vast, composite, electric *Democratic Nationality*.

—WHITMAN, 1872 preface

The reconstruction of the nation during and after the war years began for Whitman with the act of reconstructing his poems. In 1867, at the time the radical republicans in Congress were passing their series of reconstruction acts over the veto of President Andrew Johnson, Whitman brought out the fourth edition of *Leaves of Grass*. The volume included only six new poems, all minor (an inscription poem "Small the Theme of My Chant," "The Runner," "Tears," "Aboard at a Ship's Helm," "When I Read the Book," "The City Dead-House"). But Whitman extensively revised, reshaped, deleted, and added to the body of poems published in the 1860 *Leaves*. These revisions, most of which were made during the war in his own blue book copy of the 1860 *Leaves*, once again reflect the poet's attempt to bind up the nation's wounds.[1] Although Whitman retained the controversial *Children of Adam* and *Calamus* groupings, the *Chants Democratic* grouping was, like the national union that was its theme, broken up and extensively revised.

A copy of Whitman's blue book *Leaves* was found in his desk at the Department of the Interior, which led to his being fired from his position by Secretary James Harlan. But it is important to bear in mind that Whitman made the bulk of his revisions before his friend O'Connor published his inspired defense of the poet in 1866 and thus before he emerged in his more public role as "the good gray poet." What his extensive work of revision, excision, and addition indicates is not his desire to retreat from the radical posture of his earlier *Leaves* but, rather, the renewed sense of artistic calling he had gained during the Civil War.

Most of his revisions work toward an artistic tightening and sharpening of word, image, phrase, and unit. Behind his artistic reshaping is the overall impulse to mend the split in the 1860 *Leaves* between private and public, lover and poet as a means of affirming, or reaffirming, the principle of organic union in self, nation, and cosmos. This act of poetic reconstruction is particularly evident in his 1867 revisions of "Starting from Paumanok" (originally "Proto-Leaf"). If in 1860 the poem was divided by the polarities of lover and poet, in the 1867 version Whitman seeks to fuse the poet and lover by toning down the Calamus motif and sharpening the focus on democratic union.

Whitman's decision to drop three of his more confessional *Calamus* poems in the 1867 *Leaves* was probably also motivated by his desire to fuse the poet and the lover in a single national persona who would project the unitary figure of a reconstructed self and a reconstructed nation. Although he may have sought through revision and excision to cover over, at least poetically, the sense of fracture and self-doubt, guilt and shame associated with his Calamus feelings in the prewar years, there is no reason to assume that he deleted *Calamus* poems nos. 8, 9, and 16 in order to erase their personal homoerotic signature. In fact, he retained several *Calamus* poems that are equally personal (including "When I Heard at the Close of the Day" and "Trickle Drops"). And to *Calamus* no. 39 ("Sometimes with One I Love"), he added a parenthetical remark that disclosed in a straightforward manner the homosexual subject of his *Calamus* poems: "(I loved a certain person ardently and my love was not return'd,/Yet out of that I have written these songs.)"[2]

Presenting himself in his new role as the poet of the "throes of Democracy," he added to the 1867 "As I Sat Alone by Blue Ontario's Shore" ("By Blue Ontario's Shore") a series of parenthetic interpolations that project the image of the democratic mother under siege. These interpolations culminate in a reaffirmation of the democratic dream, but it is a dream perpetually deferred by the weight of contradiction in the sociopolitical landscape of post–Civil War America:

> (Mother! bend down, bend close to me your face!
> I know not what these plots and wars and deferments are for;
> I know not fruition's success—but I know that through war and peace
>     your work goes on, and must yet go on.)
>
> (*LG: Variorum*, I, 209)

As the plots and deferments increased in the postwar period, the democratic mother became an increasingly central figure in Whitman's poetic iconography.

Even before the war, "The Mother of These States" was in Whitman's view the appropriate emblem of the American republic. "None of the emblems of the classic goddess—nor any feudal emblems—are fit symbols for the republic," he wrote in a notebook,[3] and in another entry he resolved in his poems to "bring in the idea of Mother—the idea of the mother with numerous children—all, great and small, old and young, equal in her eyes—as the identity of America" (*CW*, IX, 11). Whitman's move away from his earlier emphasis on the poet as a creator of strong individuals to an emphasis on the poet as a creator of national unity corresponds to a shift in his poems from a primary identification with the male to a primary identification with the female dimensions of the universe. Amid the disintegrative forces of the postwar period, the figure of the Union as democratic mother projected the corporate identity of the states, and the familial image of the mother with many children underlined the organic and egalitarian nature of the compact among individuals and states. As an emblem of regenerative potency, the democratic mother also came to symbolize the creative and democratizing force of history itself.

But in her primary aspect, the democratic mother represented a return to a prewar, republican model of power. In one of his parenthetical invocations, Whitman presents the mother as a figure of the *E Pluribus Unum* in which power is dispersed among individuals rather than centralized in the state:

> (Mother! with subtle sense—with the naked sword in your hand,
> I saw you at last refuse to treat but directly with individuals.)
>                                          (*LG: Variorum*, I, 206)

As America began to move toward a corporate, centralized, male-identified model of power, Whitman stressed the democratic mother as a federated model of governance in which power, whether political or economic, was diffused among the many rather than concentrated in the hands of the few. Embodying the values of sympathy and love, the democratic mother represented a benignant model of female power that Whitman used to counteract the centralized administration of the state and the aggressively capitalist and male-powered ethos of the new market economy.

Just as in the post–Civil war period Whitman came to see the war as a pivotal event in the evolution toward the democratic future, so he came to see his war poems as pivotal to the thematic and structural organization of

*Leaves of Grass.* "The whole Book," he said, "revolves around that Four Years' War, which, as I was in the midst of it, becomes, in *Drum-Taps*, pivotal to the rest entire" (*LGC*, p. 752). In his 1889 essay "A Backward Glance o'er Travel'd Roads," he went even further in asserting the central- ity of the war experience to *Leaves of Grass:* "Without those three or four years and the experiences they gave, 'Leaves of Grass' would not now be existing" (*LGC*, p. 570). Whitman, of course, lies, or perhaps he experi- ences a willed lapse of memory. In the prewar years he had already published the 1855, 1856, and 1860 editions of *Leaves of Grass*, volumes that in the opinion of many constitute his major contribution as a poet. What he probably meant was that without the experience of the war, *Leaves of Grass*, like the republic itself, would not exist in its current form; the political tensions that split the nation in the prewar years and led to the armed conflict of the Union War were at the foundation of his book. But Whitman's words on the pivotal event of the war also have the effect of refashioning *Leaves of Grass* into a poetic justification of the Civil War, making the war seem somehow implicit and genetically coded in the formation of book and nation.

In 1867, however, the war appears as a rupture in Whitman's recon- structed *Leaves*. Despite his attempt to bind up the nation's wounds by presenting an image of union—personal, political, and artistic—the 1867 *Leaves* is Whitman's most chaotic edition: it exists in several versions, with different annexes tagged on with separate pagination. The fact that *Drum- Taps* and *Sequel* are added on to the volume with separate pagination is itself a sign that Whitman had not yet integrated the war into a coherent artistic or national design. The disorder of the 1867 *Leaves* seems to mirror the personal and public chaos wrought by the Civil War. Signifi- cantly, too, the 1867 edition was the first to appear without a picture of the poet on its frontispiece. Although Whitman continued to emphasize the unity of self and nation, he had still not found an appropriate image of himself as poet of postwar America to present to the public.

The Civil War marked the end of one phase of American development and the beginning of a new phase. The evolution of *Leaves of Grass* in the postwar period corresponds to this new phase of national development. As the nation entered the era of radical reconstruction, Whitman came to see *Leaves of Grass* as substantially complete and made plans to write a new volume stressing the theme of national union.[4]

The main changes in his next edition of *Leaves of Grass*, which was published in 1871–72, are reflected not in the twenty-four new poems but in Whitman's effort to reconstruct the volume in accordance with the "programme of culture" he had set forth in *Democratic Vistas*. The *Drum-Taps* poems are integrated into the body of his new volume, so that the war is now rationalized both structurally and thematically as part of a saving national vision. The war was still central to Whitman's conception of *Leaves of Grass*, as indicated by his adding two new groupings entitled *Marches Now the War Is Over* and *Bathed in War's Perfumes*. Among the poems that affirmed the advance toward the democratic future in the *Marches* grouping, he included "Respondez" as a contrapuntal sign that the war had not cured the ills of the republic; in fact, he added lines that specifically responded to the widespread corruption of the Gilded Age:

> (Stifled, O days! O lands! in every public and private corruption!
> Smother'd in thievery, impotence, shamelessness, mountain-high;
> Brazen effrontery, scheming, rolling like ocean's waves around and upon
>     you, O my days! my lands!
> For not even those thunderstorms, nor fiercest lightnings of the war,
>     have purified the atmosphere;)
>
> *(LG: Variorum, I, 261)*

Registering some of the same doubts about democracy he had expressed in *Democratic Vistas*, Whitman suggests that if anything, the war had given a new birth not to freedom but to fraud.

To keep alive the spirit of revolutionary struggle, Whitman added a more radical grouping to the 1871 *Leaves: Songs of Insurrection*. This cluster was formed in response to the corruption of power and authority in the Gilded Age, as indicated by a note in the original manuscript of the poem. Under the heading *Songs of Insurrection*, Whitman wrote:

> Not only are These States the born offspring of Revolt against mere over-weening authority—but seeing ahead for Them in the future a long, long reign of Peace with all the growths corruptions and tyrannies & formalisms of Obedience, (accumulating, vast folds, strata, from the rankness of continued prosperity and the more and more insidious grip of capital) I feel to raise a note of caution (perhaps unneeded alarm) that the ideas of the following cluster will always be needed, that it may be worth while to keep well up, & vital, such ideas and verses as the following. *(WWW, p. 229)*

This theme is announced in the opening poem, "Still Though the One I Sing," the only new poem in the cluster:

> Still though the one I sing,
> (One, yet of contradictions made,) I dedicate to Nationality,
> I leave in him revolt, (O latent right of insurrection! O quenchless,
>     indispensable fire!)
>
> *(LG: Variorum, III, 632)*

In *Songs of Insurrection*, the new emphasis on nationhood, in both Whitman's work and the country at large, is counterbalanced by the theme of revolutionary struggle. He includes "To a Foil'd Revolutionaire," "France, the 18th Year of These States," "Europe," and "Walt Whitman's Caution," which all were written in response to political tyranny at home and abroad. At a time when the national government was encroaching on rights traditionally enjoyed by state and municipality, "Walt Whitman's Caution"— "*Resist much, obey little*"—reminded American citizens of the doctrine of local and state sovereignty. But the poems are not only, or even mainly, a response to "overweening" government authority. Rather, *Songs of Insurrection* is a response to a new form of economic tyranny, the "insidious grip of capital," that took hold of America in the post–Civil War period.

To some editions of *Leaves of Grass* (1871–72) Whitman added two annexes: *Passage to India* and *After All Not to Create Only*, both of which were published as separate pamphlets in 1871. These annexes are the first sign of his attempt to inaugurate a new work, different from *Leaves of Grass*, emphasizing the themes of democratic nationality and spiritual union that he had stressed in *Democratic Vistas*.

The *Passage to India* cluster, which includes seventy-five poems, all except twenty-three culled from his earlier work, is linked by the thematic thread of death in sections that move from the postwar context of *Ashes of Soldiers* and *President Lincoln's Burial Hymn* to *Whispers of Heavenly Death*, *Sea-Shore Memories*, and a final leap into the spiritual sea in a section entitled *Finale to the Shore*. That a cluster of poems inaugurated by a new phase of American development in the post–Civil War period should have death as its pervasive focus is only one of the troubling ironies of this volume.

The cluster opens with the title poem "Passage to India," which is structured around a series of trinities: Past/Present/Future, Time/Space/Death, Man/Nature/God, Ancient/Medieval/Modern, Asia/Europe/America, Train/Cable/Telegraph, Explorer/Inventor/Poet. These trinities are rooted in what Whitman called the "triplicate process" of Hegelian evolution: "First the Positive, then the Negative, then the product of the mediation between them; from which product the process is repeated and so goes on without end" (*CW*, IX, 173). This "triplicate process" of history is the poem's organizing principle and thematic focus.

Like the trinities of unit, image, and theme from which the poem is constructed, "Passage" incorporates three separate unpublished poems in its final artistic ordering: "Fables" (lines 18–29), "Thou Vast Rondure Swimming in Space" (lines 81–115), and "O Soul, Thou Pleaseth Me" (lines 182–223).[5] These poems, which stress the theme of spiritual union, are incorporated into the celebration of technological union that is the occasion for the poem; appearing at the beginning, middle, and end, they become a kind of spiritual axis for the technological theme.

Whitman begins in an epic voice, singing the three great technological achievements of his age: the successful completion in 1866 of the Atlantic cable, which he had earlier celebrated in a Brooklyn *Times* article "The Moral Effect of the Cable"; the opening of the Suez Canal on November 17, 1869; and the joining of the Union Pacific and the Central Pacific railroads at Promontory Point on May 10, 1869:

> Singing my days,
> Singing the great achievements of the present,
> Singing the strong light works of engineers,
> Our modern wonders, (the antique ponderous Seven outvied,)
> In the Old World the east the Suez canal,
> The New by its mighty railroad spann'd,
> The seas inlaid with eloquent gentle wires.
>
> (*LGC*, p. 411)[6]

Like the radical reconstructionist Thaddeus Stevens, Whitman looked upon advances in science, technology, and industry as part of the march of humanity out of the feudal past and toward the democratic future. His vision of a world united in a common democratic culture by means of modern advances in communication and transportation was shared by his age; in fact, even the "Passage to India" refrain has sources in the popular rhetoric of the time. Hailing the completion of the transcontinental railroad, a reporter for the New York *Evening Post* wrote: "This is the way to India, telegraphed the directors of the Pacific Railroad, yesterday, from the point where the last rail had just been laid. But, give us only a half century of peace and freedom; give us only the unshackled development of all our faculties and energies, for fifty years, and we shall not cry out, This is the way to India; we shall stick up on our shores the sign, *This is India.*"[7]

Whitman shared the fascination of his age with the opening of trade relations between East and West. In fact, in 1868 he was present at the president's reception of the Chinese embassy staff, occasioned by the opening of trade and diplomatic relations with China.[8] But it was not Whitman's purpose to hail the commercial "passage to India" made possible by advances in modern technology. The technological achievements he praises at the beginning of the poem are effaced in the course of the poem; these become, like the poem itself, a "Passage to more than India." The completion of the Atlantic cable, the transcontinental railroad, and the Suez Canal is the occasion for the poem, but its subject is the relation between vision and history. Like the section "Orbic Literature" in *Democratic Vistas*, "Passage to India" is a poem about the role of the poet in history, and it is a poem about faith, providence, and spiritual grace.

"There's more of me, the essential ultimate me, in that than in any of the poems," Whitman said of "Passage to India" (*WWC*, I, 56). The "ultimate me" to which Whitman refers is the vision of history from which his poems emerged. "Passage" is his "ultimate" poetic representation of the essentially religious role he conceived for himself as both actor and initiator in the unfolding drama of democratic history. "Passage to India" and its cluster, Whitman observed, gives "freer vent and fuller expression to what, from the first . . . lurks in my writings, underneath every page, every line, everywhere" (*PW*, II, 465).

Although Whitman did not intend the poem to present a "philosophy, consistent or inconsistent," he did intend to present an essentially Darwinian vision of history: "the burden of it is evolution," he said, "the one thing escaping the other—the unfolding of cosmic purposes" (*WWC*, I, 156–57). This evolutionary pattern is evident in the opening sequence as the poet turns from his celebration of the "modern wonders" of technology toward a vision of the past as a "projectile" giving rise to the present. "Passage" moves backward and forward in time and space, forming a pattern of descent and ascent, which, along with the recurrent figures of rondure and circumnavigation, projects the vision of history as a gradually rising spiral that is the poem's evolutionary "burden." Like *Democratic Vistas*, the poem is Hegelian in vision and form, moving toward a dialectical union of opposites—past and present, East and West, religion and science, spirit and matter, fiction and fact—as the poet moves toward union with God.

The physical passage to India made possible by achievements in modern technology initiates a spiritual passage backward in time to the myths and fables of the past. In section 2, initially the "Fables" poem, the "facts" of modern science are presented in apparent opposition to the "fables" of ancient civilization:

> Passage O soul to India!
> Eclaircise the myths Asiatic, the primitive fables.
>
> Not you alone proud truths of the world,
> Nor you alone ye facts of modern science,
> But myths and fables of eld, Asia's, Africa's fables,
> The far-darting beams of the spirit, the unloos'd dreams,
> The deep diving bibles and legends,

The daring plots of the poets, the elder religions;
.    .    .    .    .    .    .
O you fables spurning the known, eluding the hold of the known,
   mounting to heaven!
.    .    .    .    .    .    .
You too with joy I sing!

                                       (*LGC*, p. 412)

The return to these myths and fables becomes the basis for a poetic and essentially religious rereading of Darwin. The voyage of the *Beagle* and the "facts" of modern science are made safe by presenting history not only as a physical evolution but also as a spiritual unfolding of God's purpose. This spiritual unfolding is enunciated, indeed propelled by, the "daring plots of the poets," the "far-darting beams of the spirit," and "unloos'd dreams" in the "deep diving bibles and legends" of the past.

As in Hart Crane's "The Bridge," in which the poet seeks, through the vaulting "curveship" of the bridge, to "lend a myth to God," Whitman envisions the technological spanning of the globe made possible by the cable, the railroad, and the canal as a sign of spiritual grace:

Passage to India!
Lo, soul, seest thou not God's purpose from the first?
The earth to be spann'd, connected by network,
The races, neighbors, to marry and be given in marriage,
The oceans to be cross'd, the distant brought near,
The lands to be welded together.

                                       (*LGC*, p. 412)

It is no coincidence that in celebrating technological progress, Whitman chose to focus on achievements that extended the possibilities of national and international communion: the Suez Canal linking the East to Europe, the Atlantic cable linking Europe with America, and the transcontinental railroad linking America with Asia. In this image of the earth literally "spann'd" by advances in technology, Whitman found a fitting symbol for his vision of history as a unitary and spiritually infused process evolving toward a global democratic community.

At the end of section 2 the poet announces a "worship new" that collapses the troubling opposition between science and religion, fact and fiction, poet and history:

A worship new I sing,
You captains, voyagers, explorers, yours,
You engineers, you architects, machinists, yours,

> You, not for trade or transportation only,
> But in God's name, and for thy sake O soul.
>
> (*LGC*, p. 412)

In Whitman's "worship new"—peopled in the past by trinities of captains, voyagers, and explorers and in the present by engineers, architects, and machinists—the fictions of the poet are reconciled with the facts of the machine age. Material progress is represented as a sign of God's will and, by implication, a realization in history of the "daring plots of the poets."

Scanning, winding, threading, crossing, and ascending the American continent, Whitman's persona merges with the transcontinental railroad, realizing through technological progress the voyage of national fusion—over, under, around, and through the country—that he had been able to take in imagination only in "Song of Myself." Rather than presenting the steam engine as a threat to the American landscape, Whitman, like the popular magazines and journals of his time, mythologizes the train as a dramatic means of opening the Mississippi valley to cultivation and eventually to realizing the democratic dream of America. Winding in "duplicate slender lines" across America, linking Europe and the East, the transcontinental railroad realizes in history the "unloos'd dreams" of another visionary—Columbus:

> (Ah Genoese thy dream! thy dream!
> Centuries after thou art laid in thy grave,
> The shore thou foundest verifies thy dream.)
>
> (*LGC*, p. 414)

Making real Columbus's dream of a "Passage to India," the train becomes a sign of the potency of vision, of dreams "verified" by history.

In the figure of Columbus, Whitman arrives at the real subject of his poem. The "spinal idea," he wrote in a notebook on "Passage to India," is that "the divine efforts of heroes, & their ideas, faithfully lived up to, will finally prevail, and be accomplish'd however long deferred . . . a brave heroic thought or religious idea faithfully pursued, justifies itself in time, not perhaps in its own way but often in grander ways."[9] Illustrating the process whereby vision is translated into history, Columbus is a prototype of Whitman's poet and the ultimate hero of "Passage to India."

Poetically enacting the stage theory of history enunciated in *Democratic Vistas*, in which the material is followed by a spiritual phase of development, "Passage to India" presents the poet as the spiritual counterpart of the explorers through whom "All these separations and gaps shall be taken up and hook'd and link'd together" (*LGC*, p. 415). Like the Emersonian poet, Whitman's poet will connect the "ME" with the "NOT ME." As the "true son of God," the poet replaces the old Christ, providing a saving and empowering vision in which man, nature, and God are compacted:

> Trinitas divine shall be gloriously accomplish'd and compacted by
>     the true son of God, the poet,
> (He shall indeed pass the straits and conquer the mountains,
> He shall double the cape of Good Hope to some purpose,)
> Nature and Man shall be disjoin'd and diffused no more,
> The true son of God shall absolutely fuse them.
>
> (*LGC*, pp. 415–16)

Moving from present to past and back again in a circular pattern that inscribes both the gradually rising spiral of history and the ascending circular movement of the poem, Whitman traces a sequence of historical events that begins in the valley of the Euphrates and culminates in the "world of 1492, with its awaken'd enterprise" and the rise of a new swell of humanity symbolized by the figure of Columbus:

> And who art thou sad shade?
> Gigantic, visionary, thyself a visionary,
> With majestic limbs and pious beaming eyes,
> Spreading around with every look of thine a golden world,
> Enhuing it with gorgeous hues.
>
> (*LGC*, p. 417)

With his "pious beaming eyes," Whitman's Columbus is both an explorer and a religious prophet; as an emblem of America (Columbia) he is also the initiator of the democratic phase of history. Erasing the more sordid facts of commerce and conquest that impelled Columbus's quest, Whitman mythologizes Columbus as the "chief histrion" of the phase of democratic history that culminated in the year 1869 with the realization of his dream of a passage to India.

"Visionary, thyself a visionary," Whitman says, insisting on the term to stress the image of Columbus as a prototype of the poet. "Behold his dejection, poverty, death," he adds, figuring his own fate and the fate of his book in post–Civil War America. But as "History's type of courage, action, faith," Columbus is also a model of democratic personalism and a figure of heroic possibility:

> (Curious in time I stand, noting the efforts of heroes,
> Is the deferment long? bitter the slander, poverty, death?
> Lies the seed unreck'd for centuries in the ground? lo, to God's
>     due occasion,

> Uprising in the night, it sprouts, blooms,
> And fills the earth with use and beauty.)
>
> (*LGC*, pp. 417–18)

Columbus is, in effect, a figure of the poet of *Leaves of Grass*, planting the seeds of a democratic golden world that, like the dream of a passage to India and a world in round, might bloom in some future transformation of vision into history.

The phase of physical exploration that culminates in the realization of Columbus's dream gives rise to a new phase of spiritual exploration in which the "chief histrion" is the poet. In the poem's final movement, Whitman emerges as the spiritual heir of Columbus and a kind of poetic founder of America:

> O soul, repressless, I with thee and thou with me,
> Thy circumnavigation of the world begin,
> Of man, the voyage of his mind's return,
> To reason's early paradise,
> Back, back to wisdom's birth, to innocent intuitions,
> Again with fair creation.
>
> (*LGC*, p. 418)

Like Thoreau drinking in the sacred waters of the Ganges, Whitman returns to the spiritual wisdom—the "primal thought"—embodied in the "budding bibles" of the past. Just as the physical exploration and progress of the world led backward to the East, so Whitman's spiritual exploration leads backward beyond the fallen world of biblical legend to the "innocent intuitions" of the East, where vision spurns the "separations and gaps" of history.

In his mystical journey toward the spiritual center of the universe, Whitman affirms his faith in the linked processes of life and death. His spiritual quest becomes, in the final sequence of the poem, a journey toward death:

> Reckoning ahead O soul, when thou, the time achiev'd,
> The seas all cross'd, weather'd the capes, the voyage done,
> Surrounded, copest, frontest God, yieldest, the aim attain'd,
> As fill'd with friendship, love complete, the Elder Brother found,
> The Younger melts in fondness in his arms.
>
> (*LGC*, p. 420)

Even though Whitman uses the imagery of democratic brotherhood, the union of his soul with God is not a union of equals. As a "moral, spiritual fountain," God is not equal to but "Greater than stars or suns"; the soul "yields" to God; and the body is lost in a bath of spirit. Whitman has lost the balance of his early poems. Whereas in "Song of Myself" the spirit comes down to merge with the body in a balance of body and soul, in "Passage to India" the soul rises out of the body and melts in the arms of God.

The God of "Passage to India" is a more orthodox figure than the pantheistic god of Whitman's earlier verse, and "Passage to India" is a more orthodox poem. Whereas God is evoked at the poem's outset as a dynamic, spiraling energy in the world, by the end of the poem, the movement becomes linear rather than circular, as the poet mounts toward a God who exists as a terminus beyond, rather than a power within, the world. Even Whitman's language becomes antiquated and mismatched as he seeks an appropriate religious vocabulary to describe this "transcendent" figure: His address is cluttered with thee and thou forms; his syntax is garbled in phrases such as "Waitest not haply for us somewhere there the Comrade perfect?" and the soul matest, smilest, fillest, swellest, copest, frontest, and yieldest in its journey to God.

Having "reckoned" the spiritual union with God that is at the end of his quest, the poet prepares his soul to set sail for this "Passage to more than India":

> Passage, immediate passage! the blood burns in my veins!
> Away O soul! hoist instantly the anchor!
> Cut the hawsers—haul out—shake out every sail!
> Have we not stood here like trees in the ground long enough?
> Have we not grovel'd here long enough, eating and drinking like
>     mere brutes?
> Have we not darken'd and dazed ourselves with books long enough?
>
> Sail forth—steer for the deep waters only,
> Reckless O soul, exploring, I with thee, and thou with me,
> For we are bound where mariner has not yet dared to go,
> And we will risk the ship, ourselves and all.
>
>                                            (*LGC*, p. 421)

The poet's leap into the sea is an affirmation of faith in death, the future, and the unknown. Resisting the pessimistic mood of his time and the course of history itself, Whitman, like some reckless Ahab, sets sail in a universe where the seas are "all the seas of God." But there is something rather desperate about the rhetorical posturing in this final passage. Whitman delivers his message of spiritual faith by waving exclamation points and blowing loud trumpet blasts into the ears of his readers. It is as if he is still not certain whether the question he asks at the end of the poem is rhetorical or real: "O daring joy, but safe! are they not all the seas of God?"

"Passage to India" is a manifesto poem that adumbrates the religious and spiritual orientation of Whitman's later work. The *Passage to India* cluster begins and ends with a leap toward spiritual transcendence and death. This increased emphasis on the spiritual world was in part a reaction against Gilded Age America. As the material conditions in post–Civil War America seemed to deny the theory of America and thus the validity of his life and work, Whitman turned increasingly toward the spiritual world to valorize and justify his democratic vision.

Ironically, however, his push away from a land populated by groveling "brutes" is also a push away from American democracy. His leap into the sea represents a push away from the land that was the putative and much-vaunted source of American character and institutions. After the war, as the passage westward led not toward democratic realization but a repeat of the past, Whitman sought a saving national vision by leaping out of history toward spiritual grace. He became, in effect, a prophet without a land. Like the ancient fables he describes in "Passage," he, too, came to spurn and elude the "hold of the known" in order to climb out of time and history and ascend to heaven.

In his later period, more than in his early years, Whitman became in full earnest Emerson's "Poet." Whereas in his early work Whitman characteristically stood on the shore, questioning and doubting, in his later work, as he pushed away from the land, he lost touch with the stubborn particularity of the physical world that had been a source of resistance and tension, paradox and wit, elegiac pathos and lyric exuberance in his early poems. As in the light-show ending of "Passage to India," Whitman tends in his later poems to affirm rather than question, declare rather than suggest, exclaim rather than explain. There is a rhetorical weightlessness in his later poems: His language is abstract and more traditionally poetic, and there is a corresponding tendency to regard the bodily self and the physical world as unreal. The "Dead" says Whitman in "Pensive and Faltering" are "the only living, only real/And I the apparition, I the spectre" (*LGC*, p. 455).

The best poems of Whitman's later period are poems like "A Noiseless, Patient Spider" and "Sparkles from the Wheel," which are based on the common and local. In "A Noiseless, Patient Spider," which is actually a revision of a *Calamus* poem of 1862–63, Whitman projects himself in the figure of a spider tirelessly unreeling filaments in search of an anchor or bridge in the "vacant vast surrounding" of the universe.[10] In "Sparkles

from the Wheel," Whitman communicates something of the anomie of a
modern city in the figure of a knife grinder displaying a craftsmanship
rapidly being replaced by wage labor and the assembly line values of
speed, profit, and efficiency.

If the *Passage to India* annex is characterized by a thrust away from the
material conditions of America, the second annex of the 1871 *Leaves, After
All Not to Create Only* (later, "Song of the Exposition") is more rooted in
the new era of expansion in industry, agriculture, and transportation fostered by the national government during the post–Civil War era. In this
poem, which Whitman read at the opening of the annual exhibition of the
American Institute in New York on September 7, 1871, and which was
published simultaneously in several newspapers, he seeks both to draw on
and to stimulate the national democratic culture he had envisioned in
*Democratic Vistas.* Using the rising glory rhetoric of the revolutionary period and the corresponding image of civilization's translation westward,
Whitman invites the muse to "migrate from Greece and Ionia" to America. The cultural products of the past are effaced as the muse joins hands
with Columbia and ensconces herself amid the gadgetry of industrial age
America:

> By thud of machinery and shrill steam-whistle undismay'd,
> Bluff'd not a bit by drain-pipe, gasometers, artificial fertilizers,
> Smiling and pleased, with palpable intent to stay,
> She's here, install'd amid the kitchen ware!
>
> (*LG: Variorum,* III, 615)

Although Whitman opposes the "great cathedral industry" to the cultural
monuments of the past, the poem's focus is not the products of industry
but the laborers who make that production possible.

> Materials here, under your eye, shall change their shape, as if by magic;
> The cotton shall be pick'd almost in the very field,
> Shall be dried, clean'd, ginn'd, baled, spun into thread and cloth,
>     before you:
> You shall see hands at work at all the old processes, and all the new
>     ones;
> You shall see the various grains, and how flour is made, and then bread
>     baked by the bakers;
> You shall see the crude ores of California and Nevada passing
>     on and on till they become bullion;
> You shall watch how the printer sets type, and learn what a composing-
>     stick is;

> You shall mark, in amazement, the Hoe press whirling its cylinders,
>   shedding the printed leaves steady and fast;
> The photograph, model, watch, pin, nail, shall be created before you.
>                                   (*LG: Variorum*, III, 617–18)

At a time when skilled labor and the traditional handcrafts were losing value and significance under the press of machine technology, Whitman stresses the creative and magically transforming hand of the worker in "all the old" and "all the new" processes of industrial production. For all Whitman's desire to install his muse amid the thud and steam of machine age America, his poetic representation of an ideal community of industrial workers is in fact a criticism of the actual conditions of city and factory life.

As Roy Harvey Pearce says in a discussion of "Song of the Exposition," Whitman "was discovering what Marx called alienation, the alienation of the laborer from the product of his labor, in that mass-industrial society in which all men willy-nilly become laborers."[11] Not only is Whitman concerned with humanizing the potentially dehumanizing effects of machine production. He is also concerned with naturalizing the essentially denaturalized landscape of modern industry: The industrial palaces that rise "tier on tier, with glass and iron facades" wear nature's colors. "Gladdening the sun and sky," they are "enhued in cheerfulest hues,/ Bronze, lilac, robin's-egg, marine and crimson" (*LG: Variorum*, III, 617). Like Louis H. Sullivan, who envisioned his skyscrapers "rising from the earth as a unitary utterance," Whitman represents the high-rising architecture of the modern age as a creation within rather than against nature.

With the massive devastation of the Civil War still fresh in his mind, Whitman is particularly concerned with celebrating the productivity of a peacetime economy in which armies of laborers replace the armies of war:

> Away with themes of war! away with War itself!
> Hence from my shuddering sight, to never more return, that show of
>   blacken'd, mutilated corpses!
> That hell unpent, and raid of blood—fit for wild tigers, or for
>   lop-tongued wolves—not reasoning men!
> And in its stead speed Industry's campaigns!
> With thy undaunted armies, Engineering!
> Thy pennants, Labor, loosen'd to the breeze!
>                                   (*LG: Variorum*, III, 619)

The passage registers Whitman's own turn away from themes of war to celebrate the values of reconciliation and rehabilitation, labor and productivity. And yet as in other postwar poems, the poet protests too much. His

strident, overbearing rhetoric is itself a sign of the psychic wound left by
the war and the vigorous effort Whitman had to make in the postwar
period to overcome the war's disruptive "show of blacken'd, mutilated
corpses," of "hell unpent," and "not reasoning men."

By stressing the value of productive labor and then the value of the
democratic Union that enables that production, Whitman opposes the
mass destruction of the war at the same time that he in some sense
rationalizes that destruction as a means of saving the Union. Overarching
the community of workers and the poem itself is the figure of the Union,
evoked once again as a sacred and infinitely bountiful Mother:

> All thine, O sacred Union!
> Ship, farm, shop, barns, factories, mines,
> City and State—North, South, item and aggregate,
> We dedicate, dread Mother, all to thee!
>
> Protectress absolute, thou! Bulwark of all!
> For well we know that while thou givest each and all (generous as God,)
> Without thee, neither all nor each, nor land, home,
> Nor Ship, nor mine—nor any here, this day, secure,
> Nor aught, nor any day, secure.
>
> (*LG: Variorum,* III, 623)

As emblem and protectress of the *Novus Ordo Seclorum* and the *E Pluribus
Unum* that were part of the revolutionary seal of the American republic,
Whitman's Union as mother is not only a source of material abundance;
she is a spiritually invested being—"generous as God"—who is the ultimate source of security for self and world.

The Union as a sacral value, as "Protectress absolute" against the
forces of darkness in the individual and the world, is the final lesson of
Whitman's song of the 1871 exhibition: "Think not our chant, our show,
merely for products gross, or lucre—it is for Thee, the Soul in thee,
electric, spiritual!" (*LG: Variorum,* III, 624). This transcendent and maternal vision of the Union became the saving faith of Whitman's final years,
protecting and securing him against the "products gross" of American
history. As a spiritually charged Mother, the Union was a democratic
bonding of individuals for the freedom and good of all; she was a bulwark
against the demons released by the war, of "hell unpent" and "not reasoning men"; she was a sign of "rightful rule" as the underlying law of the
universe and a personal sign for Whitman that his own life and work had
meaning.

It was in his 1872 preface to a sheaf of seven poems entitled *As a Strong Bird on Pinions Free* that Whitman first fully articulated the break with his earlier work that came in the postwar period. Having already given "published expression" to his "epic of Democracy" in *Leaves of Grass,* he said, "the present and any future pieces from me are really but the surplusage forming after that volume, or the wake eddying behind it." Perhaps he felt he had nothing more to say: With no interest in "business pursuits and applications usual in my time and country," he said, "it may be that mere habit has got dominion of me, when there is no real need of saying any thing further" (*PW,* II, 458–59).

Turning away from the war's scars and mounded graves, Whitman stressed the role his own work was to play in reconstructing the nation: "I say the life of the present and the future makes undeniable demands upon us each and all, south, north, east, west. To help put the United States (even if only in imagination) hand in hand, in one unbroken circle in a chant—to rouse them to the unprecedented grandeur of the part they are to play, and are even now playing—to the thought of their great future, and the attitude conform'd to it . . . these, as hitherto, are still, for me, among my hopes, ambitions" (*PW,* II, 463). The parenthetical "even if only in imagination" is—as is common with Whitman's parenthetical remarks—the key phrase here; it is the poet's concession to the sordid facts of post–Civil War America. Accepting the conflict between the ideals of the republic and the reality of Gilded Age America as a given, Whitman dedicates himself to creating an imaginary United States. It is in the gap between ideology and reality—in the absence of his true America of the mind—that Whitman found his poetic siting in the postwar period. Rather than singing an already existing world, he would sing America into being.

As part of his attempt to bring this America of the imagination into existence, Whitman turned in *As a Strong Bird* and future works toward an increased emphasis on democratic nationality: " 'Leaves of Grass,' already publish'd, is, in its intentions, the song of a great composite *democratic individual,* male or female. And following on and amplifying the same purpose, I suppose I have in my mind to run through the chants of this volume, (if ever completed,) the thread-voice, more or less audible, of an aggregated, inseparable, unprecedented, vast, composite, electric *democratic nationality*" (*PW,* II, 463). His greater emphasis on the aggregate was in part a response to the fact that the myth of Jeffersonian individualism was no longer adequate to the centrifugal force of a modern capitalist economy: Rampant individualism was at the very source of the unbridled greed, acquisitiveness, and materialism that characterized the Gilded Age, and it was also at the root of the labor–capital strife that intensified in the last quarter of the nineteenth century. There was a need for a new myth, a myth of communality and control, to counter the runaway force of laissez-faire individualism. Rather than proposing more government authority, Whitman encouraged an ethos of social responsibility by evoking a mythos of national union, a religiomystical and essentially matriarchal vision of the Union as a political ordering bound not by law or imposition from above but spiritually by a sense of common values and shared traditions.

Aside from his general desire to create a cohesive image of the nation fused by a commonality of beliefs and values, what Whitman meant specifically by a "thread-voice" of democratic nationality is unclear. He was aware of the new presence of the masses as a particularly modern, democratic phenomenon both in America and abroad, and he was interested in expressing this modern aggregate in his work. An entry in one of his notebooks preparatory to writing *Leaves of Grass* indicates that he was exploring new ways of expressing solidarity and the en masse in literature. Under the heading "idea for Novel or Play," he wrote:

> Work of some sort—instead of sporadic characters—introduce them in large masses, on a far grander scale—armies—twenty-three full-formed perfect athletes. . . . Nobody appears upon the stage simply—but all in huge aggregates nobody speaks alone—Whatever is said, is said by an immense number. Bring in whole races, or castes, or generations, to express themselves . . . voice of generations of slaves—of those who have suffered—voice of Lovers—of Night—Day—Space—the stars—the countless ages of the Past—the countless ages of the future—Do not fancy that I have come to descend among you, gentlemen. —I encompass you all.[12]

The essence of this note was translated into the "many long dumb voices" of Whitman's composite persona in "Song of Myself" (*LG*, 1855, p. 48, lines 509–20). Like D. W. Griffith in *Birth of a Nation* (1915) and John Dos Passos in *USA Trilogy* (1938), Whitman was working toward a wholly new concept of artistic expression, an art originating not in the particularities of individual experience but in the collective consciousness of the masses, in which nobody speaks or holds the stage alone.

In the post–Civil War period in particular, Whitman made the collectivity, the en masse, a focus of his work. Whereas in his early poems he imagined himself regenerating the republic by planting seeds of individual growth, in the preface poem to *As a Strong Bird*, "One Song, America, Before I Go" (later incorporated into "Thou Mother with Thy Equal Brood"), he imagines himself sowing seeds of "endless Nationality." Revising his earlier borrowing from Edmund Spenser, Whitman imagines himself fashioning not the individual, but the nation: "I'd fashion thy Ensemble, including Body and Soul; I'd show, away ahead, thy real Union, and how it may be accomplish'd" (*LG: Variorum*, III, 633). Whitman's democratic ensemble, like his democratic individual, is a construct. Unable to penetrate the "blank of the future," in the title poem "As a Strong Bird on Pinions Free" ("Thou Mother with Thy Equal Brood"), Whitman authors an imaginary script of the future nation: "I but thee name—thee prophecy—as now!/I merely thee ejaculate!" (*LG: Variorum*, III, 638). Fusing male and female metaphors of reproduction, Whitman imagines the poet as a man mother, simultaneously ejaculating and giving birth to the democratic future by naming its possibility.

Whitman was never able to carry out his plan to write a new volume of poems centered on the theme of democratic nationality. On January 23, 1873, he suffered a paralytic stroke that left him virtually immobilized for a few weeks and crippled for the rest of his life. In May he made a trip to Camden, New Jersey, to see his mother a few days before her death on May 23, 1873. "I feel that the blank in life & heart left by the death of my mother is what will never to me be fill'd," he wrote to his friends John and Ursula Burroughs (*Corr.*, II, 225). Physically handicapped, as well as emotionally depressed by his mother's death, Whitman was unable to return to his job in Washington. Although he hired a substitute for a time, on July 1, 1874, his services as a clerk in the attorney general's office were officially terminated. He spent the remainder of his life in Camden, first at his brother George's house and finally, beginning in 1884, in his own home at 328 Mickle Street (Figure XI).

Referring to his physical disability as the "war-paralysis," Whitman attributed his stroke to his service in the hospitals during the war years. Although he may not literally have inhaled poisons in the war hospitals, as he sometimes claimed, his stroke was at least partly a result of the psychic demons that came to haunt him during and after the war years. According to Whitman, his Washington physician Dr. Drinkard told him that "it was the result of too extreme bodily and emotional strain continued at Washington and 'down in front,' in 1863, '4 and '5" (*PW*, II, 704–5). Immediately after the war, Whitman appears to have suffered from a kind of shell shock that manifested itself in physical symptoms and psychic stress.

This stress intensified in 1870 when he suffered another emotional crisis, probably related to his love relationship with a horse-car conductor named Peter Doyle (Figures XII and XIII). In a notebook entry dated July 15, 1870, Whitman's public image as the good gray poet struggles with his personal desire as a closeted gay poet, leading him to resolve "TO GIVE

FIGURE XI. Whitman's house in Camden, New Jersey. Courtesy of the Walt Whitman House, Camden, New Jersey.

FIGURE XII. Photograph of
Peter Doyle, 1868. Courtesy of
the Sixsmith Collection, The
John Ryland University Library
of Manchester.

FIGURE XIII. Photograph of
Whitman and Doyle, 1865,
Washington, D.C. Courtesy of
The Library of Congress.

UP ABSOLUTELY & *for good, from this present hour,* this FEVERISH,
FLUCTUATING, *useless undignified pursuit of 164—too long, (much too
long)* persevered in, —so humiliating—*It must come at last* & had better
come now—(*It cannot possibly be a success*)." He comments on the need to
suppress his love for men: "Depress the adhesive nature[.] It is in
excess—making life a torment All this diseased, feverish disproportionate
*adhesiveness*" (*UPP*, II, 96). During this same period Whitman carried on
an extensive and loving correspondence with Doyle.[13] Although he does
not mention Doyle by name in his notebooks, it is commonly assumed that
164 is a code for the initials PD, corresponding to letters 16 and 4 in the
alphabet. Whereas during the war years, Whitman's homosexual desire
became a source of sustenance and utopian vision, in the postwar period,
as the more conservative sexual ideology of the new bourgeois order took
hold, his love relationships with men became a heightened source of self-
torment and self-doubt.

There is a rather tragic irony in the fact that the democratic poet of
bodily health and free-wheeling mobility should become in his later years
a "half-paralytic." And yet there had always been a curious correspondence between Whitman's body and the body politic of America: His body
seemed at times a kind of national seismograph, registering disturbances
in the political sphere. In the postwar period in particular, his physical
paralysis and the corresponding lack of vitality in his work seemed to
reflect the diseased condition of the political republic. Whitman frequently described his physical state in a way that suggested its connection
with the state of the Union in the postwar years. "It is singular how much
nervous disease there is—and many cases of paralysis & apoplexy," he
wrote his mother after his own stroke. "I think there is something in the
air" (*Corr.*, II, 220). He referred to the physical problems of his later years
as "bequests of the serious paralysis at Washington, D.C., closing the
Secession war—that seizure indeed the culmination of much that pre-
ceded, and real source of all my woes since" (*PW*, II, 736). Later on his
first trip west in 1879, Whitman, perhaps confronted with the conflict
between his dream vision of the West and the reality of rough conditions
and "plenty of hard-up fellows," became physically ill and had to return to
the East (*Corr.*, III, 168).

In a prefatory note to *Leaves of Grass* (1889), Whitman traced the
course of his bodily ills in such a way as to connect them with the socioeconomic ills of America in the latter half of the nineteenth century: "The
perfect physical health, strength, buoyancy . . . which were vouchsafed
during my whole life, and especially throughout the Secession War period, (1860 to '66,) seem'd to wane after those years, and were closely

track'd by a stunning paralytic seizure, and following physical debility and inertia, (laggardness, torpor, indifference, perhaps laziness,) which put me low in 1873 and '4 and '5—then lifted a little, but have essentially remain'd ever since" (*PW*, II, 736). The physical and emotional debility that put Whitman "low" between 1873 and 1875 coincided with the worst economic depression in American history. "There is an awful amount of want & suffering, from no work, hereabout," Whitman wrote to Pete Doyle of conditions in New Jersey in 1874 (*Corr.*, II, 275). By 1875, 500,000 workers were unemployed nationwide. During the same period, the worst scandals of the Grant administration were also exposed. The Credit Mobilier scandal, implicating several congressmen in a railroad fraud, broke in the fall of 1872; in February and March the "salary grab" act caused public outrage; in the summer, the secretary of the treasury exposed several treasury officials who had taken bribes from the whiskey-ring conspiracy of distillers; and five of Grant's cabinet members were implicated in illegal financial dealings.

Whitman's low mood in the period initially following his stroke is evident in "Prayer of Columbus," which appeared in *Harper's Magazine* in March 1874. In his notes for the poem Whitman wrote under the heading "Portraiture of Columbus": "pourtray [*sic*] him as a mystic he *was a mystic.*"[14] He had already included a "portraiture" of Columbus as a figure of himself in "Passage to India," but in "Prayer of Columbus" he makes that identification complete. "As I see it now," he wrote Ellen O'Connor, "I shouldn't wonder if I have unconsciously put a sort of autobiographical dash in it" (*Corr.*, II, 272). In the voice of Columbus ailing and ship-wrecked on the island of Jamaica, Whitman utters his own woe and the woefulness of his times:

> A batter'd, wreck'd old man,
> Thrown on this savage shore, far, far from home,
> Pent by the sea, and dark rebellious brows, twelve dreary months,
> Sore, stiff with many toils, sicken'd, and nigh to death,
> I take my way along the island's edge,
> Venting a heavy heart.
>
> (*LG: Variorum*, III, 661–62)

Physically paralysed and politically disillusioned, Whitman moves in "Prayer of Columbus," as in "Passage to India," toward a more traditional religious faith. Whereas the early Whitman had consistently railed against those religion systems that postulated a divine authority outside the self, in "Prayer of Columbus," he utters his own prayer, yielding the authority of self and the command of the democratic ship to the divine "Steersman" in the sky. The gesture measures the distance between the early and late Whitman and the extent of his disillusionment with America's experiment in democracy.

Wavering, like Columbus, between the hope that his work was part of a divine plan for "newer, better worlds" and his "mocking" suspicion that he was the "raving" victim of some cosmic joker, Whitman struggled to accommodate the scandals of the Grant administration and the rankness of the Gilded Age in a saving national vision. Grant was, in Whitman's view, "nothing heroic, as the authorities put it—and yet the greatest hero." His rise from the son of a tanner to general and president was the essence of the American success myth, illustrating "the capacities of that American individuality common to us all."[15] To Whitman, Grant's acts as both general and president were instrumental in preserving and solidifying the Union; and his reelection in 1872, despite the challenge to his Southern policy and his corrupt administration by the liberal Republicans, confirmed both the Union victory and the policies of radical reconstruction.

In a diary entry for March 4, 1874, Whitman noted that Grant's election "confirmed for the second time, the Principle of Nationality, as the principle dominating all others in American politics." His overwhelming victory established "the Reconstruction measures and the 13th, 14th, and 15th Amendments to the Constitutions [*sic*] as organic and immutable elements of the Constitution through the time to come."[16] Although Whitman had disagreed with his friend William Douglas O'Connor over the wisdom of immediately enfranchising black men, he did support the Reconstruction policies and constitutional amendments that would give to black persons, if not the reality then at least the promise of, equal civil and political rights.[17]

To give Grant a "moment's diversion from the weighty stream of official and political cares," in February 1874 Whitman sent him a copy of his war memoranda, "'Tis But Ten Years Since," which had appeared in the New York *Daily Graphic.* "You of all men can best return to them, in the vein in which they are composed," the poet wrote. "I am not sure whether you will remember me—or my occasional salute to you in Washington" (*Corr.,* II, 280). Whitman regarded the corruption of the Grant administration and the failure of moral energy in the nation as passing symptoms rather than as signs of terminal illness in the democratic body of America. The loss of idealism in the nation was symbolized for some by the death in 1874 of Charles Sumner, who had led the move in Congress for a civil rights bill to ensure the full equality of black freemen. Responding to Ellen O'Connor's anxiety about "political & public degradation—Sumner's death & inferior men &c. being rampant &c," Whitman assured her: "I look on all such states of things exactly as I look on a cloudy & evil state of weather, or a fog, or long sulk meteorological—it is a natural result of things, a growth of something deeper, has its uses, & will hasten to exhaust itself, & yield to something better—" (*Corr.*, II, 289).

The scandals of the Grant administration are the subject of "Nay, Tell Me Not To-day the Publish'd Shame," which was published in the *Daily Graphic* during the winter session of Congress in 1873:

> Nay, tell me not to-day the publish'd shame,
> Read not to-day the journal's crowded page,
> The merciless reports still branding forehead after forehead,
> The guilty column following guilty column.
>
> To-day to me the tale refusing,
> Turning from it—from the white capitol turning,
> Far from the swelling domes, topt with statues,
> More endless, jubilant, vital visions rise
> Unpublish'd, unreported.
>
> (*LGC*, p. 578)

Whitman's initial *Nay* is significant, for the poem marks his refusal to read or hear the "publish'd shame" of the time, which included the recent exposure of the Credit Mobilier scandal and the "salary grab" act. He turns away from the "swelling domes" of the Capitol to an "unpublish'd, unreported" vision of the country, but the vision he evokes is at once "unpublished" and unreal:

> Through all your quiet ways, or North or South, you Equal
>     States, you honest farms,
> Your million untold manly healthy lives, or East or West, city or
>     country,
> Your noiseless mothers, sisters, wives, unconscious of their good,
> Your mass of homes nor poor nor rich, in visions rise—
>     (even your excellent poverties,)
> Your self-distilling, never-ceasing virtues, self-denials, graces,
> Your endless base of deep integrities within, timid but certain.
>
> (*LGC*, p. 578)

Refusing the journalistic accounts of his time, Whitman "unimagines" the "publish'd shame" of America in a willfully idealized vision of "honest farms" and virtuous, self-denying citizens inhabiting a land-based Jeffersonian republic in which even poverty is excellent.

Whitman's desire to unmake the present in dreams of the democratic future is evident in the two-volume edition of his works that he published in conjunction with the Centennial Exhibition in Philadelphia in 1876. "O how different the moral atmosphere amid which I now revise this Volume, from the jocund influences surrounding the growth and advent of LEAVES OF GRASS" (*LGC*, p. 749), Whitman wrote in a note on his birthday in 1875. In a later note on his personal state when he prepared the centennial edition of his works, he wrote: "I was seriously paralyzed from the Secession war, poor, in debt, was expecting death . . . and I had the books printed during the lingering interim to occupy the tediousness of glum days and nights" (*PW*, II, 699). The moral atmosphere and gloom to which he refers allude specifically to his paralysis and his continued sadness over his mother's death, but the atmosphere also includes the dark days of the nation to which Whitman repeatedly refers in his new volume: "Thee, seated coil'd in evil times, my Country, with craft and black dismay—with every meanness, treason thrust upon thee" (*LG: Variorum*, III, 669). He feared that the "Pathology" of the present would enter into his new work, and indeed it did, both in the recurrent image of the "time's thick murk" and in the failure of inspiration evident in the centennial edition.

The first volume of the centennial edition was a reprint of the 1871–72 *Leaves of Grass*, with a few minor "Intercalations" added at the end of some editions. His second volume, entitled *Two Rivulets*, consisted primarily of previously published poetry and prose: *Democratic Vistas, As a Strong Bird on Pinions Free, Memoranda During the War,* and *Passage to India* all are reprinted from original plates with separate pagination.

In his preface to *Two Rivulets,* Whitman avows his earlier intent to turn the *Passage to India* cluster into a new volume of poems: "It was originally my intention, after chanting in LEAVES OF GRASS the songs of the Body and Existence, to then compose a further, equally needed Volume, [exhibiting] the problem and the paradox of the same ardent and fully appointed Personality entering the sphere of the resistless gravitation of Spiritual Law, and with cheerful face estimating Death." Unable to carry out his original plan, he decided to conclude his work with the *Passage to India* cluster and "thoughts, or radiations from thoughts, on Death, Immortality, and a free entrance into the Spiritual world" (*LGC*, p. 748).

*Two Rivulets* is, as its title suggests, double—split between politics and death. "I have not hesitated to embody in, and run through the Volume," Whitman says, "two altogether distinct veins, or strata—Politics for one, and for the other, the pensive thought of Immortality. Thus, too, the prose and poetic, the dual forms of the present book" (*LGC*, p. 748). This duality of perspective—between life and death, real and ideal, body and soul, present and future—structures the volume. The preface is itself divided between reflections on democratic nationality in the body of the essay and reflections on the purport of *Leaves of Grass* in a sequence of discursive footnotes. The contents of the volume is divided thematically between pieces on democracy, such as *Democratic Vistas* and the preface to *As a Strong Bird*, and pieces on death, such as *Memoranda During the War* and *Passage to India*. Just as this dualism of perspective is bound in a single volume, so in his epigraph Whitman establishes the idea of a union of opposites as the controlling metaphor of *Two Rivulets*: "*For the Eternal Ocean bound,/These ripples, passing surges, streams of Death and Life*" (*LG: Variorum*, III, 655). The differences between *Leaves of Grass* and *Two Rivulets*, the first representing his earlier emphasis on the "Body and Existence" and the second representing his later emphasis on "Death and the Spiritual World," are also, the poet says, "One in structure," part of "an interpenetrating, composite, inseparable Unity" (*LGC*, p. 751).

In the poems of *Two Rivulets* Whitman accommodates "the bad majority—the varied, countless frauds of men and States" in an Hegelian scheme in which "Only the Good is universal" ("Song of the Universal," *LG: Variorum*, III, 680). But by locating the drama of democracy in the "Spiritual World" outside the "Body and Existence" of his haughty, electric, contradictory, and frequently vulnerable persona, Whitman loses the vitality and specificity of his early verse. His poetry begins to limp with the hollow abstractions of democracy. Although Whitman continued to grope toward a more communal form, he lacked the stamina to complete such a work. Reflecting on the products of labor on display at the Centennial Exhibition in Philadelphia, he wrote: "The glory of Labor, and the bringing together not only representatives of all the trades and products, but, fraternally, of all the Workmen of all the Nations of the World, (for this is the Idea behind the Centennial at Philadelphia,) is, to me, so welcome and inspiring a theme, that I only wish I were a younger and a fresher man, to attempt the enduring Book, of poetic character, that ought to be written about it" (*LGC*, p. 751).

If in his early poetry Whitman struggled with and against the paradoxes and contradictions of democratic self and nation, in his centennial poems he seems content to be the poet of public policy. Rather than engage the political contradiction between clearing the West for settlement and developing a race "proportionate to Nature," in "Song of the Redwood-Tree" he celebrates the felling of trees as part of an "unseen moral essence" or "hidden national will" that molds the New World, "unswerv'd by all the passing errors, perturbations of the surface" (*LG: Variorum*, III, 674–79). Rather than explore the contradiction between the advance of "broad humanity" and the extermination of native Americans carried out by government agency, he sounds a "trumpet-note" for General George A. Custer, representing his last stand as a "lightning flash" of heroism amid the "dark days" of the present ("A Death Sonnet for Custer"; later, "From far Dakota's Cañons," *LG: Variorum*, III, 653–54).

Stiffened into his public pose as the good gray poet of democracy, Whitman seems no longer willing to give voice to his questions and fears about the future of America. But these questions and fears nevertheless come out, if only as in "Passage to India," indirectly and subterraneously. In "To a Locomotive in Winter" he celebrates the locomotive as a "Type of the modern! emblem of motion and power! pulse of the continent!" The poem registers the public fascination with the railroad which, as the key to western settlement, mass production and consumption, and the binding of the nation, dominated the economics, politics, and imagination of late-nineteenth-century America. Despite the poem's upbeat tone, however, the poet's language of celebration is fraught with alarm. The "black cylindric body" of the locomotive, with its "great protruding headlight, fix'd in front" and its smoke stacks out-belching "dense and murky clouds" as it pants and roars and "shrieks" through the countryside, bears traces of an anxiety about machine technology that anticipates Frank Norris's train as Iron Monster, grasping the land and the people in the stranglehold of "The Octopus" (*LG: Variorum*, III, 666–67).

Like the republic itself in 1876, Whitman's centennial volume is full of centrifugal impulses, held together not organically from within but by binding from without. The volume is a patchwork, with each of the different sections paginated separately. For a volume intended to honor the centennial of the republic, there is also something foreboding about the pairing of politics and death, or as Whitman says of his songs, "Strands of Patriotism and Death." Death in these songs is no longer balanced with life as part of an ongoing process; rather, as in "Lilacs," death becomes a deliverance from life, a "last impregnable retreat—a citadel and tower" ("In Former Songs," *LG: Variorum*, III, 672–73). Politically, Whitman celebrates the future, but his mind, like *Two Rivulets* itself, is turned toward thoughts "on Death, Immortality, and a free entrance into the Spiritual world" (*LGC*, p. 748). Concluding with the "Passage to India" cluster, Whitman's centennial volume marks the course of the American republic itself, leaping not toward the democratic future but toward death.

For all Whitman's effort to celebrate the national birthday, his centennial volume *had* absorbed the "pathology" of a country "coil'd in evil times." While America displayed its material wealth and industrial prowess at the Centennial Exposition in Philadelphia, thousands of workers remained unemployed. On July 4, 1876, a skirmish between black militiamen and white civilians led to an armed confrontation that was settled by the intervention of federal troops. Occurring on the very day of the republic's centennial birthday, the conflict was a reminder of unresolved political tensions within the union. Whatever reconstruction had taken place in the South was largely the result of the threat or actual exercise of federal military power.

The presidential election of 1876 revealed the growing power and resistance of the Democratic party in the South. Samuel Tilden, the Democratic candidate, won a majority of the popular vote against the Republican candidate Rutherford B. Hayes. But Tilden lacked one electoral vote to win the election. The political stalemate, which lasted several months, resulted in the Compromise of 1877: In exchange for the votes of southern Democrats, Hayes agreed as president to withdraw the last remaining troops from Louisiana and South Carolina, to include Democrats in his government, and to support policies favorable to southern whites. Returning the nation to sectional compromise and the South to home rule, the Compromise of 1877 marked the official end of Reconstruction. For almost a century, racism and segregation were upheld by the courts and institutionalized as the official policy of the nation, North as well as South.[18]

As the nation began to move toward economic recovery and a new political regime under the presidency of Rutherford B. Hayes (1877–81), Whitman, too, began to move toward a renewed sense of bodily health. He was aided in his recovery by articulating his war experiences in *Memoranda During the War*, which was published in 1875. He was also invigorated by the visits to the New Jersey farm of Susan and George Stafford that he began making in 1876. Among the Staffords, whom he met through his affectionate relationship with their son Harry, he retrieved some of his own familial past in rural Long Island; and on the banks of Timber Creek, he practiced a rigorous regime of mud baths, scrubbing, and nude sunbathing that shocked the neighbors but restored his physical strength and vigor.

After 1876, there was also a turn in Whitman's literary fortunes. It is perhaps one of the ironies of poetic history that his centennial *Leaves*, which in some ways marked a low point in the life of both poet and nation, also marked an upward swing in his reputation as a poet. Under the impetus of a subscription campaign on his behalf carried out by his admirers in England and America, the centennial *Leaves* became his first book to sell well nationally and internationally. Several prominent British writers, including William Michael and Dante Gabriel Rossetti, Edward Dowden, Alfred Tennyson, John Ruskin, George Saintsbury, and Ford Madox Brown, bought copies of Whitman's books, sometimes paying double and triple the price. "Severely scann'd, it was perhaps no very great or vehement success," Whitman wrote, "but the tide had palpably shifted at any rate, and the sluices were turn'd in my own veins and pockets. That emotional, audacious, open-handed, friendly-mouth'd just-opportune English action, I say, pluck'd me like a brand from the burning, and gave me life again, to finish my book, since ab't completed" (*PW*, II, 699–700).

The restoration of balance is reflected in Whitman's final ordering of his poems in the 1881 edition of *Leaves of Grass*. In this edition, he integrated all of his annexes, thus signifying his abandonment of the plan to write a new volume of poems centered on the theme of democratic nationality and spiritual union. Unlike Whitman's postwar volume of *Leaves of Grass*, the 1881 volume once again includes a picture of the poet, but it is not, as in earlier volumes, a current picture of himself. Rather, he returns to the 1855 daguerreotype engraving of himself, which appears opposite his longest poem, now retitled "Song of Myself." The new title confirms the move toward the recreation of himself in the image of his book that had intensified during and after the war years. But the 1855 daguerreotype is also a sign of difference. Like James Fenimore Cooper returning to the youth of both hero and country in *The Deerslayer* (1841), Whitman's return to his 1855 portrait is another leap away from the present, signifying his desire to identify himself, his book, and the nation not with the "half-paralytic" of the postwar period but with the healthy and robust democrat of the prewar years.

And yet, while this idealized Jacksonian common man stands at the head of the 1881 *Leaves*, there are signs that Whitman has lost the revolutionary fire that marked his early period. Although he does not renounce the political radicalism of his early years, his deletions, additions, and changes reveal a quest for stability and balance and a corresponding falling off of the moral passion, political commitment, and struggle to resolve contradiction that characterized his earlier verse. At the very time that the Knights of Labor and the Farmer's Alliance were beginning to gain adherents, Whitman weakened his assault on the "grip of capital" by dropping the *Songs of Insurrection* cluster and distributing the poems in this grouping through *Leaves of Grass*. He also deleted "Respondez," the poem that had over the years registered the ironic gap between ideal and reality in the American republic. He deleted his comradely invocation to workmen and workwomen at the outset of "A Song for Occupations." And in "The Sleepers," he suppressed two of his most radical utterances: his erotically charged "O hot-cheek'd and blushing" sequence and his power-ful attack on slavery in the black "Lucifer" passage. He also suppressed several passages and a few poems that revealed his doubts about the future of self and nation: He deleted "Solid, Ironical, Rolling Orb" and the "chaos" passage in "Out of the Cradle" that begins "O a word! O what is my destination? (I fear it is henceforth chaos;)" (*LGC*, p. 639). These changes had the effect of removing sites of historical struggle from the poems and reinforcing the image of national growth as natural growth.

Whitman added only twenty new poems to the 1881 *Leaves*, the best of which is "The Dalliance of the Eagles," a poem depicting in violent and graphic language the copulation of two eagles in flight. The other 1881 poems are minor, occasional poems, including poems on Whitman's trip west in 1879 ("Italian Music in Dakota," "The Prairie States," and "Spirit That Form'd This Scene"), a poem on Grant's triumphant return from his world tour in 1879 ("What Best I See in Thee"), a poem on the death of his mother ("As at Thy Portals Also Death"), and a poem on the death of President James Garfield, who was shot on July 2, 1881, in the dispute over spoils and patronage that followed the election of 1880.

Whitman was depressed by the shooting of Garfield, whom he had known when the latter was a congressman in Washington. Coming only a few days before the Fourth of July, the shooting revived his Carlylean doubts about the prospects of democracy in the New World: "We had the most horrible *celebration* here I ever knew," he wrote of the national birthday; "ruffians yelling, crackers, and all the old guns & pistols of all Jersey, with all the bad elements of humanity completely let loose & making the most infernal din possible to conceive for over thirty hours" (*Corr.*, III, 232–33).

Garfield's death on September 19, 1881, made him the second presi-dent to be assassinated in less than twenty years. And yet "The Sobbing of the Bells," like Whitman's Lincoln elegy, effaces the violent circum-stances of Garfield's death. In six short lines that approximate the conven-tional language and sentiment of Longfellow and Whittier, Whitman evokes the "heart-beats of a Nation" sobbing in unison with the bells that toll on the occasion of Garfield's death. Ironically, the shooting of presi-dents has become, along with war, the most effective means of bringing the country together in a truly national union.

The most significant change in the 1881 *Leaves* is not the addition of new poems but Whitman's restructuring of the entire volume into a final coherent form. Introduced by a series of inscriptions that reach from the "One's Self" of the poet to the "Thou" of the reader, the poems repeat this outward motion, developing not chronologically but thematically from a focus on self in "Starting from Paumanok" and "Song of Myself," to a focus on the relation of self to other in the amative theme of *Children of Adam* and the adhesive theme of *Calamus*, and then toward the national and international focus of poems such as "Salut au Monde!," "Crossing Brooklyn Ferry," "Song of the Exposition," and "A Song for Occupations." These songs are followed by three new clusters: *Birds of Passage*, which focuses on the evolutionary advance of democracy; *Sea-Drift*, which collects several of the seashore poems, including "Out of the Cradle" and "As I Ebb'd"; and *By the Roadside*, a miscellany of vignettes of the poet's life and times, including "A Boston Ballad," "Europe," and "To the States, to Identify the 16th, 17th, or 18th Presidentiad."

The poems of the war and its aftermath are now included in a single *Drum-Taps* grouping. The change signifies the move of both poet and nation away from a central preoccupation with the war in the post-Reconstruction period. The war is still the axis of *Leaves of Grass*, but the poems are not distributed over three separate groupings, as in 1871 and 1876. No longer collected in a separate volume, Whitman's Lincoln poems logically follow the *Drum-Taps* poems in a grouping entitled *Memories of President Lincoln* (still the only place in which Lincoln is specifically named as the subject of the poems). The poems on the death of Lincoln are followed by the clusters *Autumn Rivulets*, *Whispers of Heavenly Death*, and *From Noon to Starry Night*. Like the poems collected in the *Passage to India* annex, these clusters concentrate on the themes of death, immortality, and the spiritual world. The *Songs of Parting* cluster still closes the volume with a leap toward the sea in "Now Finale to the Shore" and a parting kiss to the reader in "So Long!"

These clusters radiate in ever-widening concentric circles from a focus on self, life, body, light, day, and the social world toward a focus on the cosmos, death, soul, darkness, night, and the spiritual world. At the same time, the clusters and the poems they include continually fold back on one another chronologically and thematically, temporally and spatially, in a manner that suggests the image of ensemble—of "form and union and plan"—that is the final design and desire of *Leaves of Grass*.

# 12

## *Representing America*

> The final culmination of this vast and varied Republic will be
> the production and perennial establishment of millions of com-
> fortable city homesteads and moderate-sized farms, healthy
> and independent, single separate ownership, fee simple, life in
> them complete but cheap, within reach of all. Exceptional
> wealth, splendor, countless manufactures, excess of exports,
> immense capital and capitalists, the five-dollar-a-day hotels
> well fill'd, artificial improvements, even books, colleges, and
> the suffrage—all, in many respects, in themselves, (hard as it is
> to say so, and sharp as a surgeon's lance,) form, more or less, a
> sort of anti-democratic disease and monstrosity, except as they
> contribute by curious indirections to that culmination—seem to
> me mainly of value, or worth consideration, only with reference
> to it.
>
> —WHITMAN, "Our Real Culmination"

Whitman's major work during the 1870s and 1880s was not in poetry, but
in prose—in the political pamphlet *Democratic Vistas,* his *Memoranda During the War,* and his personal reflections on his life and times in *Specimen
Days,* which was published in 1882. Based on his Civil War notebooks
(1861–65) and a series of notes he took during his trips to Timber Creek
between 1875 and 1881 and his trips to the West, Canada, Boston, and
New York between 1879 and 1881, *Specimen Days* is, as Whitman said, a
"*prose* jumble" (*Corr.,* III, 301). At the beginning of the volume, in a note
written a few days before Independence Day 1881, Whitman resolved "to
go home, untie the bundle, reel out diary-scraps and memoranda, just as
they are, large or small, one after another, into print-pages, and let the
melange's lackings and wants of connection take care of themselves. . . .
May-be, if I don't do anything else, I shall send out the most wayward,
spontaneous, fragmentary book ever printed" (*PW,* I, 1). Like his Civil

War *Memoranda,* however, Whitman's *Specimen Days* is no more "reeled out" than was the spontaneous bop prose and poetry of his literary descendants, Jack Kerouac and Allen Ginsberg. If anything, his years of paralysis had given him more time to work and rework his notes into a deliberate design.

Whitman described *Specimen Days* as "an autobiography after its sort," with Montaigne and Rousseau as "sort o' synonyms"; it was a "gathering up, & formulation, & putting in identity of the wayward itemizings, memoranda, and personal notes of fifty years, under modern & American conditions" (*Corr.,* III, 308). *Specimen Days* is autobiography written in grand cipher; by collapsing the events of his personal life and fusing "the past in a single identity," Whitman fashions himself into an emblem of the nation's history in the latter half of the nineteenth century. "I suppose I publish and leave the whole gathering," he wrote, "first, from that eternal tendency to perpetuate and preserve which is behind all Nature, authors included; second, to symbolize two or three specimen interiors, personal and other, out of the myriads of my time, the middle range of the Nineteenth century in the New World; a strange, unloosen'd, wondrous time" (*PW,* I, 3).

The narrative has four sections: a brief account of his youth and manhood, which Whitman wrote for Richard Maurice Bucke, who was planning an "official" biography of the poet; an account of the Civil War, largely a reprint of *Memoranda During the War;* a series of meditations on nature based on Whitman's Timber Creek notes; and a final sequence of reflections on social and literary matters, including an extensive account of his trip West in 1879. Like Benjamin Franklin's *Autobiography,* which was begun before the Revolution and completed afterwards, Whitman's autobiography fuses the the pre– and post–Civil War years in a seemingly unitary pattern of history that tells a "wondrous" story of personal and national success.

*Specimen Days* may be as near as Whitman ever came to writing the work of democratic nationality that he proposed in 1871. Like the prose counterpart to the poems he included in *Two Rivulets,* he regarded *Specimen Days* as a companion volume to *Leaves of Grass.* Referring to them as his Adam and Eve in a publicity note he wrote for the volume, he remarked: "It is understood that Whitman himself considers 'Specimen Days' the exponent and finish of his poetic work 'Leaves of Grass,' that each of the two volumes is indispensable in his view to the other, and that both together finally begin and illustrate his literary schemes in the New World" (*Corr.,* III, 309). Placing the self in time in a chronological account that spans America in the pre– and post–Civil War period, *Speci-*

*men Days* is a personal and historical counterpart to Whitman's poetic
embodiment of the nation in *Leaves of Grass;* but the self that emerges in
his autobiographical account of his life and times is no less invented than
is the poetic persona who tramps through *Leaves of Grass.*

Like *Leaves of Grass, Specimen Days* turns on the themes of balance and
union, and a primary concern with reconciling the paradox of democracy
in America. Like Whitman's poetic persona, the narrator of *Specimen Days*
shuttles back and forth between self and aggregate, nature and city, soli-
tude and society, matter and spirit. There is a similar dualism or balance
of voices, moving between intimate addresses to the "reader dear" in the
nature notes and the more public, oratorical voice of the pieces on the
sociopolitical landscape of America. As in *Leaves,* the overarching design
of *Specimen Days* is to deliver "a soul-sight of that divine clue and unseen
thread which holds the whole congeries of things, all history and time, and
all events, however trivial, however momentous, like a leash'd dog in the
hand of the hunter" (*PW*, I, 258).

Whitman's hunter image is apt, for it suggests the almost physical
strain he experienced in attempting to plot the "unseen thread" that
would hold his "strange, unloosen'd time," like a "leash'd dog," within a
coherent narrative of personal and national regeneration. Moving from his
personal reminiscence of his early years toward the more public engagement of the Civil War, he begins in the third section an account of the
restoration he found on the banks of Timber Creek. The aftermath of the
war is summarized in a few brief sentences:

> I continued at Washington working in the Attorney-General's department
> through '66 and '67, and some time afterward. In February '73 I was stricken
> down by paralysis, gave up my desk, and migrated to Camden, New Jersey,
> where I lived during '74 and '75, quite unwell—but after that began to grow
> better; commenc'd going for weeks at a time, even for months, down in the
> country, to a charmingly recluse and rural spot along Timber creek. . . . And
> it is to my life here that I, perhaps, owe partial recovery (a sort of second
> wind, or semi-renewal of the lease of life) from the prostration of 1874–'75.
> (*PW*, I, 118)

By essentially eliminating the period of Reconstruction and the public
and private dis-ease associated with those years, Whitman moves his
narrative directly from the tragedy of the war to the restoration of an
eternalized nature that bears no sign of political struggle and the wounds
of history. After the war memoranda, the narrative resumes in the spring
of 1876: "Dear, soothing, healthy, restoration-hours—after three confin-
ing years of paralysis—after the long strain of the war, and its wounds and

death" (*PW*, I, 120). Still shaken by the war's "show of blacken'd, mutilated corpses" and "not reasoning men," Whitman explains and contains
the war within a regenerative pattern of history at the same time that he
links his own restoration hours at Timber Creek with the political and
economic revival of the nation on the occasion of its centennial.

Whitman's Timber Creek notes are a mediating act, designed to restore self, book, and world to "natural" equilibrium after the "*convulsiveness*" of the war. The atmosphere shifts from dark to light, from the
interiors of hospitals to the open air: "I restore my book to the bracing and
buoyant equilibrium of concrete outdoor Nature," Whitman says, "the
only permanent reliance for sanity of book or human life" (*PW*, I, 120).
But the narrator's message of natural balance is at odds with the narrative
move from the strain of war to the restoration of nature, which occurs as
an unnatural rupture in the story, a sign of discord rather than equilibrium
in the text of book and world.

Whitman's desire to restore his book to the "buoyant equilibrium" of
concrete nature is executed as a narrative break away from the diseased
"concrete" of post–Civil War America, and it is only through such a break
that he is able to find medicine for his "sick body and soul" in the waters
of Timber Creek and the flora and fauna of the surrounding countryside.
This medicine of nature he passes on to the ailing nation as a saving
narrative of restoration. "If the notes of that outdoor life could only prove
as glowing to you, reader dear, as the experience itself was to me" (*PW*, I,
118–19). Avoiding the precision of either scientist or stylist, Whitman
affects the uncultivated voice of nature, imitating an old cedar tree whose
"wild and free and somewhat acrid" plums grow even in unfriendly cli-
mates. He gabs and loiters like the ambling persona in "Song of Myself,"
delivering what becomes his last will and testament to his readers. His
notes are casual, garrulous, high-spirited, and full of the good humor of a
man who has reached the end of the road and knows it. Lugging his
portable chair and his slow-moving, half-paralytic body from place to
place along the creek, the poet of the open road wryly observes: "My
range is quite extensive here, nearly a hundred rods" (*PW*, I, 151).

Whitman chats intimately with nature as he would with an old chum,
making her speak the language of democracy—of equity, balance, and
union—that had been silenced by the politics of the new market economy.
To communicate this democratic "soul-sight" to the reader, Whitman,
like Thoreau in *A Week* and *Walden*, organizes his nature notes into a
seasonal rhythm, with the high points of faith usually occurring in spring
and summer. Whether sinking his feet into the ooze of a mud bath,
counting and measuring the leaves of a mullein plant, or joining in the

flights of the kingfishers, Whitman is concerned with demonstrating, and indeed embodying, the relation between the Me and the Not Me. In a sequence of apparently random pictures, he depicts again and again the presence of self in nature, nature in self. Whereas Thoreau's meditations on nature are aphoristic and continually spill over into philosophy and parable, Whitman's reflections are descriptive and interactive. He wants to create "interchange." Exercising with trees as a "natural gymnasia, for arms, chest, trunk-muscles," he says: "I can soon feel the sap and sinew rising through me, like mercury to heat. I hold on boughs or slender trees caressingly there in the sun and shade, wrestle with their innocent stalwartness—and *know* the virtue thereof passes from them into me. (Or may-be we interchange—may-be the trees are more aware of it all than I ever thought.)" (*PW*, I, 153).

Beyond his purely physical interchange with nature, Whitman plots the spiritual interchange that fuses and infuses the created world. As in his Civil War memoranda, his moments of mystical transcendence characteristically come on clear nights as he gazes upward at the constellations of stars in a moonlit sky. In "Hours for the Soul," he records the lesson of one such night in which the "orbs thick as heads of wheat in a field" declared the glory of God: "As if for the first time, indeed, creation noiselessly sank into and through me its placid and untellable lesson, beyond—O, so infinitely beyond!—anything from art, books, sermons, or from science, old or new. The spirit's hour—religion's hour—the visible suggestion of God in space and time—now once definitely indicated, if never again" (*PW*, II, 174).

Whitman's message of physical restoration and spiritual transcendence is both a message to and a reaction against the overcrowded and overcultivated cities of the East. "What a contrast from New York's or Philadelphia's streets!" he exclaims, as he steps through a field of flowering horsemint, boneset, and wild bean. Like the vagabond of the open road in *Leaves of Grass,* Whitman's nature lover exhorts his readers to escape the constrictions of society and culture:

> Away, from curtain, carpet, sofa, book—from "society"—from city house, street, and modern improvements and luxuries—away to the primitive winding, aforementioned wooded creek, with its untrimm'd bushes and turfy banks—away from ligatures, tight boots, buttons and the whole cast-iron civilizee life—from entourage of artificial store, machine, studio, office, parlor—from tailordom and fashion's clothes—from any clothes, perhaps, for the nonce, the summer heats advancing, there in those watery, shaded solitudes. Away, thou soul, (let me pick thee out singly, reader dear, and talk in perfect freedom, negligently, confidentially,) for one day and night at least,

returning to the naked source-life of us all—to the breast of the great silent
savage all-acceptive Mother. (*PW,* I, 121–22)

The passage draws on a long tradition of antiurban sentiment, a tradition
that was strengthening in America with the rapid growth of industrialization in the late nineteenth century. But Whitman's antiurban rhetoric also
bears the erotic impress of his sexual politics. The societal undressing he
proposes is not metaphoric, as it was in "Song of Myself" and "Song of
the Open Road." What he proposes is a literal undressing in the watery
solitudes of the great mother. Like the breast of nature that beckons
toward the dream of America at the end of F. Scott Fitzgerald's *The Great
Gatsby,* the breast of the "all-acceptive Mother" registers the longing of an
entire culture to return to a mythical green place, associated with the
idealism of the founding moment and the landscape of an undefiled
America. Whitman's desire to return to the "naked-source life" of the
mother is ultimately a form of resistance to a male-powered order associated with the "cast-iron civilizee life" of city and marketplace.

"Am much of the time in the country & on the water," Whitman wrote
to the British poet Frederick Locker-Lampson in the spring of 1880; "yet
take deep interest in the world & all its bustle, (though perhaps keeping it at
arm's length)" (*Corr.,* III, 179). What Whitman intended at Timber Creek
was not to forget the world but to "keep" it in perspective—"at arm's
length." But as he came to see the dis-ease of his time as a necessary part of
some divine scheme, he succumbed to a kind of joyful determination that
sapped his desire for struggle and change. In a note on "November 8, '76,"
he commented on the hotly contested election between Tilden and Hayes
that resulted in a political stalemate of several months and the virtual end of
Reconstruction: "As I hobble down here and sit by the silent pond, how
different from the excitement amid which, in the cities, millions of people
are now waiting news of yesterday's Presidential election, or receiving and
discussing the result—in this secluded place, uncared-for, unknown" (*PW,*
I, 135). Rolling with the seasonal rhythms of the universe, Whitman appears to advocate not withdrawal from but positive indifference to the
political struggles of his time; he appears to have lost his belief in an active,
public-spirited, and engaged citizenry, which he had formerly regarded as
the safeguard of the American republic.

Unlike Thoreau's *Walden,* however, in which the self withdraws from
society to pursue a solitary, inward course of travel, in the narrative pattern
of *Specimen Days,* Whitman's "restoration-hours" in the wooded solitudes
of Timber Creek prepare for a renewed advance on the social world. In
"Human and Heroic New York," he records his impressions of the city

after an absence of many years. Although he is disturbed by the "corpse-like" faces of a new conspicuously consuming leisure class he sees "careering" through Central Park, he still finds among the common people of New York an apparently undefiled source of democratic possibility:

> . . . the current humanity of New York gives the directest proof yet of success-ful Democracy, and of the solution of that paradox, the eligibility of the free and fully developed individual with the paramount aggregate. In old age, lame and sick, pondering for years on many a doubt and danger for this republic of ours—fully aware of all that can be said on the other side—I find in this visit to New York, and the daily contact and rapport with its myriad people, on the scale of the oceans and tides, the best, most effective medicine my soul has yet partaken. . . . (*PW*, I, 172)

After this paean to the American people, *Specimen Days* moves back and forth between the country and the city, between specimens of nature and specimens of the farmers, sailors, firemen, conductors, and day laborers whom Whitman regarded as the leaven of the democratic future.

If Whitman transformed his days at Timber Creek into a myth of national restoration, and his return to New York after the war into an emblem of democratic possibility, he translated his trip West into a proph-ecy of the American future. He was initially invited to speak at the Kansas Quarter Centennial Celebration in Lawrence, Kansas; he never delivered the speech, but he included a copy of it in *Specimen Days*. In this speech and the notes on the West that follow it, Whitman invokes what Henry Nash Smith called the Myth of the Garden: the idea of the West as a rural Eden where through hard work and a virtuous yeomanry the democratic ideals of the American republic would be realized.[1] Whitman was particu-larly impressed with the vastness and range of the prairies, which seemed to promise not only a regenerated human race but also a truly democratic literature: "This favor'd central area of (in round numbers) two thousand miles square seems fated to be the home both of what I would call America's distinctive ideas and distinctive realities" (*PW*, I, 208).

Whitman's account of his western travels is followed immediately by the essay "Edgar Poe's Significance," a pairing that is not completely random. It is in the prairies that Whitman finds the democratic measure and future siting of American literature, a literature that will fashion the "infinite and paradoxical variety" of the land into a single compact iden-tity: "Subtler and wider and more solid . . . than the laws of the States, or the common ground of Congress or the Supreme Court, or the grim welding of our national wars, or the steel ties of railroads, or all the kneading and fusing processes of our material and business history, past or present, would in my opinion be a great throbbing, vital, imaginative work, or series of works, or literature, in constructing which the Plains, the Prairies, and the Mississippi river . . . should be the concrete background" (*PW*, I, 223–24).

Judging the works of the "whole crowd" of American writers by the boundless, horizontal vistas of the plains and prairies, Whitman finds that the works of his literary contemporaries fall short. "Poe's verses," he observes, "illustrate an intense faculty for technical and abstract beauty, with the rhyming art to excess, an incorrigible propensity toward nocturnal themes, a demoniac undertone behind every page—and, by final judgment, probably belong among the electric lights of imaginative literature, brilliant and dazzling, but with no heat" (*PW*, I, 231). Whitman wanted for poetry "the clear sun shining, and fresh air blowing—the strength and power of health, not of delirium, even amid the stormiest passions—with always the background of the eternal moralities." But his evaluation of Poe's work is shrewd. He was not so quick as later critics were to dismiss Poe as un-American. He recognized that Poe's verse represented a significant "sub-current" in American literary culture, that the "morbidity, abnormal beauty," and "sickness" of Poe's verse was a sign of pathology beneath the "perennial and democratic concretes at first hand" (*PW*, I, 232). Poe was the "demoniac" brother of the democratic son of America; and in Whitman's later work, as he too began pushing out of space, out of time, and away from the landscape of democratic America, he began moving, willy-nilly, closer to Poe.

Whitman's trip to Boston in the spring of 1881 to deliver his annual lecture the "Death of Abraham Lincoln" and his return to Boston in the summer to prepare *Leaves of Grass* for publication became the occasion for him to reflect in the final section of *Specimen Days* on his relationship with the New England literary establishment. Although he writes tributes to Emerson, Longfellow, Whittier, and Bryant, his evaluation of their literary achievements is qualified and reserved. Emerson, who is the subject of three essays, is a commanding and "cheery" presence in the closing passages of *Specimen Days*. But in a comment that may veil Whitman's disappointment at not being included in Emerson's anthology *Parnassus* (1872), he describes his verse as "rhym'd philosophy" (*PW*, I, 267). Longfellow is linked with the spirit of European literature: "He is not revolutionary, brings nothing offensive or new, does not deal hard blows" (*PW*, I, 285). Whittier's verse is likened to the "measur'd step of Cromwell's old veterans," reflecting the "splendid rectitude and ardor of Luther, Milton, George Fox," at the same time that it suggests their "narrowness and wilfulness" (*PW*, I, 267). Only in the nature lyrics of William Cullen

Bryant, whom Whitman addresses as "citizen and poet," does he find the "first interior verse-throbs" of an autochthonous national literature.[2]

At the outset of his nature notes, Whitman worried that his "invalidism" would infect his narrative: "Doubtless in the course of the following, the fact of invalidism will crop out, (I call myself *a half-Paralytic* these days, and reverently bless the Lord it is no worse,) between some of the lines— but I get my share of fun and healthy hours, and shall try to indicate them. (The trick is, I find, to tone your wants and tastes low down enough, and make much of negatives, and of mere daylight and the skies.)" (*PW*, I, 119). Despite Whitman's effort to "make much of negatives"—to shape his life into an exemplum of personal and national progress—the fact of disease did "crop out" in the narrative of *Specimen Days*.

His reminiscences of the prewar years are steeped in the pain of nostalgia and indelible loss. This is particularly evident in his account of his return to the "Maternal Homestead" in West Hills, Long Island. Comparing the past with the present in words that echo Volney's *Ruins*, he states: "Then [1825–40] stood there a long rambling, dark-gray, shingle-sided house, with sheds, pens, a great barn, and much open road-space. Now of all those not a vestige left; all had been pull'd down, erased, and the plough and harrow pass'd over foundations, road-spaces and every-thing, for many summers; fenced in at present, and grain and clover growing like any other fine fields. Only a big hole from the cellar, with some little heaps of broken stone, green with grass and weeds, identified the place" (*PW*, I, 7). With its language of absence and erasure, holes and breaks, Whitman's evocation of the past suggests the experience of social disruption and dispossession that underlay the progressive rhetoric of the age. His vision of the maternal homestead, with its extended family gath-ered around a "vast kitchen and ample fireplace," expresses a longing to return not only to the house of his mother but also to a simpler, rural past lost in the cataclysm of the Civil War. For all his attempt to "contain" the war years in a natural and national regenerative cycle, the historical rup-ture signified by the Civil War continually breaks the thread of his narra-tive of restoration.

The dislocation spawned by the economics of post–Civil War Amer-ica is particularly evident in the note "A Specimen Tramp Family," dated June 22, 1878, in which Whitman comments on the "number of tramps, singly or in couples" he encountered on his visit to his friend John Bur-roughs in New York. Presenting a "specimen" of one such tramp family with several children crouching in the back of a rickety one-horse wagon, he focuses on the mother as an especially disturbing figure of social dislocation: "I could not see her face, in its great sun-bonnet, but some-how her figure and gait told misery, terror, destitution. She had the rag-bundled, half-starv'd infant still in her arms. . . . Eyes, voice and manner were those of a corpse, animated by electricity. She was quite young—the man she was traveling with, middle-aged. Poor woman—what story was it, out of her fortunes, to account for that inexpressibly scared way, those glassy eyes, and that hollow voice?" (*PW,* I, 169). The unvoiced tale of terror inscribed in the destitute look of the mother and child contests as it silently tells against the national success story Whitman is trying to write.

His note "The First Spring Day on Chestnut Street" suggests similarly untold miseries even in seemingly prosperous times: "Doubtless, there were plenty of hard-up folks along the pavements," he says as he walks along the crowded Philadelphia street in 1879, "but nine-tenths of the myriad-moving human panorama to all appearance seem'd flush, well-fed, and fully-provided" (*PW,* I, 188–89). As in his poetic catalogues, Whitman's panoramic vision of the abundant flow of humanity along Chestnut Street skims over the fact of "plenty of hard-up folks." But the signs of a dispossessed and impoverished urban populace crop out:

> The peddlers on the sidewalk—("sleeve-buttons, three for five cents")—the handsome little fellow with canary-bird whistles—the cane men, toy men, toothpick men—the old woman squatted in a heap on the cold stone flags, with her basket of matches, pins and tape—the young negro mother, sitting, begging, with her two little coffee-color'd twins on her lap—the beauty of the cramm'd conservatory of rare flowers, flaunting reds, yellows, snowy lilies, incredible orchids, at the Baldwin mansion near Twelfth street—the show of fine poultry, beef, fish, at the restaurants. (*PW,* I, 189)

Whitman's mood is upbeat; his description of Chestnut Street in spring is meant to affirm a hopeful national vision of progress and abundance. But his urban panorama is a vision of want amid plenty; what he sees almost inadvertently as his eyes glide over the city that gave birth to the Constitution of the United States is an "other" America heaped "on the cold stone flags," in which people blend with commodities as part of a marketplace culture of things.

In the concluding passage of *Specimen Days,* "Nature and Democracy—Morality," Whitman stresses the link between democracy, nature, and art, as well as the necessary balance between country and city that is the structuring theme of *Specimen Days.*

> I have wanted, before departure, to bear special testimony to a very old lesson and requisite. American Democracy, in its myriad personalities, in factories, work-shops, stores, offices—through the dense streets and houses of cities,

and all their manifold sophisticated life—must either be fibred, vitalized, by regular contact with out-door light and air and growths, farm-scenes, animals, fields, trees, birds, sun-warmth and free skies, or it will certainly dwindle and pale. We cannot have grand races of mechanics, work people, and commonality, (the only specific purpose of America,) on any less terms. (*PW*, I, 294–95)

The "old lesson" that Whitman teaches is the agrarian lesson of Jefferson: the land as the foundation of the American republic and the source of virtue among its citizens. But at a time when national resources were being unleashed for general exploitation and thousands of people were flocking to the city, it is a lesson uttered against the course of American history.

"Do you know what *ducks & drakes* are?" Whitman asked his friend William Douglas O' Connor. "Well, S. D. is a rapid skimming over the pond-surface of my life, thoughts, experiences, that way—the real area altogether untouch'd, but the flat pebble making a few dips as it flies & flits along—enough at least to give some living touches and contact-points—I was quite willing to make an immensely *negative* book" (*Corr.*, III, 315). What Whitman's comment suggests is that only by negating the "real area" of his life and times was he able to mold *Specimen Days* into a story of personal and national success. It was paradoxically only by writing "an immensely *negative* book" that he was able to create an immensely *positive* myth of national restoration.

If in *Specimen Days* Whitman refashions the diseased concrete of his life and times into a myth of personal and national restoration, in the *Collect* section that follows he gives full voice to his doubts about the future of democracy. In "Lacks and Wants Yet," he reflects with alarm on the signs in America of a class conflict between rich and poor: "What is more terrible, more alarming, than the total want of any such fusion and mutuality of love, belief, and rapport of interest, between the comparatively few successful rich, and the great masses of the unsuccessful, the poor? As a mixed political and social question, is not this full of dark significance? Is it not worth considering as a problem and puzzle in our democracy—an indispensable want to be supplied?" (*PW*, II, 533–34). In his later years, Whitman came to share what Henry George in *Progress and Poverty* (1883) described as the "wide-spread consciousness among the masses that there is something radically wrong in the present social organization."[3]

In "The Tramp and Strike Questions," which is part of a lecture Whitman never delivered, he looked with the eyes not of a national mythmaker but of a political realist at the economic threat to American democracy: "Beneath the whole political world, what most presses and perplexes to-day, sending vastest results affecting the future, is not the abstract question of democracy, but of social and economic organization, the treatment of working-people by employers, and all that goes along with it—not only the wages-payment part, but a certain spirit and principle, to vivify anew these relations" (*PW*, II, 527). Whitman's reflections on the increased conflict between labor and capital were probably stimulated by the railroad strike of July 1877, the first major labor strike in the nation's history. The strike resulted in massive losses of life and property along railroad lines across the country, and federal troops had to be called in to quell the violence.[4]

Along with *The Eighteenth Presidency!*, "The Tramp and Strike Questions" is another of those unpublished speeches to the American people in which Whitman reveals his roots in the revolutionary enlightenment and the political radicalism of the early nineteenth century. In considering what he calls the "Poverty Question" and the question of strikes, he does not, as does John Hay in *The Bread-Winners* (1884), treat strikes as signs of criminal lawlessness; nor does he, as does Samuel Gompers, the first president of the American Federation of Labor (founded in 1881), urge the workers to strive to gain advantages for themselves in the current economic system. Whitman's perspective is more revolutionary. Relating the ravages of the new economic order to the "rapine, murder, outrages, treachery, hoggishness" of the old order in Europe, he foresees the possible overthrow of the current political system. He links the striking workers with what he calls the "great" strike of the French Revolution. Whereas the American Revolution was political and limited, the French Revolution was economic and far-reaching: "The American Revolution of 1776 was simply a great strike, successful for its immediate object—but whether a real success judged by the scale of the centuries, and the long-striking balance of Time, yet remains to be settled. The French Revolution was absolutely a strike, and a very terrible and relentless one, against ages of bad pay, unjust division of wealth-products, and the hoggish monopoly of a few, rolling in superfluity, against the vast bulk of the work-people, living in squalor" (*PW*, II, 528).

If the Civil War had secured the political union, it had also secured a capitalist market economy that was making the political order of democracy no different from the feudal order of the past: "If the United States, like the countries of the Old World, are also to grow vast crops of poor, desperate, dissatisfied, nomadic, miserably-waged populations, such as we see looming upon us of late years—steadily, even if slowly, eating into them like a cancer of lungs or stomach—then our republican experiment, notwithstanding all its surface-successes, is at heart an unhealthy failure" (*PW*, II, 528). "The Tramp and Strike Questions" concludes with a somber note, dated February 1879, indicating that the signs of the economic slump of the 1870s were still evident in the sociopolitical landscape. "I saw to-day a sight I had never seen before—and it amazed, and made me serious; three quite good-looking American men, of respectable personal presence, two of them young, carrying chiffonier-bags on their shoulders, and the usual long iron hooks in their hands, plodding along, their eyes cast down, spying for scraps, rags, bones, &c." (*PW*, II, 528–29). As in the novelistic reflections of Howells, Norris, and Dreiser on the ironic reversals of economic democracy, Whitman is troubled by the arbitrary power of a new economic order that can undo the identities and fortunes of even the most solid-seeming American citizens. Translated into the actual landscape of late-nineteenth-century America, Whitman's mythical vagabond of the open road had become an impoverished commoner grubbing for sustenance along the backroads of the country.

But it is not only the crops of destitute and nomadic Americans bred by laissez-faire capitalism that trouble Whitman in "Notes Left Over"; he is also disturbed by the failures of democracy itself. "The Tramp and Strike Questions" is immediately followed by "Democracy in the New World," in which he reflects on the "weakness, liabilities and infinite corruptions of democracy" (*PW*, II, 529). In a passage that is closer to the bleaker view of humanity in Henry Adams's *Democracy* than it is to the usually sunny prospect of the bard of democracy, Whitman observes: "By the unprecedented opening-up of humanity en-masse in the United States, the last hundred years, under our institutions, not only the good qualities of the race, but just as much the bad ones, are prominently brought forward. Man is about the same, in the main, whether with despotism, or whether with freedom" (*PW*, II, 529). The passage marks a nadir in Whitman's reflections on democracy in America: No longer does he envision America as the site of a regenerated human race and the beacon of a transformed world. If humans are about the same "whether with despotism or with freedom," then the republican experiment of America was indeed an "unhealthy failure."

Despite Whitman's doubts about the republican experiment, however, his solutions to the problems of democracy in America were still framed by the political vision of the radical democrat: more freedom, less government, no tariff, equal rights, and a general distribution of land and wealth. In "Who Gets the Plunder?", which was originally intended for his lecture "The Tramp and Strike Questions," he returns to the locofoco rhetoric of his early years in addressing the question of protective tariffs. The profits of protection, he asserts, should go "to the masses of laboring-men—resulting in homesteads to such, men, women, children—myriads of actual homes in fee simple, in every State, (not the false glamour of the stunning wealth reported in the census, in the statistics, or tables in the newspapers,) but a fair division and generous average to those workmen and workwomen" (*PW*, II, 531). In reality, however, the "profits of 'protection' go altogether to a few score select persons—who, by favors of Congress, State legislatures, the banks, and other special advantages are forming a vulgar aristocracy, full as bad as anything in the British or European castes, of blood, or the dynasties there of the past" (*PW*, II, 532).

Although Whitman, who once said that he admired Andrew Carnegie's *Triumphant Democracy* (1886), never stopped celebrating American progress, he recognized that the monopolies and trusts being formed by Carnegie, J. P. Morgan, and John D. Rockefeller in the latter part of the nineteenth century were creating an antidemocratic leviathan that not even the poet could bring under control. As this new industrial combine tightened its hold, Whitman clung even more fiercely to the dream of an artisan republic. In "Our Real Culmination," one of the final essays in *Specimen Days and Collect*, the end he foresees for America looks backward toward the revolutionary dreams of the founding moment:

> The final culmination of this vast and varied Republic will be the production and perennial establishment of millions of comfortable city homesteads and moderate-sized farms, healthy and independent, single separate ownership, fee simple, life in them complete but cheap, within reach of all. Exceptional wealth, splendor, countless manufactures, excess of exports, immense capital and capitalists, the five-dollar-a-day hotels well fill'd, artificial improvements, even books, colleges, and the suffrage—all, in many respects, in themselves, (hard as it is to say so, and sharp as a surgeon's lance,) form, more or less, a sort of anti-democratic disease and monstrosity, except as they contribute by curious indirections to that culmination—seem to me mainly of value, or worth consideration, only with reference to it. (*PW*, II, 539)

Despite advances in wealth, technology, and education and even the right to vote, Whitman sees in the new economic order an "anti-democratic disease and monstrosity" that is carrying America away from its "real culmination" as a republic of independent laborers, in which property and wealth are generally distributed and "within reach of all."

If at the close of the nineteenth century, the future was for Whitman, as it was for America, a blank page, he spent his final years inscribing it with the republican legends of the past. But while Whitman dreamed backward toward the artisan republic of the past, he also dreamed forward toward an alternative social order. At a time when urban workers were beginning to organize and strike against the corporate monolith and, even in the agrarian South and West, farmers were organizing against the impoverishment of their lives in widespread tenancy and indebtedness, Whitman imagined a "real culmination" of the American republic in which the values of liberty and equality, shared wealth and general property ownership, productive labor and local control, independence and cooperation would offer a road to the democratic future different from the capitalist road in which the few profit at the expense of the many and at the expense finally of the republic itself.

# 13

## *"How Dangerous, How Alive"*

> Leaves of Grass is essentially a woman's book: the women do not know it, but every now and then a woman shows that she knows it: it speaks out the necessities, its cry is the cry of the right and wrong of the woman sex—of the woman first of all, of the facts of creation first of all—of the feminine: speaks out loud: warns, encourages, persuades, points the way.
>
> —WHITMAN, from Traubel, *With Walt Whitman in Camden*

On March 1, 1882, the Boston district attorney, Oliver Stevens, initiated proceedings to suppress the 1881 edition of *Leaves of Grass*. Perhaps responding to a complaint by the Society for the Prevention of Vice, the district attorney wrote to Whitman's publisher, Osgood & Company of Boston: "We are of the opinion that this book is such a book as brings it within the provisions of the Public Statutes respecting obscene literature and suggest the propriety of withdrawing the same from circulation and suppressing the editions thereof" (*Corr.*, III, 267 n).

Whitman at first agreed to remove what Osgood called "the obnoxious features," thinking that the changes involved "about ten lines to be left out, & half a dozen words or phrases" (*Corr.*, III, 267). But when he received a full list of the "lines and pages and pieces" to be "expunged," which included several passages as well as the entire text of "A Woman Waits for Me," "To a Common Prostitute," and "The Dalliance of the Eagles," he demurred. "The list whole & several is rejected by me, & will not be thought of under any circumstances" (*Corr.*, III, 270). He withdrew his book from Osgood & Co., and placed an official ban on any future attempt to suppress what he called his "sexuality odes."

Whitman had been repeatedly vilified in the American press for his obscenity, and New England had been particularly vocal in protesting his base sensuality. "The author should be kicked from all decent society as below the level of the brute," wrote the *Boston Intelligencer*, in response to the 1855 *Leaves of Grass*. Even the otherwise-sympathetic Thoreau had said: "It is as if the beasts spoke." And in 1860 Emerson himself had attempted to persuade Whitman to eliminate *Children of Adam* from *Leaves of Grass*.[1]

Whitman had already received other "pretty serious special official buffetings," including being dismissed on moral grounds from his job in the Bureau of Indian Affairs in 1865, but this was the first time that legal proceedings had been initiated against *Leaves of Grass*. The district attorney's threat of prosecution involved not only an attack on the sexual politics of *Leaves of Grass* and a violation of the First Amendment rights of freedom of speech and freedom of the press. It was also an attempt by what Osgood called "the official mind" to suppress the sexually active female body of Whitman's poems.

From the period of the founding of the British colonies in North America, a sexually unruly body had been feared as a source of social danger. In *Of Plymouth Plantation 1620–1647*, William Bradford described the outbreak of bodily wickedness that occurred shortly after the colonists' arrival: "Not only incontinency between persons unmarried, for which many both men and women have been punished sharply enough, but some married persons also. But that which is worse, even sodomy and buggery (things fearful to name) have broke forth in this land oftener than once." The scrupulous detail of Bradford's account of the trial of the servant Thomas Granger, who in the year 1642 was "detected of buggery, and indicted for the same, with a mare, a cow, two goats, five sheep, two calves and a turkey," suggests the rigor with which the government sought to regulate the political economy by controlling, in particular, the potentially dangerous bodies of the lower class. Granger was dutifully executed along with the bodies of the polluted and polluting animals.[2]

In the revolutionary and postrevolutionary period, as the egalitarian implications of the American and French revolutions began to spread, the fear of an unruly body became bound up with the fear of democracy itself. In fact, the bizarre gothic narratives of Charles Brockden Brown in such works as *Wieland* (1798) and *Edgar Huntly* (1799) might be read as signs not so much of a personal neurosis but of a more general cultural anxiety, rooted in the fear of a body and democracy run amuck.

Whitman, of course, had always courted and flaunted a sexually turbulent and unruly body as a sign of health and a badge of his democratic faith. Recognizing the link between sexual liberation and political liberation, he celebrated the body and sexual love in his poems and made an athletic, liberated, and magnetically charged female body the center of his political program. "To the movement for the eligibility and entrance of women amid new spheres of business, politics, and the suffrage, the current prurient, conventional treatment of sex is the main formidable obstacle. The rising tide of 'woman's rights,' swelling and every year advancing farther and farther, recoils from it with dismay" (*PW*, II, 494).

By the end of the nineteenth century, Whitman's sexual politics and the official politics of America were clearly at odds. As women in the late nineteenth century moved into the public sphere and, even in the role of social housekeepers, challenged the separation of spheres and the economy of the family that were the base of bourgeois hegemony, the "official mind" of America focused more and more on the regulation of the female body and female desire as a means of controlling the direction of the nation. "No matter how her mythic representations changed," says the historian Carroll Smith-Rosenberg, "from the mid-nineteenth century on woman had become the quintessential symbol of social danger and disorder."[3]

What Whitman's offending poems had in common was the naming of an active and potentially dangerous female body that those in power were attempting to legislate into an acceptable model of true womanhood. At a time when medical journals were arguing against any activity that would upset the female body in its natural reproductive capacity, Whitman's "A Woman Waits for Me" celebrated the procreative power of an athletic and sexually charged woman. At a time when middle-class moral reformers were crusading to control or silence the fact of female prostitution, which had been on the rise since the end of the Civil War, "To a Common Prostitute" named the prostitute not only as a sexual being but as a figure of sympathy and a victim of socioeconomic oppression. And at a time when female sexuality was being deployed into the biological determinism of motherhood, "The Dalliance of the Eagles" suggested, in the figure of two eagles copulating in midair, a fierce and erotically compelling female desire that was unrelated to any maternal or reproductive drive.[4]

By 1882 the female body of Whitman's poems was deemed dangerous to the public morality. Not only was *Leaves of Grass* banned in Boston, but those who attempted to publish his offending poems were persecuted. When Ezra Heywood of Princeton, Massachusetts, published "A Woman Waits for Me" and "To a Common Prostitute" in *The Word* in 1883, he was arrested, brought to trial, and finally acquitted. Whereas Bradford, living in colonial New England, scrupulously named "things fearful to name," the attempt to suppress Whitman's poems suggests that in the late nineteenth century, the "official mind" of New England had decided to suppress a sexually active female body by "un-naming" it on the level of language itself. And Massachusetts was not alone in its response. When Whitman contracted to publish *Leaves of Grass* with Rees Welsh and

Company in Philadelphia in 1882, the Society for the Prevention of Vice initiated similar proceedings to suppress his work in Pennsylvania.[5]

The irony, of course, is that while Whitman's sexually charged women and the heterosexual love poems of *Children of Adam* were singled out as a source of political danger, his love poems to men in *Calamus* went virtually unnoticed by the organs of official culture.[6] The other irony in the proceedings against *Leaves of Grass* is that while the official male mind of America sought to protect true womanhood by silencing the female text of Whitman's poems, women readers loved him and defended him passionately in letters and reviews. " 'Leaves of Grass' thou art unspeakably delicious," declared the popular writer Fanny Fern (Sara Parton) in a review of the 1855 *Leaves of Grass*. Raising a "woman's voice of praise" in opposition to the "charge of coarseness and sensuality" launched by "small critics," Fanny Fern's review, which appeared in her popular column in the New York *Ledger* in 1856, drew early and specific attention to the sexual politics of the debate over Whitman's work. "Walt Whitman, the effeminate world needed thee," she said, congratulating him for being "enamored of *women*, not *ladies*." As a woman rather than a "lady" reader, she for one had no problem with his sensuality and his candid "undraping" of the female body. "My moral constitution may be hopelessly tainted or—too sound to be tainted," she told her several hundred thousand readers, "but I confess that I extract no poison from these 'Leaves'—to me they have brought only healing. Let him who can do shroud the eyes of the nursing babe lest it should see its mother's breast." Extending the "cordial grasp of a woman's hand" to the poet, she concludes by citing Whitman's egalitarian appeal to his readers in "Song of Myself":

> The wife—and she is not one jot less than the husband,
> The daughter—and she is just as good as the son,
> The mother—and she is every bit as much as the father.[7]

The debate over *Leaves of Grass* developed in its initial stage into a battle over the propriety of naming the female body, particularly in its sexual and reproductive capacities. The battle divided male and female critics and, in a few instances, husband and wife. When Whitman's friend Henry Clapp sent a review copy of the 1860 *Leaves* to Juliette Beach, her response was intercepted by her husband, who submitted a review of his own. He compared Whitman with a "stock-breeder" who regarded "woman only as an instrument for the gratification of his desires, and the propagation of the species." The review was published under Juliette Beach's name, but she refused to be silenced by her husband's voice. In succeeding issues the *Saturday Press* printed her "Correction" and a review, signed by "A

Woman," in which she defends the "deep spiritual significance" of Whitman's work and the inevitable success of his "bold and truthful pages" in the "future of America."[8]

Women readers were particularly ardent in their defense of Whitman's poems of sex and the body.[9] At times his impassioned, tactile appeals for love—"touch me, touch the palm of your hand to my body"—elicited a strong erotic response. In 1860 he received a love letter from a Hartford, Connecticut, workingwoman named Susan Garnet Smith, who offered to bear his child: "My womb is clean and pure. It is ready for thy child my love. Angels guard the vestibule until thou comest to deposit our and the world's precious treasure" (*WWC*, IV, 312–13). Whitman scribbled "? insane asylum" on the envelope, but Susan Smith was not alone in her response to his work.

Anne Gilchrist, a woman of letters and respected member of the Rossetti circle in England, was similarly aroused by Whitman's poems. "I had not dreamed that words could cease to be words, and become electric streams like these," she wrote William Michael Rossetti after reading the unexpurgated manuscript of *Leaves* in 1869.[10] After suffering from the neurasthenia that came to characterize the Victorian "angel in the house," Gilchrist was given "new birth" by reading Whitman's poems. Like Fanny Fern and Juliette Beach, she insisted on the particularity of her female perspective. Commenting on the sex poems that Rossetti had expurgated from his 1868 edition of *Leaves of Grass,* she wrote: "I will take courage to say frankly that I find them also beautiful, and that I think even you have misapprehended them. Perhaps indeed they were chiefly written for wives!"[11]

In an "English Woman's Estimate of Walt Whitman," which appeared in the Boston *Radical* in 1870, Gilchrist singled out the sex poems of *Children of Adam* for special praise. Reflecting on the "ignominious shame brooding darkly" around female sexuality, she asks: "Do you think there is ever a bride who does not taste more or less this bitterness in her cup?" She praises Whitman for giving voice to a sexualized female body: "It was needed that this silence, this evil spell, should for once be broken, and the daylight let in." Rather than being "harmful to the woman," Whitman's poems were a source of self-realization, inciting the female to a feeling of pride in her body, her sexuality, and her creative power, "where foolish men, traitors to themselves, poorly comprehending the grandeur of their own or the beauty of a woman's nature, have taken such pains to make her believe there was none" (Gilchrist, pp. 14–17).

When Whitman failed to understand that the words of a "Woman's Estimate" were in fact "the breath of a woman's love," Gilchrist wrote him personally, offering herself as bride, wife, and eternal mate. "Dear Walt. It is a sweet & precious thing, this love; it clings so close, so close to the Soul and Body, all so tenderly dear, so beautiful, so sacred; it yearns with such passion to soothe and comfort & fill thee with sweet tender joy" (Gilchrist, p. 61). Despite her forty-two years, she, like Susan Smith, offered to bear his child: "I am yet young enough to bear thee children, my darling, if God should so bless me. And would yield my life for this cause with serene joy if it were so appointed" (Gilchrist, p. 66).

Whitman cautioned her not to confuse "an unauthorized & imaginary ideal Figure" with the "plain personage" of Walt Whitman (*Corr.*, II, 170). But Gilchrist was determined. After her mother died, she set sail for Philadelphia, where she arrived with her three children in September 1876. During her two-year residence in Philadelphia, her passion modulated into the warmth of friendship. She invited Whitman to early evening "tea-suppers," did viva voce translations of Victor Hugo, and offered him a place to stay when he wished (Figure XIV). To Whitman, Gilchrist was a model of heroic womanhood: "She was the mother of a number of children: she had done justice to her children: she had lived a real life with her husband: that was the substratum—a noble substratum, base: then on top of this she built the great scientific, intellectual, esthetic superstructure as the sort of crown to all. She was harmonic, orbic: she was a woman—then more than a woman" (*WWC*, IV, 93).

The female response to *Leaves of Grass* was, at least in part, Whitman's design. "Leaves of Grass is essentially a woman's book," he said; "the women do not know it, but every now and then a woman shows that she knows it: it speaks out the necessities, its cry is the cry of the right and wrong of the woman sex—of the woman first of all, of the facts of creation first of all—of the feminine: speaks out loud: warns, encourages, persuades, points the way" (*WWC*, II, 331). In a conversation with Traubel, he attributed the female sensibility of his book to his mother's influence: "Leaves of Grass is the flower of her temperament active in me" (*WWC*, II, 113). But the sources of his "woman's book" were deeper and more complex. The book had roots in the American revolt against patriarchal authority and in the fundamental myth of the democratic republic; in the fears for the Union in the antebellum period and the emergence of a centralized capitalist state in the post–Civil War period; in the women's rights movement and the cataclysmic socioeconomic transformations of the nineteenth century; in Whitman's desire for personal, social, and spiritual love and his own anomalous position as a homosexual in heterosexual America.

FIGURE XIV. Whitman with Anne Gilchrist in *The Tea Party*, 1882, by Herbert
Gilchrist. Courtesy of Special Collections, Van Pelt Library, University of Pennsylvania.

*Leaves of Grass* was designed to "arouse and set flowing in men's and
women's hearts, young and old, (my present and future readers,) endless
streams of living, pulsating love and friendship, directly from them to
myself, now and ever" (*LGC*, p. 753). If at times Whitman's work seems
to reinscribe the conservative sexual ideology of his time, his poems had
and still do have a galvanizing effect on women readers. The potential

danger of this effect is suggested by the poet Audre Lorde in her essay "Uses of the Erotic: The Erotic As Power":

> When we begin to live from within outward, in touch with the power of the erotic within ourselves, and allowing that power to inform and illuminate our actions upon the world around us, then we begin to be responsible to ourselves in the deepest sense. For as we begin to recognize our deepest feelings, we begin to give up, of necessity, being satisfied with suffering and self-negation, and with the numbness which so often seems like their only alternative in our society. Our acts against oppression become integral with self, motivated and empowered from within."[12]

Even in the nineteenth century, Whitman's poems aroused and set flowing erotic female energies, which, when not confined to a Victorian model of service to man, could be both politically dangerous and prodigiously empowering. In a world rigidly segregated into female and male spheres, Whitman's poems offered intimacy, companionship, sympathy, and sexual freedom among men and women. They were invitations, beckoning women readers out of domestic confinement toward an open road of equality and comradeship with men. To Eliza Farnham in her study *Woman and Her Era* (1864), Whitman was one of the pioneering feminists of his age. It was no doubt in recognition of Whitman's designs on his female readers—and his danger—that his poems became in the 1880s the target of Anthony Comstock's Society for the Prevention of Vice and ultimately the victim of government censorship.

By the late nineteenth century, as the male-driven engine of America appeared to be rushing headlong in the direction of the "fabled damned" of *Democratic Vistas*, Whitman's dream of America became increasingly matriarchal. At a time when the medical profession—with forceps in hand—was assuming increasing control of the female reproductive process, *Leaves of Grass* proclaimed the primacy and superiority of "divine maternity" in the order of creation. This emphasis on the power of motherhood was not a prescription that all women should become mothers. If Whitman celebrated the female in her reproductive capacity, he also celebrated the "female equally with the male," and his poems are full of images of women as workers and athletes who "know how to swim, row, ride, wrestle, shoot, run, strike, retreat, advance, resist, defend themselves" (*LGC*, p. 102). If he was particularly fond of mothers, his female heroes were Frances Wright, George Sand, Margaret Fuller, Anne Gilchrist, Lucretia Mott, and Delia Bacon—all women who had challenged traditional women's roles. "I expect to see the time," he said, "in Politics, Business, Public Gatherings, Processions, Excitements, when women shall not be divided from men, but shall take their part on the same terms as men" (*DN*, III, 739).

A part of the problem in dealing critically with Whitman's attitude toward women—and female critics still tend to be more ardent in defense of his women than male critics—is that his celebration of "divine maternity" inscribes a contradiction at the root not only of nineteenth-century American culture but of feminism itself. On the one hand, the mother was a limiting figure, the very symbol of what Michel Foucault has called the "deployment" of female sexuality by the bourgeois social order. On the other hand, as in Harriet Beecher Stowe's *Uncle Tom's Cabin* (1852), Louisa May Alcott's *Little Women* (1868–69), and Charlotte Perkins Gilman's *Herland* (1915), the mother was a utopian figure who represented a radical critique of the capitalist industrial order and the possibility of an alternative matriarchal economy.

In the last few years of his life Whitman wrote a poem entitled "America." In opposition to the commercial and material culture of America and the aggressively capitalist spirit of the time, he imagined the future in the egalitarian and communal image of the mother:

> Centre of equal daughters, equal sons,
> All, all alike endear'd, grown, ungrown, young or old,
> Strong, ample, fair, enduring, capable, rich,
> Perennial with the Earth, with Freedom, Law and Love,
> A grand, sane, towering, seated Mother,
> Chair'd in the adamant of Time.
>
> (*LGC*, p. 511)

The towering mother of "America" is more than an emblem of the corporate identity of the United States: "Perennial with the Earth, with Freedom, Law and Love," she bears the traces of an alternative social order.

To Whitman at the close of the nineteenth century, history and the future were female. Rather than reinforcing the cult of true womanhood and the status quo, the towering mother—"Chair'd in the adamant of Time"—challenged a political economy based on the separation of female and male, private and public, home and world, by placing the values of community, equality, creation, and love at the center rather than at the margins of democratic culture. This vision represented a nostalgic glance backward toward the revolutionary image of America as an abundant female republic and a utopian glance forward toward what Herbert Marcuse would call the "femalization" of man.[13] Like the figure of the woman who emerges as the base of a "reconstructed sociology" at the end of *Democratic Vistas*, the future that Whitman imagined—and the only future possible for America—was in the image of a divinely charged matriarch, self-poised, swinging through time.

It was this collective and female dream of the future that Whitman sought to maintain against the fact of a dying body and diminished hopes in the final years of his life. Weakened by a sunstroke in 1885 and another paralytic stroke in 1888 that "chain'd" him to a chair "as never before" (*PW*, II, 681), he wrote three brief annexes to *Leaves of Grass—Sands at Seventy* (1888–89), *Good-Bye My Fancy* (1891–92), and *Old Age Echoes* (1897). In these final poems, Whitman became the poet of old age: "old, alone, sick, weak-down, melted-worn with sweat" (*LGC*, p. 546). His crippled and disease-ridden body, old before its time, seemed once again an image of America itself at the close of the century. But despite the wreckage of his "old, dismasted, gray and batter'd ship, disabled, done" (*LGC*, p. 534), the poet continued to sing, picking over the remains of his own and the national life in poems on the Wallabout martyrs, the Washington Monument, the Civil War, Abraham Lincoln, General Grant, General Sheridan, Anne Gilchrist, Nature, and the Indians.

The Indians. After having suppressed in his previous poems the fact of the dead bodies of the Indians on which America was founded, Whitman wrote no fewer than three Indian poems in the final decade of his life. Each is sited at a crucial epoch in the "Indian wars": "Yonnondio," an Iroquois term meaning "lament for the aborigines," registers the silencing of the original inhabitants of the land—"blank and gone and still, and utterly lost"; "Osceola" is about the imprisonment of the well-known Seminole warrior Osceola, who died of "a broken heart" in Fort Moultrie, South Carolina, in 1838—at the time when Jackson's policy of "removing" the Indians west of the Mississippi was being completed; and "Red Jacket (from Aloft.)" is a tribute to the Iroquois orator, whose bones were reburied in Buffalo City on October 9, 1884, shortly after government agents had been employed to slaughter 2.5 million buffalo in order to herd the survivors onto reservations.[14] The poems appear to be merely topical, but coming as they do at the very moment when the stage was being set for the final massacre at Wounded Knee in 1890, they signal the ways in which the dead bodies of the Indians would rise out of the very ground of America, demanding to be reckoned with in any future dream of democracy.

Whitman's last poems are end-of-the-century poems that register the "cherish'd lost designs" of youth and the "ungracious glooms" of old age at the same time that they affirm "the rhythms of Birth eternal." Anticipating Robert Frost's "The Oven Bird," Whitman compares himself to "a single snow-bird merrily sounding over the desolation" from "arctic bleak and blank" (*LGC*, p. 520). If his voice died—he seemed to fear—he would die and America would die. He no longer believed in the inevitability of democracy. If democracy was going to happen, the people, urged on by the poets, would have to make it happen. And so he continued to sing.

"Behind all else that can be said," Whitman wrote in "A Backward Glance o'er Travel'd Roads" (1888), "I consider 'Leaves of Grass' and its theory experimental—as, in the deepest sense, I consider our American republic itself to be, with its theory" (*LGC*, pp. 562–630). In this essay written to conclude all future editions of *Leaves of Grass*, Whitman reasserts the active, essentially political design of his work: His poems are a democratic experiment that parallels and indeed figures the experimental politics of the American republic. He envisions *Leaves of Grass* as a form of poetic history, "an attempt, from first to last, to put *a Person*, a human being (myself, in the latter half of the Nineteenth Century, in America,) freely, fully and truly on record."[15] "But," he adds, "it is not on 'Leaves of Grass' distinctively as *literature*, or a specimen thereof, that I feel to dwell, or advance claims. No one will get at my verses who insists upon viewing them as a literary performance, or attempt at such performance, or as aiming mainly toward art or aestheticism" (*LGC*, pp. 573–74).

The difference that Whitman poses is unclear. He appears to mean to distinguish between the closure of traditional works of literature and the democratic design of *Leaves of Grass* which is, like the republic itself, open-ended. What he had in mind for "the States" was a poetry that would "express, vitalize and give color to and define their material and political success, and minister to them distinctively" (*LGC*, p. 574). To Whitman the value of poetry was not formal and aesthetic but affective and social; its value would be measured not by its conformity to universal standards of taste but by its expressive and vitalizing power in fostering a democratic change of world. But Whitman's words about literature as social practice are more than a rejection of art for art's sake; they also register the artist's resistance to the commodification of art in the age of steam-engine production. Whitman's aim is not to produce "polish'd" and finished aesthetic objects. He has designs on his readers: He wants them to be not spectators but actors in the work of democracy. His design is not merely to "record" but—with the cooperation of the reader—to "make" democratic history.

Whatever Whitman's limits as a political theorist and apologist for America in the nineteenth century, his poems continue in some sense to "make" democratic history. "Still his words leap from their pages," wrote Langston Hughes in his 1946 essay "The Ceaseless Rings of Walt Whitman": "His all-embracing words lock arms with workers and farmers, Negroes and whites, Asiatics and Europeans, serfs, and free men, beaming democracy to all."[16] In the days of Bill Haywood, Eugene Debs, and the Industrial Workers of the World, Whitman's poems were printed and distributed to workers and farmers in little blue book editions small enough to fit in an overalls' pocket. Similarly, in England, France, Germany, and Russia in the period preceding and immediately following World War I, radical thinkers and writers carried Whitman's work to the people by means of lectures, translations, group readings, and slide presentations.[17]

In an article on "Reading Whitman During the Vietnam War," the writer Patricia Hampl remembers her boyfriend going to jail in 1970, after bewildering his Illinois draft board with "a letter that explained that because Walt Whitman had said, 'Dismiss whatever insults your own soul,' he was dismissing his draft card (see enclosed, etc., etc.)." During the Vietnam War, Whitman was frequently cited in letters that young men wrote to their draft boards. But, says Hampl, it was acting on Whitman's vision, "taking the utopic idea seriously into history, and then living with the consequences, which finally gave the idea dignity."[18]

Taking Whitman's democratic poetry "seriously into history," the minister of culture in Managua, Nicaragua, included a selection of Whitman's poems in *Poesía Libre,* a collection of Nicaraguan freedom poems. Even though the American poet comes from the same country that opposes their own struggle for freedom, says the editor, his poems articulate the free voice of the people and thus might serve as a model for the new poets of Nicaragua.[19] Along with an imprint from Whitman's blue book revisions of *Chants Democratic,* the anthology includes translations of "There Was a Child Went Forth" and section 3 of "I Sing the Body Electric," which begins "I knew a man, a common farmer, the father of five sons."

"There is no other poet who is so consistently permeated by democratic ideas," wrote the Austrian proletarian poet M. R. von Stern.[20] It is perhaps because the democratic voice of Whitman's poems is so strong that the rise and fall of his literary reputation have tended to coincide with periods of revolutionary ferment and reaction at home and abroad.[21] After a brief upsurge of interest in Whitman's democratic work in the 1920s and 1930s in America, his work fell out of favor during the subsequent period of academic formalism and the Cold War. In his introduction to a selection of Whitman's poems, Robert Creeley remembers that when he was a student at Harvard in the 1940s, "it was considered literally bad taste to have an active interest in his [Whitman's] writing." Allen Ginsberg also remembers that Whitman's "work was little famous, not much read and a bit put down in the years after his death, to the point of . . . or to the situation that when I went to Columbia College in 1945, between '44 and '49, by scholars and academic poets and by professors and their ilk by the Cold War soldiers and warriors of those days, Whitman was considered some lonesome, foolish crank who'd lived in poverty and likely Bohemian dis-splendor, having cantankerous affairs with jerks of all nations, in his mind." Gay Wilson Allen encountered a similar lack of interest in Whitman's work when a major textbook publisher turned down his *Walt Whitman Handbook* in 1945.[22]

In the 1950s, Whitman's reputation had a similarly charged political dimension. "During the McCarthy period you could not write your dissertation in Minnesota on Whitman," remembers the writer Meridel Le Sueur. "The southern agrarians made him a bad word, at best a naive yokel of the mob. Their fierce animosity showed how good he was for us, how dangerous, how alive."[23] Under the impetus of the splendid centennial biographies published by Gay Wilson Allen and Roger Asselineau and Ginsberg's comic tribute "A Supermarket in California," Whitman's reputation soared during the politically tumultuous 1960s and 1970s. But during the period of political reaction that followed in the late 1970s and 1980s, interest in his work once again subsided.

"The proof of a poet is that his country absorbs him as affectionately as he has absorbed it," Whitman wrote at the end of his 1855 preface (*LG* 1855, p. 24). The irony of Whitman's literary reception is that he has never really been absorbed by the American people. Even among writers and intellectuals, his work was first absorbed abroad—mainly by French, Spanish, and Latin American writers—and then filtered back into the mainstream of American literature. Whitman as a people's poet has been kept from the people, says the poet June Jordan. "If you hope to hear about Whitman," she quips, "your best bet is to leave home."[24]

But it is not only that other countries have been the first to recognize Whitman as *the* poet of America. They also read him differently. Whereas American critics tend to emphasize Whitman's individualism, his transcendentalism, and his aesthetic revolution, foreign writers and critics tend to dwell on his collectivism, his democratic politics, and the revolutionary implications of his verse.

In an article entitled "Whitman Exilado," the poet C. W. Truesdale remembers meeting the Ecuadorian poet and political exile Miguel Donoso Pareja in Mexico in 1966. "This is my text," Parejo told him, giving him a Spanish translation of *Leaves of Grass*. "This is my only Bible. Your country is very fortunate to have had such a great poet. Without Whitman, there is no [Pablo] Neruda. Without Neruda, there is no poetry. Without poetry, there is no culture." Having been "given" Whitman by the Latin American poet, Truesdale admits his bewilderment, a bewilderment that in some sense sums up the difference between the American and the foreign response to Whitman's work: "Such a frank and open passion for any sort of poetry was something almost entirely foreign to my own experience, as was Miguel's political commitment to a revolutionary cause and his association of Whitman, and poetry, with that cause."[25]

If Whitman was the mythologizer of what Georg Lukacs has called the heroic age of bourgeois individualism, the anticipatory tendency of his work, particularly in the post–Civil War period, was collective. "One's-Self I sing, a simple separate person,/Yet utter the word Democratic, the word En-Masse," says Whitman in the opening inscription to *Leaves of Grass*, an inscription that was added immediately after the Civil War. Amid the anti-Communist politics of the Cold War years and after, we have tended to play down—or forget—that *yet*. It is a *yet* that inscribes the paradox at the core of democracy. As the poet of many and one, Whitman sang of individuality, independence, and freedom, *yet* also uttered the words *comradeship, equality,* and *solidarity*.

Marginalized groups in America and throughout the world have heard the word *en masse* at the revolutionary heart of *Leaves of Grass*. "I, too, sing America," says Langston Hughes in "I, Too," a poem that self-consciously responds to Whitman's democratic call. "I too am a descendant of Walt Whitman" echoes June Jordan in "For the Sake of a People's Poetry: Walt Whitman and the Rest of Us." "In America," she says, "the father is white: It is he who inaugurated the experiment of this republic." But within the whiteness of the founding moment, she discovers the democratic green space generated by Whitman's *Leaves:* "Within the Whitman tradition, Black and Third World poets traceably transform, and further, the egalitarian sensibility that isolates that one white father from his more powerful compatriots." In this tradition, Jordan includes feminists, Agostinho Neto, and the Chilean poet Pablo Neruda.[26]

As a kind of new world descendant of Whitman, Neruda has himself acknowledged his debt to the American poet. "I hold him to be my greatest creditor," he said in a speech delivered to P.E.N. in 1972. "I stand before you feeling that I bear with me always this great and wonderful debt which has helped me to exist." Whitman, he says, was a "lyric moralist," who "had no fear of either moralizing or immoralizing, nor did he seek to separate the fields of pure and impure poetry." His intention was "not just to sing, but to impose on others his own total and wide-ranging vision of the relationships of men and nations." It is in this "total and organic universal vision" that Neruda finds the source of Whitman's greatness: "There are many kinds of greatness, but let me say (though I be a poet of the Spanish tongue) that Walt Whitman has taught me more than Spain's Cervantes: in Walt Whitman's work one never finds the ignorant being humbled, nor is the human condition ever found offended." Neruda's conclusion is in the utopian voice of Whitman's own democratic vistas: "Man's liberation may often require bloodshed, but it always requires song—and the song of mankind grows richer day by day, in this age of sufferings and liberation."[27]

Toward the close of his life Whitman told Traubel: "The political class is too slippery for me—even its best examples: I seem to be reaching for a new politics—for a new economy: I don't know quite what, but for something" (*WWC*, I, 101). Whitman's work embodies the complexities, anxieties, and contradictions of America in the age of democratic and capitalist transformation. At the same time, it dreams forward, toward the revolutionary possibility of *an other*, as yet unrealized politics of freedom and solidarity.

In a comment on the relation between realism and myth, Maxim Gorky said:

> Myth is invention. To invent means to extract from the sum of a given reality its cardinal idea and embody it in imagery—that is how we get realism. But if to the idea extracted from the given reality we add—completing the idea by the logic of hypothesis—the desired, the possible, and thus supplement the image, we obtain that romanticism which is at the basis of myth, and is highly beneficial in that it tends to provoke a revolutionary attitude to reality, an attitude that changes the world in a practical way.[28]

Gorky's words might be read as a gloss on *Leaves of Grass*, suggesting the ways that the poems simultaneously mediate and resist their age. Whitman's art extracts from the "given reality" of nineteenth-century America its cardinal democratic idea and embodies it in imagery. But his democratic myth is not only an invention of the past and present; it is also an invention of the future. Whitman's grammar of the future is the grammar of a utopian will that resists the existing system, postulating a "desired," a "possible" that can provoke "a revolutionary attitude to reality" and transform the world in practical ways.

"In literature, all things become possible" said Carlos Fuentes in an article on the role of the writer in the modern world.[29] For Whitman the art of the possible was also the art of life's possibility. To create an internationality of people, it was his ambition to "practically start an internationality of poems." "The final aim of the United States of America," he said in 1884, "is the solidarity of the world. What fails so far, may yet be accomplished by song, radiating, clustering, concentrating from all the lands of the earth" (*Corr.*, III, 369).

Like the workers and students who scrawled *L'imagination au pouvoir* on the walls of Paris during the 1968 uprising in France, Whitman realized that the power to change the world resided in the power to imagine a revolutionary change of world. This is the lesson of "A Thought of Columbus," the last poem he wrote. It is Columbus's "thought" of a world in round, a thought that defied the belief structure of his time, that initiated the process of transforming vision into history. The "Thought of Columbus," which concludes Whitman's life and work, is really the thought of the poet's power to bring *an other* America and ultimately *an other* world of the imagination into being by naming and renaming its possibility. In the words of Hart Crane's "The Bridge":

> thy wand
> Has beat a song, O Walt,—there and beyond!

# Notes

## Chapter 1

1. Whitman, *Leaves of Grass: The First (1855) Edition*, ed. Malcolm Cowley (New York: Viking, 1959), p. 87. Subsequent references will be cited in the text as *LG* 1855.

2. Whitman, *Leaves of Grass: Comprehensive Reader's Edition*, ed. Harold W. Blodgett and Sculley Bradley (New York: Norton, 1965), p. 505. Subsequent references will be cited in the text as *LGC*.

3. Walt Whitman, *The Complete Writings*, ed. Richard Maurice Bucke et al. (New York: Putnam, 1902), vol. IX, pp. 35–36. Subsequent references will be cited in the text as *CW*.

4. *Putnam's Magazine*, September 1855, in Walt Whitman, *Leaves of Grass* (1855) (New York: Eakins Press, 1966).

5. Ibid.

6. Many critics have stressed the determining influence of Emerson. For a full consideration of the Emerson–Whitman connection, see Jerome Loving, *Emerson, Whitman, and the American Muse* (Chapel Hill, N.C.: University of North Carolina Press, 1982). Esther Shephard argues for the determining influence of George Sand in *Walt Whitman's Pose* (New York: Harcourt, Brace, 1938). Henry Binns was the first to advance the New Orleans Romance theory in *A Life of Walt Whitman* (New York: Dutton, 1905). Emory Holloway supports this theory in *Whitman: An Interpretation in Narrative* (New York: Knopf, 1928). Richard Maurice Bucke stresses Whitman's mystical origins in *Cosmic Consciousness* (Philadelphia: Innes & Sons, 1901). In the introduction to his edition of *LG* 1855, Malcolm Cowley suggests the origins of *Leaves of Grass* in a mystical experience, "essentially the same as the illuminations or ecstasies of earlier bards and prophets" (p. xii). For recent psychosexual explanations of Whitman's poetic origins, see Gay Wilson Allen, *The Solitary Singer* (1955) (New York: Macmillan, 1960); Roger Asselineau, *The Evolution of Walt Whitman* (1954) (Cambridge, Mass.: Harvard University Press, 1960–62); Edwin Haviland Miller, *Walt Whitman's Poetry: A Psychologi-*

*cal Journey* (New York: New York University Press, 1968); Stephen Black, *Whitman's Journeys into Chaos: A Psychoanalytic Study of the Poetic Process* (Princeton, N.J.: Princeton University Press, 1975); and David Cavitch, *My Soul and I: The Inner Life of Walt Whitman* (Boston: Beacon Press, 1985).

7. For recent discussions of the politics of literary value, see Terry Eagleton, "The Rise of English," in his *Literary Theory* (Minneapolis: University of Minnesota Press, 1983); Barbara Herrnstein Smith, "Contingencies of Value," *Critical Inquiry*, 10 (September 1983), 1–36; Paul Lauter, "History and the Canon," *Social Text*, 12 (Fall 1985), 94–101; Sacvan Bercovitch, "The Problem of Ideology in American Literary History," *Critical Inquiry*, 12 (Summer 1986), 631–53; and Myra Jehlen, "Beyond Transcendence," in *Ideology and Classic American Literature*, ed. Sacvan Bercovitch and Myra Jehlen (New York: Cambridge University Press, 1986), pp. 1–18.

8. Williams, *Culture and Society: 1780–1950* (1958) (New York: Columbia University Press, 1983), p. 30.

9. Bucke, *Cosmic Consciousness*, p. 187.

10. T. S. Eliot, *For Lancelot Andrewes* (Garden City, N.Y.: Doubleday, Doran, & Co., 1928), p. vii; T.S. Eliot, "Introduction," in Ezra Pound, *Selected Poems* (London: Faber & Gwyer, 1928), pp. x–xi; William Carlos Williams, "An Essay on *Leaves of Grass*," in *Leaves of Grass: One Hundred Years After*, ed. Milton Hindus (Stanford, Calif.: Stanford University Press, 1955), p. 22.

11. Ezra Pound, "What I Feel About Walt Whitman" (1909), in *Whitman: A Collection of Critical Essays*, ed. Roy Harvey Pearce (Englewood Cliffs, N.J.: Prentice-Hall, 1962), p. 8; Ezra Pound, "A Pact," in his *Personae* (New York: New Directions, 1926), p. 89.

12. Vernon Louis Parrington, "The Afterglow of the Enlightenment—Walt Whitman," in *Main Currents in American Thought: The Beginnings of Critical Realism*, vol. III (New York: Harcourt, Brace, 1930), pp. 85–86; Newton Arvin, *Whitman* (New York: Macmillan, 1938).

13. F. O. Matthiessen, *American Renaissance: Art and Expression in the Age of Emerson and Whitman* (New York: Oxford University Press, 1941), p. ix; "Only a Language Experiment," pp. 517–625.

14. Richard Chase, *Walt Whitman Reconsidered* (New York: William Sloane Associates, 1955); Charles Feidelson, Jr., *Symbolism and American Literature* (Chicago: University of Chicago Press, 1953); R. W. B. Lewis, *The American Adam: Innocence, Tragedy, and the Tradition in the Nineteenth Century* (Chicago: University of Chicago Press, 1955); Howard Waskow, *Whitman: Explorations in Form* (Chicago: University of Chicago Press, 1966).

15. Allen, *The Solitary Singer;* Asselineau, *The Evolution of Walt Whitman;* James E. Miller, *A Critical Guide to Leaves of Grass* (Chicago: University of Chicago, 1957).

16. Joseph Jay Rubin, *The Historic Whitman* (University Park: Pennsylvania State University Press, 1973); Justin Kaplan, *Walt Whitman: A Life* (New York: Simon and Schuster, 1980); Paul Zweig, *Whitman: The Making of the Poet* (New

York: Basic Books, 1984); Cavitch, *My Soul and I;* James L. Machor, "Pastoralism and the American Urban Ideal: Hawthorne, Whitman, and the Literary Pattern," *American Literature,* 54 (October 1984), 335. See also M. Wynn Thomas's Marxist study *The Lunar Light of Whitman's Poetry* (Cambridge, Mass.: Harvard University Press, 1987), which examines several of Whitman's poems in the context of an emergent capitalist economy. In stressing Whitman's roots in artisan republican culture, both of us are commonly indebted to Sean Wilentz's *Chants Democratic: New York City & the Rise of the American Working Class, 1788–1850* (New York: Oxford University Press, 1984).

17. Walter Benjamin, "The Work of Art in the Age of Mechanical Reproduction," in Walter Benjamin, *Illuminations: Essays and Reflections,* ed. Hannah Arendt and trans. Harry Zohn (New York: Schocken, 1969), p. 223.

18. Walt Whitman, *Prose Works 1892,* ed. Floyd Stovall (New York: New York University Press, 1963), vol. II, p. 733. Subsequent references will be cited in the text as *PW.*

19. William Wordsworth and Samuel Coleridge, *Lyrical Ballads, 1798,* ed. H. Littledale (1911) (London: Oxford University Press, 1959), p. 240.

20. Thomas Jefferson, 1821, cited in *The National Experience: A History of the United States to 1877,* ed. John Blum et al. (New York: Harcourt Brace Jovanovich, 1981), p. 212.

21. Sean Wilentz, *Chants Democratic: New York City and the Rise of the American Working Class (1788–1850),* p. 245.

22. Walt Whitman, *Faint Clues and Indirections: Manuscripts of Walt Whitman and His Family,* ed. Clarence Gohdes and Rollo G. Silver (Durham, N.C.: Duke University Press, 1949), p. 46.

23. Walt Whitman, *Early Poems and Fiction,* ed. Thomas L. Brasher (New York: New York University Press, 1963), p. 248. Subsequent references will be cited in the text as *EPF.*

24. *PW,* I, 14; Walt Whitman, *Uncollected Poetry and Prose,* ed. Emory Holloway (Garden City, N.Y.: Doubleday, Page & Co., 1921), vol. II, p. 247. Subsequent references will be cited in the text as *UPP.*

25. Henry May, *The Enlightenment in America* (New York: Oxford University Press, 1976), pp. 153–304.

26. Horace Traubel, *With Walt Whitman in Camden,* vol. II (New York: D. Appleton & Company, 1908), p. 205. Subsequent references will be cited in the text as *WWC,* II.

27. *The Free Enquirer,* July 31, 1830, cited in Arthur M. Schlesinger, Jr., *The Age of Jackson* (Boston: Little, Brown, 1945), p. 183.

28. Horace Traubel, *With Walt Whitman in Camden,* vol. I (Boston: Small Maynard, 1906), p. 79. Subsequent references will be cited in the text as *WWC,* I.

29. For a discussion of the relationship between Wright and Whitman, see David Goodale, "Some of Walt Whitman's Borrowings," *American Literature,* 10 (May 1938), 202–13; see also Allen, *The Solitary Singer,* pp. 138–39; and Floyd

Stovall, *The Foreground of Leaves of Grass* (Charlottesville: University Press of Virginia, 1974), pp. 19–20.

30. Frances Wright, *A Few Days in Athens* (1822) (New York: Arno Press, 1972), p. 205.

31. Constantin F. Volney, *The Ruins*, trans. Thomas Jefferson and Joel Barlow (1802) (New York: Garland, 1979), pp. 25–26. For a more detailed account of Whitman's specific borrowings from Volney, see Goodale, "Some of Walt Whitman's Borrowings," pp. 202–13; Gay Wilson Allen, *The New Walt Whitman Handbook* (New York: New York University Press, 1975), pp. 176–78; and Betsy Erkkila, *Walt Whitman Among the French: Poet and Myth* (Princeton, N.J.: Princeton University Press, 1980), pp. 14–19.

32. A volume of Leggett's *Political Writings*, which Whitman probably consulted, was published posthumously in 1840. For a discussion of Leggett's influence on Whitman see Rubin, *The Historic Whitman*, pp. 39–44; and Stovall, *The Foreground of Leaves of Grass*, pp. 41–42. For a study of Leggett as an interpreter of Jackson, see Richard Hofstadter, "William Leggett, Spokesman of Jacksonian Democracy," *Political Science Quarterly* (December 1943), 581–94.

33. *Democrat*, September 4, 1839, cited in Rubin, *The Historic Whitman*, p. 45.

34. Walt Whitman, *The Gathering of the Forces*, ed. Cleveland Rogers and John Black (New York: Putnam, 1920), vol. I, p. 218. Subsequent references will be cited in the text as *GF*.

35. Joseph J. Rubin, "Whitman in 1840: A Discovery," *American Literature*, 9 (May 1937), 239–42.

36. *Plebeian*, October 14, 1842, cited in Rubin, *The Historic Whitman*, p. 50.

37. For a discussion of the psychological workings of ideology in society, see Louis Althusser, "Ideology and Ideological State Apparatuses," in *Lenin and Philosophy and Other Essays*, trans. Ben Brewster (London: New Left Books, 1971), pp. 127–86.

38. See Edward Pessen, *Jacksonian America* (Homewood, Ill.: Dorsey Press, 1969); and Edward Pessen, *Riches, Class, and Power Before the Civil War* (Lexington, Mass: Heath, 1973).

39. Herbert Bergman and William White, "Walt Whitman's Lost 'Sun-Down Papers,' Nos. 1–3," *American Book Collector*, 20 (January 1970), 18–20.

40. Letter to Pierre de Nemours, April 24, 1816, in *The Writings of Thomas Jefferson*, ed. Paul L. Ford (New York: Putnam, 1892–99), vol. X, p. 25.

41. Traubel, *WWC*, I, 80; Walt Whitman, *Daybooks and Notebooks*, ed. William White (New York: New York University Press, 1978), vol. 3, p. 729 n.

42. Although several of Whitman's poems appeared in the 1840s, most were written earlier. His two new compositions of the 1840s, "Lesson of the Two Symbols" and "Ode," were also politically inspired: "Lesson of the Two Symbols" appeared in a working-class paper, *The Subterranean*, in 1843; and "Ode," which was written as part of the campaign to save Fort Greene, was sung at Fort Greene on July 4 1846, to the tune of the "Star Spangled Banner"; it was published in the

*Brooklyn Daily Eagle* on July 2, 1846. See *EPF,* pp. 34–35; and Elliott B. Gross, " 'Lesson of the Two Symbols': An Undiscovered Whitman Poem," *Walt Whitman Review,* 12 (December 1966), 77–80.

## Chapter 2

1. Letter to Roger C. Weightman, June 24, 1826, in *The Writings of Thomas Jefferson,* vol. X, p. 391.

2. Thomas Jefferson, *Notes on the State of Virginia,* ed. William Peden (New York: Norton, 1982), p. 165. Subsequent references will be cited in the text as *Notes.*

3. Gordon Wood, *The Creation of the American Republic* (Chapel Hill, N.C.: University of North Carolina Press, 1969), p. 54. In the last few decades there has been considerable controversy about the sources and significance of the American Revolution. Bernard Bailyn, in *The Ideological Origins of the American Revolution* (Cambridge, Mass.: Harvard University Press, 1967), and Wood, in *The Creation of the American Republic,* trace the ideological origins of the American Revolution in the "commonwealth" or "country" ideology of eighteenth-century English politics. J. G. A. Pocock's chapter on of the conflict between virtue and commerce in the founding of the American republic, in *The Machiavellian Moment: Florentine Political Thought and the Atlantic Republican Tradition* (Princeton, N.J.: Princeton University Press, 1975), challenges the standard Lockean interpretation of the American Revolution. Rather, he argues, the founding moment had roots in the classical and Renaissance tradition of civic humanism, which stressed the values of public good, civic virtue, and personal sacrifice against the corruption of commerce and the dependency it engendered. The revolution might be viewed not as the first act of the revolutionary enlightenment but as a *ridurre ai principi* and thus the last great act of the Renaissance. In *Inventing America: Jefferson's Declaration of Independence* (New York: Random House, 1978), Garry Wills argues that Hutcheson and the Scottish Enlightenment rather than Locke were the primary sources of Jefferson's thought in the Declaration of Independence. Recently, John Diggins challenged Pocock's thesis, in *The Lost Soul of American Politics: Virtue, Self-Interest, and the Foundations of Liberalism* (New York: Basic Books, 1984), arguing that the rhetoric of republican virtue masked a Lockean concern with property and commerce that was the ultimate cause of the American Revolution.

4. For a splendid discussion of the background of artisan republican sentiment in New York City in the early nineteenth century, see Wilentz, *Chants Democratic,* particularly pp. 61–106.

5. Walt Whitman, *Walt Whitman of the New York Aurora,* ed. Joseph J. Rubin and Charles H. Brown (State College, Pa.: Bald Eagle Press, 1950), p. 41. Subsequent references will be cited in the text as *Aurora.*

6. For a study of this movement, see John Stafford, *The Literary Criticism of "Young America": A Study of the Relationship of Politics and Literature, 1837–1850* (Berkeley and Los Angeles: University of California Press, 1952).

7. Georg Lukacs, *The Theory of the Novel,* trans. Anna Bostock (Cambridge, Mass.: MIT Press, 1971), pp. 144–53.

8. Wilentz, *Chants Democratic,* pp. 281–84, 306–14.

9. Gross, " 'Lesson of the Two Symbols,' " pp. 78, 80.

10. See Pessen, *Jacksonian America* (1969) and *Riches, Class, and Power* (1973).

11. Joel Barlow, *The Works of Joel Barlow* (Gainesville, Fla.: Scholars' Facsimiles & Reprints, 1970), vol. II, p. 343.

12. For a study of attitudes toward the machine in nineteenth-century America, see Leo Marx's monumental study, *The Machine in the Garden* (New York: Oxford University Press, 1964).

13. John O'Sullivan, "Annexation," *Democratic Review* (July–August 1845), 5.

14. Letter to James Madison sent as an enclosure to Uriah Forrest, December 31, 1787, in *Papers of Thomas Jefferson* (Princeton, N.J.: Princeton University Press, 1950), vol. XII, pp. 478–79.

15. Alexis de Tocqueville, *Democracy in America,* trans. Henry Reeve and Francis Bowen (New York: Random House, 1945), vol. 1, p. 195.

## Chapter 3

1. Letter to John Holmes, 1820, in *The Writings of Thomas Jefferson,* vol. X, p. 157. For studies of Jefferson and the issue of slavery, see William Cohen, "Thomas Jefferson and the Problem of Slavery," *Journal of American History,* 16 (1969–70), 503–26; and William Freehling, "The Founding Fathers and Slavery," *American Historical Review,* 77 (February 1972), 81–93. For studies of the problem of slavery in the American republic, see Winthrop D. Jordan's important study *White over Black: American Attitudes Toward the Negro, 1550–1812* (Chapel Hill, N.C.: University of North Carolina Press, 1968); D. L. Robinson, *Slavery in the Structure of American Politics, 1765–1820* (New York: Harcourt Brace Jovanovich, 1971); and Edmund S. Morgan, "Slavery and Freedom: The American Paradox," *Journal of History,* 59 (June 1972), 5–29.

2. *The Congressional Globe,* August 8, 1846. See also Chaplain W. Morrison, *Democratic Politics and Sectionalism: The Wilmot Proviso Controversy* (Chapel Hill, N.C.: University of North Carolina, 1967).

3. Cited in *The National Experience,* p. 290.

4. Emory Holloway originally argued that the materials in this notebook were written in 1847. Others have contended that all or some of the material may have been written at a later date, perhaps as late as 1854. Unfortunately, the problem is complicated by the fact that a copy of the original notebook is available only on microfilm in the Library of Congress collection; the original notebook has been lost or perhaps misplaced. See Edward Grier, "Whitman's Earliest Known Notebook," *Publications of the Modern Language Association of America (PMLA),* 83 (October 1968), 1453–56; John Broderick, "Whitman's Earliest Known Notebook: A Clarification," *PMLA,* 84 (October 1969), 1657; Esther Shephard, Comment on

"Whitman's Earliest Known Notebook: A Clarification," *PMLA*, 86 (March 1971), 266; and Shephard, "Inside Front and Backcovers of Whitman's Earliest Known Notebook," *PMLA*, 87 (October 1972), 1119–22.

5. *The Writings of Thomas Jefferson*, Inaugural Address, March 4, 1801, vol. VIII, p. 3.

6. *Advertiser*, June 23, 1848, cited in Rubin, *The Historic Whitman*, p. 206.

7. The only extant copy of the Brooklyn Weekly *Freeman*, dated September 9, 1848, is in the Trent Collection at Duke University; cited in Rubin, *The Historic Whitman*, p. 211.

8. Ibid.

9. Horace Traubel, Richard Maurice Bucke, and Thomas Harned, eds., *In Re Walt Whitman* (Philadelphia: McKay, 1893), p. 35. Whitman's notes and fragments are collected in Edward F. Grier, ed., *Notebooks and Unpublished Prose Manuscripts* (New York: New York University Press, 1984); and William White, ed., *Daybooks and Notebooks*.

10. Clifton Joseph Furness, ed., *Walt Whitman's Workshop* (New York: Russell & Russell, 1964), p. 35. Subsequent references will be cited in the text as *WWW*.

11. *CW*, IX, xvi. For other studies of Whitman and oratory, see Thomas B. Harned, "Whitman and Oratory," *CW*, VIII, 244–60; and C. Carroll Hollis's detailed study, *Language and Style in Leaves of Grass* (Baton Rouge: Louisiana State University Press, 1983).

12. Whitman's friend John Burroughs says: "Between the ages of twenty and thirty, he was variously occupied as writer and editor on the press of New York and Brooklyn, sometimes going into the country and delivering political addresses." In "Walt Whitman and his 'Drum Taps,' " *Galaxy*, 2 (1866), 606. See also Furness, *WWW*, p. 72.

13. Fetridge and Co., ed., *The Boston Riot and Trial of Anthony Burns* (Boston, 1854), p. 8; and Rubin, *The Historic Whitman*, pp. 293–96. For a discussion of "A Boston Ballad," see Stephen Malin, " 'A Boston Ballad' and the Boston Riot," *Walt Whitman Review*, 9 (September 1963), 51–57; and Edward A. Martin "Whitman's 'A Boston Ballad' (1854)," *Walt Whitman Review*, 11 (1965), 61–69.

14. Wilentz, *Chants Democratic*, p. 380.

## Chapter 4

1. Emerson, *Complete Works* (Boston: Houghton Mifflin, 1903–4), vol. III, pp. 37–38. Subsequent references will be cited in the text as *CW*.

2. For an in-depth study of the Whitman–Emerson connection, see Loving, *Emerson, Whitman, and the American Muse;* and also Stovall, *The Foreground of Leaves of Grass*, pp. 282–305.

3. "Whitman to Emerson, 1856," *LGC*, p. 739. For a considered discussion of Whitman's critical attitude toward Emerson, see Kenneth M. Price, "Whitman on Emerson: New Light on the 1856 Open Letter," *American Literature*, 56 (March 1984), 83–87.

4. John Trowbridge, *My Own Story* (Boston: Houghton Mifflin, 1903), p. 360.

5. Noah Webster, *A Grammatical Institute of the English Language* (1783) (Menston, England: Scholar Press, 1968), vol. I, p. 14.

6. Barlow, *The Columbiad,* in *The Works of Joel Barlow,* vol. II, pp. 377–78, 389.

7. Reverend Theodore Dehon, "Discourse upon the Importance of Literature to Our Country," *Monthly Anthology and Boston Review,* 4 (September 1807), 472.

8. For studies of early American critical thought, see William Charvat, *The Origins of American Critical Thought, 1810–1835* (Philadelphia: University of Pennsylvania Press, 1936); and Benjamin Spencer, *The Quest for Nationality: An American Literary Campaign* (Syracuse, N.Y.: Syracuse University Press, 1957).

9. William Ellery Channing, "Remarks on National Literature," *The Christian Examiner* (1830), in Robert E. Spiller, *The American Literary Revolution, 1783–1837* (New York: New York University Press, 1967), pp. 346, 348–72.

10. Address entitled "American Literature" delivered at Brown University in 1839; in Richard Ruland, *Native Muse* (New York: E.P. Dutton, 1972), p. 282.

11. Cited in Spencer, *The Quest for Nationality,* pp. 117–18.

12. For a discussion of Whitman's relation to French writers, see Erkkila, *Walt Whitman Among the French.*

13. For a study of Whitman and the epic tradition, see James E. Miller, *The American Quest for a Supreme Fiction: Whitman's Legacy in the Personal Epic* (Chicago: University of Chicago Press, 1979), pp. 30–49.

14. *Daybooks and Notebooks,* vol. III, p. 754. Subsequent references will be cited in the text as *DN.*

15. C. Carroll Hollis, "Whitman and Swinton: A Co-operative Friendship," *American Literature,* 30 (January 1959), 425–49. See also Edward F. Grier, ed., *Notebooks and Unpublished Prose Manuscripts,* vol. V, pp. 1624–62.

16. Leon Howard, "Whitman and the American Language," *American Speech,* 5 (August 1930), 442; William Sloane Kennedy, *The Fight of a Book for the World* (West Yarmouth, Mass: Stonecraft Press, 1926), p. 95. For insightful discussions of Whitman's language experiments, see Matthiessen, *American Renaissance;* and C. Carroll Hollis, "Whitman and the American Idiom," *Quarterly Journal of Speech,* 43 (December 1957), 408–20. James Perrin Warren analyzes Whitman's use of deverbal nouns to achieve stative and dynamic effects in "The 'Real Grammar': Deverbal Style in 'Song of Myself,'" *American Literature,* 56 (March 1984), 1–16. For other noteworthy studies of Whitman's language, see Louise Pound, "Whitman's Neologisms," *American Mercury,* 12 (February 1925), 199–201; Rebecca Coy, "A Study of Whitman's Diction," *University of Texas Studies in English,* 16 (July 1936), 115–24; Donald Kummings, "Walt Whitman's Vernacular Poetics," *Canadian Review of American Studies,* 7 (1976), 119–31; and Michael Dressman, "Whitman's Plans for the Perfect Dictionary," in *Studies in the American Renaissance,* ed. Joel Myerson (Boston: Twayne, 1979), pp. 457–74.

17. Oliver Wendell Holmes, *Works* (Boston: Houghton, 1892), vol. IV, p. 234.

18. *The Works of John Adams* (Boston, 1852), vol. VII, p. 249, in *The Beginnings of American English*, ed. M. M. Mathews (Chicago: University of Chicago Press, 1931), p. 42.

19. Noah Webster, *Dissertations on the English Language* (1789) (Gainesville, Fla.: Scholars' Facsimiles & Reprints, 1951), p. 20.

20. John Pickering, *A Vocabulary or Collection of Words and Phrases Which Have Been Supposed to Be Peculiar to the United States of America* (1816), in *The Beginnings of American English*, p. 72.

21. *DN*, III, 809. The transcription in the notebooks is incomplete; to complete it I have used the transcript made by Hollis, "Whitman and the American Idiom," pp. 419–20.

22. Sir William Craigie, "The Study of American English," *Society for the Purity of English Tracts*, no. 27 (1927), 203.

23. William Swinton, *Rambles Among Words* (New York: Scribner, 1859), p. 288.

24. See Betsy Erkkila, "Walt Whitman: The Politics of Language," *American Studies*, 29 (Spring 1984), 21–34.

25. William Blake, *Prophetic Books* (1793–1804), cited in Bliss Perry, *Walt Whitman: His Life and Works* (Boston: Houghton Mifflin, 1906), p. 89.

26. The question of whether Whitman's lines may be regularly scanned is still being debated. In *The New Walt Whitman Handbook*, Gay Wilson Allen argues that the biblical methods of parallelism and repetition are at the base of Whitman's poetics. Others have persuasively argued that rhythmic recurrence is even more fundamental to Whitman's verse than is logical parallelism. See, in particular, Basil De Selincourt, *Walt Whitman: A Critical Study* (London: Martin Secker, 1914); and Sculley Bradley, "The Fundamental Metrical Principles in Whitman's Poetry," *American Literature*, 10 (January 1939), 437–59. For more recent studies of Whitman's use of patterns of rhythmic recurrence, see Asselineau, *The Evolution of Walt Whitman*, pp. 207–52; Robert E. Cory, "The Prosody of Walt Whitman," *North Dakota Quarterly*, 28 (Summer 1960), 74–79; Milton Hindus, "Notes Toward the Definition of a Typical Poetic Line in Whitman," *Walt Whitman Review*, 9 (December 1963), 75–81; and Roger Mitchell, "A Prosody for Whitman?" *PMLA*, 84 (October 1969), 1606–12. For other considerations of Whitman's free-verse technique, see Mattie Swayne, "Whitman's Catalogue Rhetoric," *University of Texas Studies in English*, 21 (July 1941), 162–78; Walter Sutton, "The Analysis of Free Verse Form, Illustrated by a Reading of Whitman," *Journal of Aesthetics and Art Criticism*, 18 (December 1959), 241–54; and Robert Griffin, "Notes on Structural Devices in Whitman's Poetry," *Tennessee Studies in Literature*, 6 (1961), 14–24.

27. For the New Critical response to Whitman, see in particular, William K. Wimsatt, Jr., and Cleanth Brooks, *Literary Criticism: A Short History* (New York: Knopf, 1957), pp. 587, 708; Cleanth Brooks, *Modern Poetry and the Tradition* (Chapel Hill, N.C.: University of North Carolina Press, 1939), p. 76; R. P. Blackmur,

*Language As Gesture: Essays in Poetry* (New York: Harcourt, Brace, 1952), p. 503; and Yvor Winters, *In Defense of Reason* (New York: Swallow Press, 1946).

28. For a discussion of Whitman's use of "indirection" in his poems, see Thomas J. Rountree, "Whitman's Indirect Expression and Its Application to 'Song of Myself,'" *PMLA*, 73 (December 1958), 549–55; and E. H. Eby, "Walt Whitman's 'Indirections,'" *Walt Whitman Review*, 12 (March 1966), 5–16.

29. For a rich and subtle analysis of the creative interplay among the poet, the reader, and the "crowd" in Whitman's verse, see Larzer Ziff, *Literary Democracy* (New York: Viking, 1981), pp. 230–43; and Larzer Ziff, "Whitman and the Crowd," *Critical Inquiry*, 10 (June 1984), 579–91.

## *Chapter 5*

1. Whitman, *Memoranda During the War* (1875–76) (Bloomington: Indiana University Press, 1962), p. 65. Subsequent references to this volume will be cited in the text as *Memoranda*.

2. Cited in Eric Foner, *Politics and Ideology in the Age of the Civil War* (New York: Oxford University Press, 1980), p. 53.

3. James Madison, *The Federalist Papers*, ed. Clinton Rossiter (New York: New American Library, 1961), p. 80.

4. Abraham Lincoln's famous "lost speech" delivered in 1855. Cited in Carl Sandburg, "Introduction," in *Leaves of Grass* (New York: Modern Library, 1921), p. vii.

5. Carl F. Strauch, "The Structure of Walt Whitman's 'Song of Myself,'" *English Journal*, 29 (September 1939), 597–607, divides the poem into a five-part structure moving from self to superman. In his influential essay, "'Song of Myself' As Inverted Mystical Experience," in his *A Critical Guide to Leaves of Grass*, pp. 6–35, James E. Miller argues for a seven-part structure corresponding to the stages of a mystical experience. Malcolm Cowley, in his introduction to *LG* 1855, divides the poem into nine stages corresponding to a similar pattern of mystical transcendence. Roy Harvey Pearce traces a four-part structure of self-realization in *The Continuity of American Poetry* (Princeton, N.J.: Princeton University Press, 1961), pp. 69–83. F. De Wolfe Miller, in "The Partitive Studies of 'Song of Myself,'" *American Transcendental Quarterly*, 12 (Fall 1971), 11–17, finds a tripartite structure that corresponds to Whitman's emergence as a poet-messiah seeking to save the souls of his readers. Alfred S. Reid, in "The Structure of 'Song of Myself' Reconsidered," *Southern Humanities Review*, 8 (Fall 1974), 507–14, finds a pattern of birth, growth, and expansion in five major units. Some of the more interesting studies of the poem focus on the relationship between the poet and the reader: See, for example, Thomas Rountree, "Whitman's Indirect Expression and Its Application to 'Song of Myself'"; Chaviva Hosek, "The Rhetoric of Whitman's 1855 'Song of 'Myself,'" *Centennial Review*, 20 (Summer 1976), 263–77; and Donald D. Kummings, "The Vernacular Hero in 'Song of Myself,'" *Walt Whitman Review*, 23 (March 1977), 23–34.

6. For a consideration of the unity-in-diversity theme of "Song of Myself," see Ronald Beck, "The Structure of 'Song of Myself' and the Critics," *Walt Whitman Review*, 15 (March 1969), 32–38; and Michael D. Reed, "First Person Persona and the Catalogue in 'Song of Myself,' " *Walt Whitman Review*, 23 (December 1977), 147–55.

7. The initial version of "Song of Myself" was not divided into numbered sections. To facilitate discussion of the poems, I shall use the section numbers that Whitman used in his 1881 version of *LG*.

8. Frederik Schyberg, *Walt Whitman*, trans. Evie Allison Allen (New York: Columbia University Press, 1951), pp. 98–99.

9. For a discussion of the female in section 11 as spiritual unifier, see T. J. Kallsen, "The Improbabilities in Section 11 of 'Song of Myself,' " *Walt Whitman Review*, 13 (September 1967), 87–92; and O. K. Nambiar, "Whitman's Twenty-Eight Bathers: A Guessing Game," in *Indian Essays in American Literature*, ed. Sujit Mukherjee and D. V. K. Raghavacharyulu (Bombay: Popular Prakashan, 1969), pp. 129–37.

10. Cited in Clarence Gohdes, "Whitman As One of the Roughs," *Walt Whitman Review*, 8 (March 1962), 18.

11. Jefferson, January 30, 1787, *Writings*, vol. XI, p. 93.

12. See, for example, Orson S. Fowler, *Amativeness* (New York: Fowler & Wells, 1844); and R. T. Trall, *Home-Treatment for Sexual Abuses: A Practical Treatise* (New York: Fowler & Wells, 1856). For a discussion of attitudes toward the body during Whitman's time, see Harold Aspiz's excellent study *Walt Whitman and the Body Beautiful* (Urbana: University of Illinois Press, 1980). In a different context, Carroll Smith-Rosenberg links the obsession with familial and bodily disorder during the Jacksonian age with personal anxiety about social disorder and change in the political sphere; see *Disorderly Conduct: Visions of Gender in Victorian America* (New York: Knopf, 1985), pp. 77–108.

13. For a study of the sources of this episode, see D. M. McKeithan, *Whitman's Song of Myself 34 and Its Background* (Uppsala, Sweden: Lundequistska Bokhandeln, 1969).

14. See James E. Miller, *A Critical Guide to Leaves of Grass*, p. 25; Waskow, *Explorations in Form*, p. 180; John J. Belson, "Whitman's 'Overstaid Fraction,' " *Walt Whitman Review*, 17 (June 1971), 63–65; and Sholom J. Kahn, "Whitman's 'Overstaid Fraction' Again," *Walt Whitman Review*, 20 (June 1974), 67–73.

15. Library of Congress (Feinberg), Item no. 32.

16. In the final version of the poem, Whitman toned down the nationalism of this passage by eliminating the specific reference to the states; he also eliminated the specific references to Christ by changing the "growth of two thousand years" to the "growth of thousands of years," thus associating regeneration with the ancient cult of Osiris. But the federal balance between the one and the many and the republican dream of regeneration remain the same. An article Whitman wrote for the *Eagle* also provides an analogue for his treatment of Christ in this passage. Speaking of Christ's birth, Whitman says: "Then vitality started in manifold seeds

of true good which had for ages lain dormant in humanity. Ah, Thou whose office it was 'to give light to them that sit in darkness and in the shadow of death, to guide our feet into the way of Peace,' how the hearts of the children of men yet turn to Thy soothing counsels; and how refreshing to know that the same founts of consolation at which we drink, have been tasted by the now dead and past ages, and still by thousands every day!" (*GF*, II, 215–16).

17. For a discussion of "The Sleepers" as sexual-psychological drama, see, in particular, Edwin H. Miller, *Whitman: A Psychological Journey*, pp. 66–84; Chase, *Walt Whitman Reconsidered*, pp. 54–57; Black, *Whitman's Journeys into Chaos*, pp. 125–37; and Robert Martin, *The Homosexual Tradition in American Poetry* (Austin: University of Texas Press, 1979), pp. 9–15.

18. Whitman reviewed Headley's *Washington and His Generals* for the *Eagle* in 1847 (*UPP*, I, 128). Parson Weems's *The Life of George Washington* appeared in 1800. For a discussion of Washington as an antipatriarchal symbol, see Jay Fliegelman, *Prodigals and Pilgrims: The American Revolution Against Patriarchal Authority, 1775–1800* (New York: Cambridge University Press, 1982), pp. 197–226.

19. When a slave "kills his master," says Douglass, he "imitates the heroes of the revolution"; in *My Bondage and My Freedom*, ed. Philip S. Foner (New York: Dover, 1969), p. 191.

20. Richard Maurice Bucke, ed., *Notes and Fragments* (Ontario, Canada: A. Talbot & Co., 1899), p. 19.

21. Cited in Stephen Oates, *To Purge This Land with Blood* (New York: Harper & Row, 1970), p. 351. J. C. Furnas discusses the "Spartacus complex" in *The Road to Harper's Ferry* (New York: W. Sloane Associates, 1959), pp. 206–44.

22. For studies of the native American movement, see Ira M. Leonard and Robert D. Parmet, *American Nativism, 1830–1860* (New York: Van Nostrand Reinhold, 1971); and Thomas J. Curran, *Xenophobia and Immigration, 1820–1930* (Boston: Twayne, 1975).

## Chapter 6

1. Walt Whitman, *The Correspondence*, ed. Edwin H. Miller (New York: New York University Press, 1961–69), vol. I, pp. 41–42. Subsequent references will be cited in the text as *Corr*. According to Miller, Seward probably sent a copy of *Statistical View of the United States . . . A Compendium of the Seventh Census* (1854), comp. J. D. B. De Bow. For a study of Seward's role in the antislavery movement, see G. G. Van Deusen, *William Henry Seward* (New York: Oxford University Press, 1961).

2. Walt Whitman, *Leaves of Grass: Facsimile of 1856 Edition* (Norwood, Pa.: Norwood Editions, 1976), p. 188. Subsequent references to this edition will be cited in the text as *LG 1856*.

3. For a study of the transition from the Union as political experiment to the Union as absolute, see Paul C. Nagel, *One Nation Indivisible: The Union in American Thought, 1776–1861* (New York: Oxford University Press, 1964).

4. For a compelling analysis of the deployment of female sexuality by the bourgeois social order, see Michel Foucault, *The History of Sexuality: An Introduction*, trans. Robert Hurley (New York: Pantheon, 1978). In *Whitman and the Body Beautiful*, Aspiz links Whitman's electrically charged persona with the eugenic theories of Lorenz Oken, *Elements of Physiophilosophy* (London, 1847); and Russell Thacher Trall, *The Hydropathic Encyclopedia: A System of Hydropathy and Hygiene* (New York, 1853), pp. 147–48. For a discussion of Whitman in relation to the sexual ideology of the age, see Myrth Jimmie Killingsworth, "Whitman and Motherhood: An Historical View," *American Literature*, 54 (March 1982), 28–43. The best case in defense of Whitman's women is made by Kay F. Reinartz, "Walt Whitman and Feminism," *Walt Whitman Review*, 19 (December 1973), 127–37. In " 'Noble American Motherhood': Whitman, Women, and the Ideal Democracy," *American Studies*, 21 (Fall 1980), 7–25, Arthur Wrobel follows D. H. Lawrence in arguing that Whitman was interested in women only as mothers.

5. Elizabeth Cady Stanton, *Elizabeth Cady Stanton As Revealed in Her Letters, Diary and Reminiscences*, ed. Theodore Stanton and Harriet Stanton Blatch (New York: Harper, 1902), vol. II, p. 210.

6. Library of Congress (Feinberg), Item no. 39.

7. See, for example, James E. Miller, " 'Brooklyn Ferry' and Imaginative Fusion," in *A Critical Guide to Leaves of Grass*, pp. 80–89; James W. Gargano, "Technique in 'Crossing Brooklyn Ferry': The Everlasting Moment," *Journal of English and Germanic Philology*, 62 (April 1963), 262–69; Gay Wilson Allen, *A Reader's Guide* (New York: Farrar, Straus & Giroux, 1970), pp. 187–91; Paul Orlov, "Of Time and Form in Whitman's 'Crossing Brooklyn Ferry,' " *Walt Whitman Quarterly Review*, 2 (Summer 1984), 12–21. In *The Lunar Light of Whitman's Poetry*, M. Wynn Thomas also argues that "in 'Crossing Brooklyn Ferry' Whitman tries heroically to spiritualize circumstance" (p. 116).

8. Library of Congress, Notebook on Government.

9. Allen, *The Solitary Singer*, p. 220. The lecture is in the Trent Collection at Duke University. Clifton J. Furness transcribed it while it was in the collection of Dr. R. M. Bucke and published parts of it in his article "Walt Whitman's Politics," *American Mercury*, 16 (April 1929), 459–66.

10. Furness, "Whitman's Politics," p. 461.

11. For a study of the significance of the Dred Scott decision, see D. E. Fehrenbacher, *The Dred Scott Case: Its Significance in American Law and Politics* (New York: Oxford University Press, 1978).

12. Walt Whitman, *I Sit and Look Out: Editorials from the Brooklyn Daily Times* (New York: Columbia University Press, 1932), p. 43. Subsequent references will be cited in the text as *ISL*.

13. David M. Potter, *The Impending Crisis: 1848–61* (New York: Harper & Row, 1976), p. 36.

14. Abraham Lincoln, *Collected Works*, ed. Roy P. Basler (New Brunswick, N.J.: Rutgers University Press, 1954), August 21, 1858, vol. III, p. 16. For studies of racism among Free-Soil advocates, see Eric Foner, *Free Soil, Free Labor, Free*

*Men* (New York: Oxford University Press, 1970); and Saul Sigelschiffer, *The American Conscience: The Drama of the Lincoln–Douglas Debates* (New York: Horizon, 1973).

15. Fredson Bowers, ed., *Whitman's Manuscripts: Leaves of Grass (1860)* (Chicago: University of Chicago Press, 1955), p. 1. Subsequent references will be cited in the text as Bowers.

16. For a discussion of Whitman's dispute with local church authorities, see Allen, *The Solitary Singer*, pp. 214–15.

17. Fredson Bowers reprints these poems and gives a complete account of their composition history in *Whitman's Manuscripts: Leaves of Grass (1860)*.

## Chapter 7

1. Lincoln, *Collected Works*, March 4, 1861, vol. III, p. 271. For a good general study of Lincoln during the war years, see Stephen B. Oates, *Abraham Lincoln: The Man Behind the Myths* (New York: Harper & Row, 1984). In *The Poet and the President: Whitman's Lincoln Poems* (New York: Odyssey, 1962), William Coyle reprints Whitman's four Lincoln poems and documents and criticism pertaining to the Whitman–Lincoln relationship. For studies of this relationship, see Clarence A. Brown, "Walt Whitman and Lincoln," *Journal of the Illinois State Historical Society*, 47 (Summer 1954), 176–84; and F. De Wolfe Miller, "The 'Long Foreground' of Whitman's Elegies on Lincoln," *Lincoln Herald*, 58 (Spring–Summer 1956), 3–7. Allen Grossman richly and suggestively compares Whitman and Lincoln on the problem of union in "The Poetics of Union in Whitman and Lincoln: An Inquiry Toward the Relationship of Politics and Art," in *The American Renaissance Reconsidered*, ed. Walter Benn Michaels and Donald E. Pease (Baltimore: Johns Hopkins University Press, 1985), pp. 183–208.

2. Walt Whitman, *Leaves of Grass: Facsimile Edition of the 1860 Text*, ed. Roy Harvey Pearce (Ithaca, N.Y.: Cornell University Press, 1961), p. 194. Subsequent references to this volume will be cited in the text as *LG* 1860.

3. For a discussion of *Chants Democratic* as a political handbook of democracy, see Robin P. Hoople, "Chants Democratic and Native American: A Neglected Sequence in the Growth of *Leaves of Grass*," *American Literature*, 42 (May 1970), 181–96.

4. Kenneth Burke, "Towards Looking Back," *Journal of General Education*, 28 (Fall 1976), 185–86. For a good discussion of Whitman's poems of doubt, see Timothy J. Lockyer, "The Mocking Voice: Whitman's Poems of Doubt," *Walt Whitman Review*, 26 (September 1980), 101–13.

5. See, for example, Edwin H. Miller, *Whitman: A Psychological Journey*, pp. 44–47; Kaplan, *Walt Whitman: A Life*, pp. 62–63; and Zweig, *Whitman: The Making of the Poet*, p. 309.

6. Herman Melville, *Moby Dick*, ed. Harrison Hayford and Hershel Parker (New York: Norton, 1967), p. 470.

7. Richard Maurice Bucke says that Whitman's friend Helen Price remembers Whitman's reading the poem to her family as early as 1858; in Bucke, *Walt Whitman* (Philadelphia: McKay, 1883), p. 29.

.     8. Whitman told the Price family that his poem "about a mocking bird" was "founded on a real incident," but this could suggest the reality of the bird story rather than a personal experience of love and loss. See Allen, *The Solitary Singer*, p. 233. The poem has been variously interpreted as ode, opera, riddle, autobiography, romance, narrative, and oedipal drama. See Leo Spitzer, " 'Explication de Texte' applied to Whitman's 'Out of the Cradle,' " *English Literary History*, 16 (September 1949), 229–49; Robert D. Faner, *Walt Whitman and Opera* (Philadelphia: University of Pennsylvania Press, 1951), pp. 173–77; Northrop Frye, *Anatomy of Criticism* (Princeton, N.J.: Princeton University Press, 1957), pp. 123–24; Stephen Whicher, "Whitman's Awakening to Death: Toward a Biographical Reading of 'Out of the Cradle Endlessly Rocking,' " in *The Presence of Walt Whitman*, ed. R. W. B. Lewis (New York: Columbia University Press, 1962), pp. 1–27; Chase, " 'Out of the Cradle' as Romance," in *The Presence of Walt Whitman*, pp. 52–71; Paul Fussell, "Whitman's Curious Warble: Reminiscence and Reconciliation," in *The Presence of Walt Whitman*, pp. 28–51; Waskow, *Whitman's Explorations in Form*, pp. 115–29; and Edwin H. Miller, *Whitman: A Psychological Journey*, pp. 171–86.

9. In his essay "Poet of Death: Whitman and Democracy," in *Literary Democracy*, Ziff argues that Whitman's "great poems of democracy are poems of death" (p. 250).

10. James E. Miller argues that spiritual love is the central focus of *Calamus* in " 'Calamus': The Leaf and the Root," in *A Critical Guide to Leaves of Grass*, pp. 52–79. For the best discussions of Whitman's radical homosexual vision in *Calamus*, see Joseph Cady, "Not Happy in the Capitol: Homosexuality and the Calamus Poems," *American Studies*, 19 (Fall 1978), 5–22; and Martin, *The Homosexual Tradition in American Poetry*, pp. 47–89. In *Gay American History: Lesbians and Gay Men in the U.S.A.* (New York: Crowell, 1976), Jonathan Katz discusses Whitman's central place in the quest for identity and freedom among nineteenth-century male homosexuals.

11. See Nagel, *One Nation Indivisible*, pp. 252–80.

12. The term *homosexual* was probably introduced into English by John Addington Symonds, who referred to "homosexual instincts" in *A Problem in Modern Ethics* (1891). It was not until the 1930s that the term received wide currency in English. See John Boswell, *Christianity, Social Tolerance, and Homosexuality* (Chicago: University of Chicago Press, 1980), pp. 41–45.

13. See Clifton Furness, "Walt Whitman Looks at Boston," *New England Quarterly*, 1 (July 1928), 353–70; and Emory Holloway, *Whitman: An Interpretation in Narrative* (New York: Knopf, 1928), pp. 164–65.

14. Allen Ginsberg, *Howl and Other Poems* (San Francisco: City Lights, 1956), p. 24.

## Chapter 8

1. Hopkins to Levi Hart, January 29, 1788, in Samuel Hopkins Papers, New York Historical Society, cited in May, *The Enlightenment in America*, p. 100.

2. Jefferson, *Notes*, p. 163; see also Jefferson to Jean Nicholas Demeunier, June 26, 1786, in *Papers*, vol X, p. 63.

3. Lincoln, *Collected Works*, March 4, 1865, vol. VIII, p. 333.

4. Ibid., vol. II, p. 276. For studies of Lincoln's relationship with the revolutionary founders, see Edmund Wilson, *Patriotic Gore: Studies in the Literature of the American Civil War* (New York: Oxford University Press, 1962); George B. Forgie, *Patricide in the House Divided* (New York: Norton, 1979); Dwight G. Anderson, *Abraham Lincoln, the Quest for Immortality* (New York: Knopf, 1982); and Diggins, *The Lost Soul of American Politics*.

5. Alexis de Tocqueville, *Democracy in America*, vol. I, p. 210.

6. Whitman to Horace Traubel, cited in Daniel Aaron, *The Unwritten War: American Writers and the Civil War* (New York: Knopf, 1973), p. 70.

7. Whitman, *Walt Whitman and the Civil War: A Collection of Original Articles and Manuscripts*, ed. Charles I. Glicksberg (Philadelphia: University of Pennsylvania Press, 1933), p. 174. Subsequent references will be cited in the text as Glicksberg.

8. Allen, *The Solitary Singer*, pp. 270–71.

9. Lincoln, *Collected Works*, March 4, 1861, vol. III, p. 268.

10. Whitman, *Notes and Fragments*, p. 196.

11. Asselineau, *The Evolution of Walt Whitman*, I, 170.

12. Holloway, *Whitman*, p. 185.

13. Horace Traubel, *With Walt Whitman in Camden*, vol. III (New York: Mitchell Kennerly, 1914), p. 581. Subsequent references will be cited in the text as *WWC*, III.

14. John Burroughs, *The Heart of Burroughs' Journals* (Boston: Houghton Mifflin, 1928), p. 237. Whitman's emergence in a female role is particularly evident in the letters that he regularly wrote to his mother during the war. At times he appears to experience the war from her point of view.

15. For a discussion of the Civil War and the death of the old republican order, see in particular, Schlesinger, *The Age of Jackson;* John Kassan, *Civilizing the Machine: Technology and Republican Values in America 1776–1900* (New York: Grossman, 1976); and Foner, *Politics and Ideology*.

16. Lincoln, "Gettysburg Address," *Collected Works*, November 19, 1863, vol. VII, p. 23.

17. Whitman published the following articles on his war experiences: "The Great Army of the Sick," *New York Times*, February 26, 1863; "Washington in the Hot Season," *New York Times*, August 16, 1863; "Our Wounded and Sick Soldiers," *New York Times*, December 11, 1864; "The Soldiers," *New York Times*, February 28, 1865; and " 'Tis But Ten Years Since," a series of six articles in the

New York *Weekly Graphic,* January 24, February 7, 14, 21, 28, and March 7, 1874.
See *PW,* I, 22–23.

18. For a discussion of Whitman's vision of the Civil War as a struggle
between democratic and antidemocratic forces, see M. Wynn Thomas, "Whitman
and the American Democratic Identity Before and During the Civil War," *Journal
of American Studies* (April 1981), 73–93.

19. See Richard Slotkin's important study *Regeneration Through Violence: The
Mythology of the American Frontier, 1600–1860* (Middletown, Conn.: Wesleyan
University Press, 1973).

20. Henry Adams, *The Education of Henry Adams* (Boston: Houghton Mifflin,
1961), p. 472.

21. During the war years, Whitman also had a copy of the *Odyssey of Homer
with Hymns, Epigrams, and Battles of the Frogs and Mice,* trans. T. A. Buckley, 1863.
On the flyleaf he wrote: "Possess'd by me from 1868 to 1888 and read by me
during those times— . . . Often in camp or Army Hospitals." See Glicksberg, pp.
82–83.

22. Walt Whitman, *Drum-Taps (1865) and Sequel to Drum-Taps (1865–6): A
Facsimile Reproduction,* ed. F. De Wolfe Miller (Gainesville, Fla.: Scholars' Facsimiles & Reprints, 1959), p. 58. Subsequent references to *Drum-Taps* will be
cited in the text as *DT.*

23. For a comparison of the war poetry of Whitman and Melville, see John Mc
Williams, " 'Drum Taps' and *Battle Pieces:* The Blossom of War," *American Quarterly,* 23 (May 1971), 181–201; and Vaughn Hudson, "Melville's *Battle Pieces* and
Whitman's Drum-Taps: A Comparison," *Walt Whitman Review,* 19 (September
1973), 81–92.

24. For an excellent general consideration of the effects of the Civil War on
Whitman's work, see Aaron, "Whitman: The 'Parturition Years,' " in *The Unwrit-
ten War,* pp. 56–74. See also George M. Fredrickson, *The Inner Civil War* (New
York: Harper & Row, 1968), pp. 90–97; and James Cox, "Walt Whitman, Mark
Twain, and the Civil War," *Sewanee Review,* 69 (April–June 1961), 185–204.
James E. Miller considers the overall themes of the *Drum-Taps* volume in *A
Critical Guide to Leaves of Grass,* pp. 219–26; and Denise T. Askin discusses the
dialectic between prophetic and private voice in " 'Retrievements Out of the
Night': Prophetic and Private Voices in Whitman's 'Drum-Taps'," *American Tran-
scendental Quarterly,* 51 (Summer 1981), 211–23.

25. Whitman, *DT and Sequel to DT,* p. 23. Subsequent references to *Sequel to
Drum-Taps* will be cited in the text as *Sequel.*

26. Ezra Pound, "A Few Don'ts by an Imagiste," *Poetry,* 1 (March 1913), 200–
206; William Carlos Williams, *Collected Earlier Poems* (New York: New Directions,
1951), p. 233.

27. Samuel Chester Reid's firsthand account of the battle of Kennesaw
Mountain bears a striking resemblance to the site of battle in "The Veteran's
Vision": "Rank after rank went down, but still on they pressed only to meet the

same fate, until impeded by the piles of their own dead." See J. Cutler Andrews, ed., *The South Reports the War* (Princeton, N.J.: Princeton University Press, 1970), p. 449.

28. In "*Drum-Taps* and Nineteenth-Century Male Homosexual Literature," in *Walt Whitman: Here and Now,* ed. Joann P. Krieg (Westport, Conn.: Greenwood Press, 1985), pp. 49–60, Joseph Cady challenges the standard critical assumption that Whitman's war poems are a sublimation of his homosexuality. Instead, he argues that the motif of wartime comradeship gave Whitman a self-protective context in which to express his homosexual feeling.

## Chapter 9

1. Lincoln, *Collected Works,* March 4, 1865, vol. VIII, p. 333.

2. For a discussion of the printing history of *Drum-Taps and Sequel,* see F. De Wolfe Miller's introduction to his facsimile edition of *Drum-Taps,* pp. vii–lix.

3. Richard P. Adams discusses Whitman's use of the conventions of pastoral elegy in "Whitman's 'Lilacs' and the Tradition of Pastoral Elegy," *PMLA,* 72 (June 1957), 479–87.

4. "I remember," Whitman said of the day of Lincoln's assassination, "the season being advanced, there were many lilacs in full bloom. By one of those caprices that enter and give tinge to events without being at all part of them, I find myself always reminded of the great tragedy of that day by the sight and odor of these blossoms. It never fails" (*PW,* II, 503). In Whitman's imagination, the western star was also bound up with the portentousness of the war years. "The heavens, the elements, all the meteorological influences, have run riot for weeks past," he said of the spring of Lincoln's second inaugural address in 1864: "But there have been samples of another description. Nor earth nor sky ever knew spectacles of superber beauty than some of the nights lately here. The western star, Venus, in the earlier hours of evening, has never been so large, so clear; it seems as if it told something, as if it held rapport indulgent with humanity, with us Americans" (*PW,* I, 94).

5. In "Whitman's Lilacs and the Grammars of Time," *PMLA,* 97 (January 1982), 31–39, Mutlu Konuk Blasing argues just the opposite: "In 'Lilacs,' Whitman moves beyond taking refuge in conventions or in the authority of authorship, which is formalized in poetic usage and syntax. In the final stanza, his rejection of the orders of a sentence indicates his psychic mastery of the event that occasioned the poem." "Psychic mastery" is not a term I would use to characterize Whitman's state of mind in the aftermath of the war and Lincoln's death. However, Blasing does concede that Whitman's "fears of chaos" are more evident in the unrevised version of the poem, which reveals "the attractions that formal music had for Whitman" (p. 37).

*Chapter 10*

1. Herman Melville, *The Battle Pieces of Herman Melville,* ed. Hennig Cohen (New York: T. Yoseloff, 1963), p. 200.

2. *WWW,* p. 57. The note appears in a Civil War notebook dated 1863.

3. For useful studies of the politics of Reconstruction, see Kenneth M. Stampp, *The Era of Reconstruction: 1865–1877* (New York: Knopf, 1966); Rembert W. Patrick, *The Reconstruction of the Nation* (New York: Oxford University Press, 1967); and W. E. B. DuBois, *Black Reconstruction in America* (New York: Atheneum, 1969). Harold M. Hyman focuses on constitutional issues in *A More Perfect Union: The Impact of the Civil War and Reconstruction on the Constitution* (New York: Knopf, 1973).

4. Melville, *Battle Pieces,* p. 40.

5. Noam Chomsky, *Turning the Tide: U.S. Intervention in Central America and the Struggle for Peace* (Boston: South End Press, 1986). See also Edward C. Kirkland, *Industry Comes of Age: Business, Labor, and Public Policy, 1860–1897* (New York: Holt, Rinehart & Winston, 1961).

6. Letter of James Harlan to William P. Dole, cited in Jerome Loving, "Whitman and Harlan: New Evidence," *American Literature,* 48 (May 1976), 221.

7. Carlyle, "Shooting Niagara: And After?," originally published in *Macmillan's Magazine,* 16 (1867), 319–336, and reprinted in Horace Greeley's *New York Tribune,* August 16, 1867, and in *Works of Thomas Carlyle* (London: Chapman & Hall Limited, 1869–72), vol. XXX, pp. 1–48. For studies of the political views of Whitman and Carlyle, see Gregory Paine, "The Literary Relations of Whitman and Carlyle with Especial Reference to Their Contrasting Views of Democracy," *Studies in Philology,* 36 (July 1939), 550–63; Joseph Jones, "Carlyle, Whitman, and the Democratic Dilemma," *English Studies in Africa,* 3 (September 1960), 179–97; and Linden Peach, "The True Face of Democracy?: Carlyle's Challenge to Whitman's Idealism," in Peach's *British Influence on the Birth of American Literature* (New York: St. Martins Press, 1982), pp. 162–93.

8. Carlyle, *Works,* vol. XXX, p. 5.

9. Ibid., pp. 6–7.

10. The *Galaxy,* which was founded in New York in 1866 by William and Frank Church to rival the Boston-based *Atlantic Monthly,* was an appropriate place for Whitman to launch his attack on genteel culture as it was represented by the New England literary establishment; see Edward F. Grier, "Walt Whitman, the *Galaxy,* and *Democratic Vistas,*" *American Literature,* 23 (November 1951), 332–50.

11. Whitman's notes on Hegel were drawn from Joseph Gostwick's *German Literature* (1854). For studies of the relation between Whitman and Hegel, see Mody C. Boatright, "Whitman and Hegel," *Texas Studies in English,* 9 (July 1929), 134–50; W. B. Fulghum, Jr., "Whitman's Debt to Joseph Gostwick," *American Literature,* 7 (January 1941), 491–96; Robert P. Falk, "Walt Whitman and German Thought," *Journal of English and Germanic Philology,* 40 (July 1941), 315–30; Olive W. Parsons, "Whitman the Non-Hegelian," *PMLA,* 58 (December 1943),

1073–93; and Paschal Reeves, "The Silhouette of th⟨
Hegelian or Whitmanian," *Personalist*, 43 (Summer 196⟨

12. In "A Mediated Vision, a Measured Voice: ⟨
Whitman's Prose," *Walt Whitman Quarterly Review*, 2 (S⟨
Robert L. Pincus argues that the "bifocal" vision of *Democra..*
new critical vision and voice foreshadowed by Whitman's early ⟩u⟨
"thwarted in the process of forging the persona of *Leaves of Grass.*" Actually,
have seen, a contrapuntal method was also at the root of Whitman's poetic strategy
in *Leaves of Grass;* here, too, the critical perspective of the political realist struggles
with the visionary perspective of the poet.

13. Mark Twain and C. D. Warner, *The Gilded Age: A Tale of To-day* (1873).

14. For an interesting consideration of the ways that the Abolition movement
deflected attention away from labor problems, see Foner, "Abolitionism and the
Labor Movement in Ante-bellum America," in *Politics and Ideology in the Age of the
Civil War.*

15. Matthew Arnold, *Culture and Anarchy*, ed. J. Dover Wilson (New York:
Cambridge University Press, 1932), p. 82.

16. Susan Sontag discusses the ambitious scope and inevitable failure of
Whitman's "Great American Cultural Revolution" in *On Photography* (New York:
Farrar, Straus & Giroux, 1975), pp. 27–48. For a recent discussion of the ideologi-
cal power of American democracy as a cultural system, see Bercovitch, "The
Problem of Ideology in American Literary History."

## Chapter 11

1. Arthur Golden edited a two-volume facsimile and textual analysis of Whit-
man's blue book *Leaves*. See *Walt Whitman's Blue Book: The 1860–61 Leaves of
Grass Containing His Manuscript Additions and Revisions* (New York: New York
Public Library, 1968).

2. Walt Whitman, *Leaves of Grass: A Textual Variorum of the Printed Poems*, ed.
Sculley Bradley et al. (New York: New York University Press, 1980), vol. II, p.
404. Subsequent references to this three-volume edition will be cited in the text as
*LG: Variorum.*

3. Cited in Aspiz, *The Body Beautiful*, p. 229.

4. In July 1871 he wrote to William Rossetti, who had published an "expurgated" edition of *Leaves of Grass* in England in 1869: "My 'Leaves of Grass' I
consider substantially finished, as in the copies I sent to you. To 'Democratic
Vistas' it is my plan to add much, if I live" (*Corr.*, II, 131).

5. These lines correspond to the final version of the poem in *LGC*, pp. 411–
21. For a discussion of the manuscript of "Passage," see B. R. McElderry, Jr.,
"The Inception of 'Passage to India,'" *PMLA*, 71 (September 1956), 837–39;
and Fredson Bowers, "The Earliest Manuscript of Whitman's 'Passage to India'
and Its Notebook," *Bulletin of the New York Public Library*, 61 (1957), 319–52. In
"Passage to Less Than India: Structure and Meaning in Whitman's 'Passage to

*MLA*, 88 (October 1973), 1095–1103, Arthur Golden argues that man never effectively integrated these sections into the substance of his poem. Like Newton Arvin, Richard Chase, Roy Harvey Pearce, and Edwin Miller, Golden regards "Passage to India" as a failed poem and symptomatic of Whitman's decline in his later period. For studies defending the poem's artistic power, see especially Stanley Coffman, "Form and Meaning in Whitman's 'Passage to India,'" *PMLA*, 70 (June 1955), 337–49; John Lovell, Jr., "Appreciating Whitman: 'Passage to India,'" *Modern Language Quarterly Review*, 21 (June 1960), 131–41; James S. Leonard, "The Achievement of Rondure in 'Passage to India,'" *Walt Whitman Review*, 26 (December 1980), 129–38; and Martin K. Doudna, "The Essential Me: Whitman's Achievement in 'Passage to India,'" *Walt Whitman Quarterly Review*, 2 (Winter 1984), 1–9.

6. Because Whitman made only minor changes in "Passage to India," I am using the final version of the poem in *LGC.*

7. *New York Evening Post*, May 11, 1869, p. 2.

8. See *Corr.*, II, 34 n.

9. "Passage to India" notebook, in Bowers, "The Earliest Manuscript of Whitman's 'Passage to India' and Its Notebook," pp. 349–50.

10. Library of Congress, Harned Notebook, 1862–1863; in *UPP*, II, 93.

11. Roy Harvey Pearce, "Whitman and Our Hope for Poetry," in his *Historicism Once More* (Princeton, N.J.: Princeton University Press, 1969), p. 348.

12. Library of Congress (Feinberg), Item no. 39.

13. In "What's in a Title? Whitman's 'Calamus' and Bucke's *Calamus*," *Studies in the American Renaissance*, ed. Joel Myerson (Boston: Twayne, 1979), Artem Lozynsky contends that Richard Maurice Bucke's publication of Whitman's letters to Peter Doyle in 1897 was intended to deal with the problem of Whitman's homosexuality by demonstrating that his relationship to Doyle was spiritual and chaste (pp. 475–88).

14. Library of Congress (Feinberg), Item no. 39. The idea for the poem came from his reading an article on "The Last Days of Columbus," published in *The Irish Republic* (May 1869). Across the margin of the article, he wrote: "Poems— Columbus—(? that name for piece)—make the poem an utterance of Columbus— there on Jamaica Island (read first *Ulysses* by Tennyson)." The article was abstracted from Sir Arthur Phelps's *The Spanish Conquest in America.*

15. "The Silent General," *PW*, I, 226–27; and "Rulers Strictly Out of the Masses," *PW*, II, 534–35.

16. Library of Congress (Feinberg), Item no. 32.

17. For a discussion of Whitman's dispute with O'Connor, see Jerome Loving, *Walt Whitman's Champion: William Douglas O'Connor* (College Station: Texas A&M Press, 1978), pp. 94–102.

18. For a discussion of the Compromise of 1877, see Kenneth M. Stampp, *The Era of Reconstruction;* Rembert W. Patrick, *The Reconstruction of the Nation;* and Keith J. Polakoff, *The Politics of Inertia: The Election of 1876 and the End of Reconstruction* (Baton Rouge: Louisiana State University Press, 1973). For studies of

racial attitudes during and after Reconstruction, see especially C. Vann Woodward, *American Counterpoint: Slavery and Racism in the North–South Dialogue* (Boston: Little, Brown, 1971); George M. Fredrickson, *The Black Image in the White Mind;* and Eric Foner, *Nothing But Freedom: Emancipation and Its Legacy* (Baton Rouge: Louisiana State University Press, 1983).

## Chapter 12

1. Henry Nash Smith, *The Virgin Land: The American West As Symbol and Myth* (Cambridge, Mass.: Harvard University Press, 1950); see also Edwin Fussell, *Frontier: American Literature and the American West* (Princeton, N.J.: Princeton University Press, 1965), pp. 397–442.

2. See "Death of William Cullen Bryant," *PW*, I, 165–67; "My Tribute to Four Poets," *PW*, I, 266–67; "A Visit, at the Last, to R. W. Emerson," *PW*, I, 278–80; "Boston Common—More of Emerson," *PW*, I, 290–91. For a discussion of Whitman's essays on literary figures in *Specimen Days*, see James Bristol, "Literary Criticism in *Specimen Days*," *Walt Whitman Review*, 12 (March 1966), 16–19; and Kenneth M. Price, "Whitman on Other Writers: Controlled 'Graciousness' in *Specimen Days*," *Emerson Society Quarterly*, 26 (1980), 79–87.

3. Henry George, *Progress and Poverty* (1883), cited in Foner, *Politics and Ideology in the Age of the Civil War*, p. 195.

4. See Melvyn Dubofsky, *Industrialism and the American Worker* (Arlington Heights, Ill.: AHM Press, 1975); Herbert Gutman, *Work, Culture and Society in Industrializing America* (New York: Knopf, 1976); and David Montgomery, *Workers' Control in America: Studies in the History of Work, Technology, and Labor Struggles* (New York: Cambridge University Press, 1979).

## Chapter 13

1. Boston *Intelligencer* cited in Asselineau, *The Evolution of Walt Whitman*, vol. I, p. 72; Thoreau letter to Harrison Blake, December 7, 1856, in *Letters to Various Persons*, ed. R.W.E. [Ralph Waldo Emerson] (Boston: Ticknor & Fields, 1865), p. 146. Whitman comments in *PW*, II, 494, on Emerson's "vehement arguments" against including "Children of Adam" in the 1860 *LG*.

2. William Bradford, *Of Plymouth Plantation 1620–1647* (New York: Random House, 1981), pp. 351, 355.

3. Smith-Rosenberg, *Disorderly Conduct*, p. 181.

4. For studies of the new wave of reactionary propriety in the 1870s and 1880s, see Smith-Rosenberg, *Disorderly Conduct;* William Leach, *True Love and Perfect Union* (New York: Basic Books, 1980); and Carl Degler, *At Odds* (New York: Oxford University Press, 1980). For more general studies of attitudes toward women in the nineteenth century, see Foucault, *The History of Sexuality;* Barbara Welter, *Dimity Convictions* (Athens: Ohio University Press, 1976); Barbara Harris, *Beyond Her Sphere* (Westport, Conn.: Greenwood Press, 1978); and

Mary Ryan, *Womanhood in America: From the Colonial Times to the Present* (New York: Franklin Watts, 1983).

5. Asselineau, *The Evolution of Walt Whitman*, vol. I, pp. 241–42.

6. It was not really until Allen Ginsberg published his comic tribute "In a Supermarket in California" that Whitman the homosexual poet came fully out of the closet, at least in America; see Ginsberg, *Howl and Other Poems*, pp. 23–24.

7. In Emory Holloway and Ralph Adimari, eds., *New York Dissected* (New York: Rufus Rockwell Wilson, 1936), pp. 162–65. Whitman may have borrowed the florid design of the 1855 *Leaves of Grass* from Parton's book *Fern Leaves from Fanny's Portfolio* (1853).

8. Material from the *Saturday Press* is cited in Allen, *The Solitary Singer*, pp. 260–62. According to Allen, there is some evidence that Beach may have been in love with Whitman and that "Out of the Rolling Ocean" was addressed to her. This was not the first time that Whitman would come between husband and wife. Later, Ellen O'Connor's intense feeling for Whitman would be at least partly responsible for the feud between Whitman and her husband that lasted several years. See Loving, *Walt Whitman's Champion: William Douglas O'Connor*.

9. In addition to the review by Juliette H. Beach, the *Saturday Press* carried a review of the 1860 *Leaves* by Mary A. Chilton, who defended Whitman's daring portraits of the "nude form" and the "functions of the human body," no matter how "brutal or degrading"; and the New York *Sunday Mercury* carried a laudatory review by the Bohemian actress and poet Adah Isaacs Menken (Allen, *The Solitary Singer*, pp. 262–63). For an examination of Whitman's special appeal to the female reader, see Lottie L. Guttey, "Walt Whitman and the Woman Reader," *Walt Whitman Review*, 22 (September 1976), 102–10.

10. Thomas Harned, ed., *The Letters of Anne Gilchrist and Walt Whitman* (New York: Doubleday, Page & Co., 1919), pp. 3–4. Subsequent references will be cited in the text as Gilchrist.

11. Cited in Schyberg, *Walt Whitman*, p. 218.

12. Audre Lorde, *Sister Outsider: Essays and Speeches* (Trumansburg, N.Y.: Crossing Press, 1984), p. 58.

13. Herbert Marcuse, *Counterrevolution and Revolt* (Boston: Beacon Press, 1972), pp. 74–78.

14. A copy of George C. Catlin's portrait of Osceola hung on the wall of Whitman's Camden room; see Edgely W. Todd, "Indian Pictures and Two Whitman Poems," *Huntington Library Quarterly* (November 1955), 1–11. For an excellent study of native American policy, see Richard Drinnon, *Facing West: The Metaphysics of Indian Hating and Empire Building* (Minneapolis: University of Minnesota Press, 1980).

15. In emphasizing the historic nature of *Leaves of Grass*, Whitman may have been influenced by a reading of Hippolyte Taine's *History of English Literature*. His article on Taine was published by Roger Asselineau, "Un inédit de Walt Whitman: Taine's *History of English Literature*," *Études Anglaises*, 10 (April–June 1957), 128–38. For a discussion of Taine's influence on Whitman, see Erkkila, *Walt Whitman Among the French*, pp. 41–45.

16. Langston Hughes, "The Ceaseless Rings of Walt Whitman," in *I Hear the People Singing: Selected Poems of Walt Whitman,* ed. Langston Hughes (New York: International Publishers, 1946), p. 9.

17. See Gay Wilson Allen's discussion of "Walt Whitman and World Literature" in *The New Walt Whitman Handbook,* pp. 249–327; Gay Wilson Allen, ed., *Walt Whitman Abroad* (Syracuse, N.Y.: Syracuse University Press, 1955), which includes critical essays from Germany, France, Scandinavia, Russia, Italy, Spain, and Latin America; Harold Blodgett, *Walt Whitman in England* (Ithaca, N.Y.: Cornell University Press, 1934); Anne Jacobson, "Walt Whitman in Germany Since 1914," *Germanic Review,* 133 (April 1926), 132–41; and Erkkila, *Whitman Among the French.*

18. "The Mayflower Moment: Reading Whitman During the Vietnam War," in *Walt Whitman: The Measure of His Song,* ed. Jim Perlman, Ed Folsom, and Dan Campion (Minneapolis: Holy Cow! Press, 1981), pp. 308–9.

19. Julio Valle-Castillo, ed., *Poesía Libre* (Managua: Ministerio de Cultura, 1981), p. 15.

20. Cited in Allen, *The New Walt Whitman Handbook,* p. 301.

21. When Whitman's works were first introduced to the French reading public during the reign of Emperor Napoleon III, Louis Étienne presented him as a democratic "rowdy." It was not until the Third Republic that Whitman's reputation began to rise. At that time, the conservative critic Henri Cochin saw in *Leaves of Grass* the very symbol of "democracy run wild, a form of insanity and megalomania" ("Un Poète Américain: Walt Whitman," *Le Correspondant,* November 25, 1877, p. 634). Translations of Whitman's work were suppressed in Austria and Hungary in the period immediately following World War I. And in Germany, after an almost cultish interest in Whitman's works during the years of Social Democracy, his reputation declined with the rise of Fascism in the 1920s and 1930s.

22. Robert Creeley, "Introduction to *Whitman Selected by Robert Creeley,*" in his *Was That a Real Poem & Other Essays* (Four Seasons Foundation, 1979), p. 62; Allen Ginsberg, "Allen Ginsberg on Walt Whitman Composed on the Tongue" (recorded for Centre Films, 1980), in *Whitman: The Measure of His Song,* p. 252.

23. Meridel LeSueur, "Jelly Roll," in *Whitman: The Measure of His Song,* p. 356.

24. June Jordan, "For the Sake of a People's Poetry: Walt Whitman and the Rest of Us," in her *Passion* (Boston: Beacon Press, 1980), p. xii.

25. C. W. Truesdale, "Whitman Exilado," in *Whitman: The Measure of His Song,* p. 335.

26. Langston Hughes, *Selected Poems* (New York: Vintage, 1959), p. 275; Jordan, "For the Sake of a People's Poetry," pp. xxiv, ix, xv.

27. Pablo Neruda, "We Live in a Whitmanesque Age," *New York Times,* April 14, 1972, p. 39.

28. Cited in Raymond Williams, *Culture and Society,* p. 279.

29. Carlos Fuentes, "When Don Quixote Left his Village, the Modern World Began," *New York Times Book Review,* March 23, 1986, p. 15.

# Index

*Note:* Page references to illustrations are italicized.